Communications
in Computer and Information Science   287

Andrzej Dziech   Andrzej Czyżewski (Eds.)

Multimedia Communications, Services and Security

5th International Conference, MCSS 2012
Krakow, Poland, May 31–June 1, 2012
Proceedings

 Springer

Volume Editors

Andrzej Dziech
AGH University of Science and Technology
Department of Telecommunications
Krakow, Poland
E-mail: adzie@tlen.pl

Andrzej Czyżewski
Gdansk University of Technology
Multimedia Systems Department
Gdansk, Poland
E-mail: indect@sound.eti.pg.gda.pl

ISSN 1865-0929   e-ISSN 1865-0937
ISBN 978-3-642-30720-1   e-ISBN 978-3-642-30721-8
DOI 10.1007/978-3-642-30721-8
Springer Heidelberg Dordrecht London New York

Library of Congress Control Number: 2012938500

CR Subject Classification (1998): K.6.5, K.4.4, H.5.1, H.4, D.4.6, C.2

Typesetting: Camera-ready by author, data conversion by Scientific Publishing Services, Chennai, India

Printed on acid-free paper

Springer is part of Springer Science+Business Media (www.springer.com)

Preface

The 5th Conference on Multimedia Communications, Services and Security, MCSS 2012, reflected the growing number of applications of innovative ICT technologies in the area of public security.

Of special importance are the technologies and applications which help to increase the privacy of citizens and to protect sensitive data. The conference continues to present research and development activities contributing to theoretical and experimental aspects of the considered domain.

The following areas were of main interest at MCSS 2012:

- Privacy and data protection using digital watermarking and other technologies
- Object and threat detection
- Data protection and distribution
- Human—centric multimedia analysis and synthesis
- Cybercrime—threats detection and counteracting

Similarly to the previous editions of the conference, oral sessions were accompanied by demonstration and poster sessions. Demonstration sessions at MCSS are a very good opportunity to present and discuss the outcomes and solutions of the performed research and to experience in practice the emerging implementations and research prototypes.

The conference was again an occasion for the exchange of knowledge among specialists involved in multimedia and security research.

May 2012   Andrzej Dziech

Organization

The International Conference on Multimedia Communications, Services and Security (MCSS 2012) was organized by the AGH University of Science and Technology in the scope of and under the auspices of the INDECT project

Executive Committee

General Chair

Andrzej Dziech — AGH University of Science and Technology, Poland

Committee Chairs

Andrzej Dziech — AGH University of Science and Technology, Poland

Andrzej Czyżewski — Gdansk University of Technology, Poland

Technical Program Committee

Emil Altimirski — Technical University of Sofia, Bulgaria

Alexander Bekiarski — Technical University of Sofia, Bulgaria

Fernando Boavida — University of Coimbra, Portugal

Eduardo Cerqueira — Federal University of Para, Brazil

Ryszard Choras — University of Technology and Life Sciences

Marilia Curado — University of Coimbra, Portugal

Andrzej Czyżewski — Gdansk University of Technology, Poland

Jacek Dańda — AGH University of Science and Technology, Poland

Jan Derkacz — AGH University of Science and Technology, Poland

Marek Domański — Poznan University of Technology, Poland

Andrzej Duda — Grenoble Institute of Technology, France

Andrzej Dziech — AGH University of Science and Technology, Poland

Andrzej Głowacz — AGH University of Science and Technology, Poland

Czesław Jędrzejek — Poznan University of Technology, Poland

Nils Johanning — InnoTec Data, Germany

Jozef Juhar — Technical University of Kosice, Slovakia

Marek Kisiel-Dorohnicki — AGH University of Science and Technology, Poland

Jaroslav Zdralek	VSB - Technical University of Ostrava, Czech Republic
Tomasz Zieliński	AGH University of Science and Technology, Poland
Mariusz Ziółko	AGH University of Science and Technology, Poland

Organizing Committee

Jacek Dańda	AGH University of Science and Technology, Poland
Jan Derkacz	AGH University of Science and Technology, Poland
Sabina Drzewicka	AGH University of Science and Technology, Poland
Andrzej Głowacz	AGH University of Science and Technology, Poland
Michał Grega	AGH University of Science and Technology, Poland
Piotr Guzik	AGH University of Science and Technology, Poland
Magdalena Hrynkiewicz-Sudnik	AGH University of Science and Technology, Poland
Paweł Korus	AGH University of Science and Technology, Poland
Mikołaj Leszczuk	AGH University of Science and Technology, Poland
Andrzej Matiolański	AGH University of Science and Technology, Poland
Piotr Romaniak	AGH University of Science and Technology, Poland
Krzysztof Rusek	AGH University of Science and Technology, Poland

Sponsoring Institutions

- European Commission, Seventh Framework Programme (FP7)
- Institute of Electrical and Electronics Engineers (IEEE)
- Intelligent information system supporting observation, searching and detection for security of citizens in urban environment (INDECT Project)
- AGH University of Science and Technology, Department of Telecommunications

Table of Contents

Web-Based Knowledge Acquisition and Management System Supporting Collaboration for Improving Safety in Urban Environment⋆

Weronika T. Adrian, Przemysław Ciężkowski, Krzysztof Kaczor,
Antoni Ligęza, and Grzegorz J. Nalepa

AGH University of Science and Technology
al. A. Mickiewicza 30, 30-059 Krakow, Poland
{wta,kk,ligeza,gjn}@agh.edu.pl

Abstract. Effectiveness and popularity of an information system is nowadays significantly influenced by its social aspects and the benefits it provides for the user community. This paper presents a progress report of recent developments within the Task 4.6. of the INDECT project aimed at development of a threat monitoring system with social features and GIS integration. The main focus is on the improved prototype system for distributed knowledge acquisition and management. The prototype is a result of a thorough analysis of previous attempts to develop the system, and a reworked design and implementation with several important improvements. In the paper, the concepts and requirements of the system are briefly recalled, selected features of the enhanced system are highlighted and a practical use case example is described. The presented scenario illustrates how citizens and police officers can cooperate with high efficiency and low cost using the system.

Keywords: security, citizens, GIS, knowledge management, INDECT.

1 Introduction

Building efficient tools for supporting Knowledge Acquisition and Knowledge Management is a challenging topic. In modern information systems, web technologies facilitate massive, distributed knowledge acquisition. Examples of such social phenomena as Wikipedia constitute a working proof of high potentials incorporated in the synergy of human and web interaction. Another trend consists in using Geographical Information Systems (GIS) for providing meaningful services and tools. GIS is a wide term encompassing various systems aimed at capturing, storing, analysis, management and presentation of geospatial data. It

---

⋆ The research presented in this paper is carried out within the EU FP7 INDECT Project: "Intelligent information system supporting observation, searching and detection for security of citizens in urban environment" (http://indect-project.eu).

A. Dziech and A. Czyżewski (Eds.): MCSS 2012, CCIS 287, pp. 1–12, 2012.

includes solutions for different applications, from touristic mapping to strategic military and emergency planning solutions.

The aim of the Task 4.6. of the INDECT project is to develop a system for distributed knowledge acquisition, gathering and organizing knowledge on threats of various nature. The system should leverage the social networks concepts and GIS component to support communities in improving safety of citizens in urban environments. The focus of the project is on tools to process the information provided by citizens via a specialized website. In fact, a Web System software for citizen provided information, automatic knowledge extraction, knowledge management and GIS integration is to be developed.

Principles and a conceptual model of the system have been described in [4]. The general idea of the system can be observed in Figure 1.

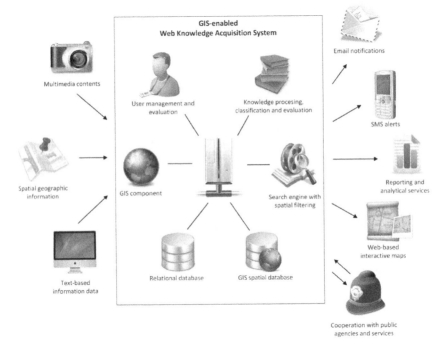

Fig. 1. Conceptual model of the system [3]

The input data, in general, may be composed of: a text description of a threat, its spatial location, and multimedia documentation. The data, stored in a relational database equipped with spatial features, should be presented to the audience in a combined visual and textual form. The system should provide means for searching, filtering, aggregation and grouping of data for final users, according to their preferred form and level of detail. The threats should be presented in a convenient and transparent way as icons on the map, in reports or notifications.

The basic assumptions for the system have been presented in [5]. These include: (i) willingness of citizens to share information about dangers with others, (ii) collaborative rating of the credibility, usefulness and importance of the threats entered into the system, (iii) possibility of entering information in various formats, and (iv) geographic data presentation.

Architecture, requirements and use cases have been then described in [5]. The functional requirements defined for the system can be summarized as follows: (i) threat data submission and management (adding, editing, deleting, validating, voting and commenting), (ii) map-based information sharing and visualization, (iii) data analysis using predefined reports, (iv) advanced search capabilities, (v) multiple categories of users, (vi) user evaluation framework to assess credibility of submitted threats, (vii) a notification system, including alerts for significant dangers submitted by credible users, (viii) a newsletter, with attribute- and location-based customization, (ix) a rule-based engine to facilitate data administration and report generation, (x) integration with existing social network applications, (xi) user authentication with open authentication frameworks, (xii) support for mobile platforms.

Based on the definition of requirements, several prototypes have been developed [3], [8]. The systems offer slightly different functionalities and have been developed with use of different technologies. Based on the experiences with these systems, a new, improved prototype has been developed [2]. It serves as a base for the final deliverable within the INDECT project.

In this paper, we summarize the major observations and changes in the new system in Sect. 2, highlight selected features of the system in Sect. 3 and present an illustrative use case example in Sect. 4. The paper is summarized in Sect. 5.

2 System Evolution and Lessons Learned

To enhance the automated knowledge processing of the system, semantic technologies for GIS were analyzed and discussed in [6]. The semantic research thread led to the development of a prototype described in [8] which in greater details investigates the integration issues of databases and ontologies. In the ontology, the general categories of threats were stored, whereas in the database, the actual data about selected areas in particular time were located. This prototype provided interesting insights and ideas for future investigations. However, for the INDECT purposes, more lightweight semantics and reasoning has been chosen. Three systems, referenced in [3] use lightweight reasoning and metadata annotations of threat such as simple tags. In the newest prototype, codenamed Social Threat Monitor (STM), only basic semantics is added with use of tags and categories (see Section 3).

Based on the experiences and tests on the prototype systems, the focus of the new improved one has been placed on Web 2.0 features, such as managing comments, voting etc. Moreover, the assumption that people want to cooperate is valid only when the system interface is simple, entering information is not time-consuming, and the user can immediately see the results. It is desirable that users can see their contribution and the impact of it (other people can evaluate and confirm it). Further, the knowledge must be processed automatically (cost elimination), and intuitive graphical interface is necessary.

The reworked design of the recent implementation emphasizes the social aspects of the system. It defines the application as a platform to exchange information about dangers and their location among citizens and with public services. The expected result of the application is to help civilians and public services to improve safety in urban areas and detect dangers faster. Users of the system are able to browse threats in selected area and fill up information about a threat by adding comments and photos; public services are able to inform people about dangers and to monitor threats and decide if an intervention is needed.

The core functionalities, operation and user interface have been slightly extended and modified. The alteration consists mainly in reorganization of the User Interface (UI), so that access to specific actions is more intuitive. The UI is described in more details in Section 3.1. The improvements include the following areas: map (viewing, filtering the threats, geopositioning), left menu (searching, selecting area, defining threats), top menu (account and application management), social features (comments, voting, banning users by a moderator etc.) and administrating the application (categories, users and groups management).

Use cases have been reviewed and a more thoughtful users hierarchy has been developed (for a thorough analysis of the use cases see [2]). The apparent problem with the previous systems was that a guest user had no or very limited scope of rights to use the system. This discouraged potential users who wanted to try the system before actual registration. In the STM, the following groups of users are defined, with gradually increasing capacitites:

1. **Guest** is a user with the anonymous web account. He is able to use basic features of the application. In order to gain more privileges, a guest need to register and log in into the system.
2. **Member** is a user with registered account in the system. With this account user can add threats, manage his own threats, comment and vote threats of other users and edit his profile.
3. **Services User** is a special account with features helping threats monitoring.
4. **Moderator** is a user with full access to threat records, able to ban users.
5. **Administrator** is a user with full access to the application.

The system has been implemented using widely-accepted, cost-free Web technologies: HTML, CSS, JavaScript, jQuery, and Google Maps API version 3 and the Django framework (for details see [2]). It has been deployed on a dedicated server and is available at: `http://vigil.ia.agh.edu.pl`.

3 Selected Features of the System

Based on beta tests of the prototypes described in [3] and analysis done in [2], the usability of the system has been significantly improved. In this section, selected improvements introduced to the system are presented.

3.1   User Interface

Main Panel / View. Main part of application is the map (see Fig. 2). It has been identified that it should cover all space user have and resize immediately when needed. Elements such as menu and header should be small and possible to hide. A header should contain a small logo, a place for status messages and account management links (login, profile edit and logout), as well as help and about links. Menu should be placed on the left, with a hide and show button. This button, should be near menu - the user should easily guess what it is for.

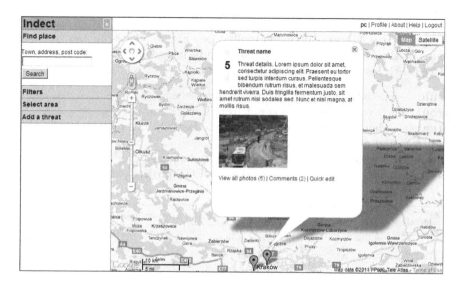

Fig. 2. Improved User Interface of the system

Interactive Map. In order to ensure interactive map, AJAX technology – Asynchronous JavaScript and XML – has been used. AJAX provides easy and quick reloading as few times as possible. Because the URL does not change when only partial reloading is done, there is a need to identify the state of the map (visible location or threat). This is done with hashing. Every time map position is changed or other action is performed, the hash changes. Moreover, the system uses appropriate JavaScript libraries to simplify catching a hash change event.

Another important thing, when it comes to interactive map, are the editing modes. In default mode, clicking on the map only shows details about a threat or gives an option to drag a position. When a user is in the editing mode (e.g. adding a threat), the map changes mode to edit: clicking on map creates a marker or move already created. Moreover, the map locates the user's position. If the user saved page to bookmarks with location, the map is set on this location. The map can be moved (with mouse or keyboard) and zoomed (with mouse scroll or scroll on the left on the map).

Viewing the Details. In the previous INDECT prototypes, viewing threat details required redirection to another page, where the user could find full description, gallery, tags etc. If a user clicked on threat icon on the map, only small speech balloon was shown, containing only name, short description and a link. In STM, four options (voting, one picture if available, number of comments and quick edit link) have been added. Voting is available only for registered users, but anonymous are able to see how many votes the threat got. Full gallery, list of tags and all comments are available on new page.

Enhanced Top Menu. Top menu toolbar (see Fig. 3) allows the user to toggle options of the login panel. If a user is already logged in, the login link is be replaced with his account name and five options:

Profile – where the user can change account settings.
Logout – where the user can log out from the system.
Threats – where the user can list all threats, last added threats, top reliability threats and most dangerous threats.
Language – which allows user to change site language.
About Indect – which is a link to the INDECT project page[1].

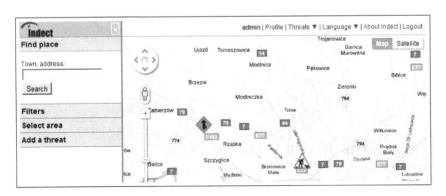

Fig. 3. Top menu for logged in user with visible account name and options

Interactive Left Menu. Left menu options are responsible for map interaction. Each section can be shown by clicking its name in menu (see Fig. 4). Menu can be toggled and resized.
Find place. This option allows user to find a place on the map. The system can look for town names, addresses and names of popular places. A user types the name in the field, click on the **Search** button and the system shows information when searching is done.

---

[1] http://www.indect-project.eu/

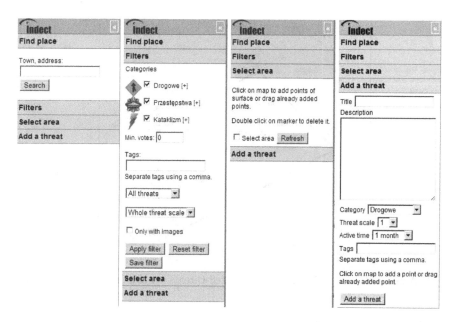

Fig. 4. Left menu: Find place, Filters, Select area and Add a threat

Filters. User can apply different filters on threats. First, filters should be selected, then the **Apply filter** button needs to be clicked. To remove filters, **Reset filter** should be clicked. Filters are defined by: (i) `Categories` – show only threats from selected categories, (ii) `Min. votes` – show threats with at least number of votes, (iii) `Tags` – show only threats for given tags, (iv) `Date added` – show only threats for selected date or period of time, (v) `Threat scale` – show only threats for selected severity of danger, and (vi) `Only with images` – show only threats with at least one image. There is also option **Save filter** which allows user to save selected filter to database. If the user is logged in and has at least one filter saved, select box with filter names is shown. Filter can be deleted with the **Delete filter** button.

Select area. With this feature user is able to select area on the map, where threats will be loaded. When Select area section is visible, every click on map creates a point which will be vertex of polygon (see Fig. 5). At least three points need to be created. Each vertex can be dragged (push button mouse down and move cursor) or deleted (double click). When polygon is ready, **Select area** checkbox need to be checked and the **Refresh** button clicked to apply area on map.

Add a threat. When a user is logged in, he or she is able to add a threat on the map. When this section is visible, every click on the map creates (or moves if is already created) a marker. It can be dragged and it indicates location of a new threat. After threat position is set, a form should be filled with threat details. Clicking on the **Add a threat** button will create a new threat.

Fig. 5. A map with selected area. Newly created threats are located inside the polygon.

3.2   Managing the Threats

Categories. Categories constitute a hierarchical way of describing threats. One threat has one category, but categories can contain many threats. Categories are organized into a tree structure. Common parent (root) of the categories is not visible on site. There is no children limit for categories.

As relational databases are not designed for storing hierarchical data, this brings many problems, especially performance issues. SQL is a procedural language, so retrieving category and its all parents is complicated. Also, fetching the whole tree requires additional operations after database query execution. To solve this problems in STM, Nested Set Model was implemented through `django-mptt` module. It provides very efficient way to retrieve categories from database. Modifying the tree is more complicated, thus slower, but it is only possible for admin user and it is rarely executed.

Tags. A tag is non-hierarchical keyword or term that describes an object, specifically a threat. One threat can be assigned by many tags. They help searching and categorizing threats. Threats should be tagged while adding by user. Well tagged threats will be more reliable for other users and should get more votes.

Groups. Groups allows users to publish threats for specified users. When adding a threat, the user (who is at least in Services group) can decide if the threat will be public or visible only for groups he or she chooses. This solution allows special groups have their own threats, that will never be published for all portal users. An examplary use case is given in the following section.

4   Use Case Example of the System

Using STM system, an ordinary user can follow the threats in his city/area. However, the system can be also used in more sophisticated situations, e.g. police investigations. Let us assume that there are three police centers, each in different city: Warszawa, Kraków, Gdańsk. Each center conducts its own investigation: Gdańsk: *Investigation 1*, Kraków: *Investigation 2*, and Warszawa: *Investigation 3*. The important aspects related to investigations, like important localizations, can be stored in STM. For the sake of preserving confidentiality of inquiries, the information should not be disclosed to ordinary users. Moreover, even not every police center must be aware of information stored by other centers. This use case presents steps that should be taken to achieve threat encapsulation in STM system and increase the safety of sensitive data.

4.1   Assigning Categories

The first step is to create an appropriate hierarchy of categories. The information related to particular investigation is stored in one category (or its subcategories). This facilitates information search. In this use case, one main category called *Investigations* is created. This category contains subcategories, where each subcategory is related to particular investigation. The result is depicted in Fig. 6.

Fig. 6. Hierarchy of categories

4.2   Defining Groups of Users

In the second step, the appropriate groups of users must be created. Each created group corresponds to one investigation. In this use case, three example groups called *Investigation1*, *Investigation2* and *Investigation3* were created. There is no obligatory coincidence between the names of the groups and categories.

4.3 Users Configuration

The users must be added to the system, either by the administrator in administration panel or by user registration form. In our use case, three user accounts were added: (1) *PolicemanGdansk* – a user from Gdańsk police center, (2) *PolicemanKrakow* – a user from Kraków police center, (3) *PolicemanWarszawa* – a user from Warszawa police center.

The administrator then has to configure the users' accounts, i.e. to assign the user account to group or groups corresponding to investigations in which user takes a part. In our use case: (1) The user *PolicemanGdansk* is assigned to group: *Investigation1*. (2) The user *PolicemanKrakow* is assigned to group: *Investigation2*. (3) The user *PolicemanWarszawa* is assigned to groups: *Investigation1*, *Investigation2* and *Investigation3*.

4.4 Threats Encapsulation

Threats encapsulation allows for hiding threats from the users which have no access to confidential information. The user who adds a threat to the system decides who can see it. The **Add threat** form provides field *Only for groups* which allows the user to assign threat to appropriate group or groups. Choosing one of the groups corresponding to investigation cause that the added threat is visible only for users that are involved in this investigation.

For safety reasons, not every user can add a threat for any (existing) group. The STM system provides several simple policies concerning groups, users and threats: (i) The registration automatically assigns user to the *Member* group. (ii) In order to add a threat for any group, a user has to belong to other group than *Member*. (iii) A user can add a single threat to more that one group. (iv) A user can add a threat only for the groups he or she belongs to. (v) A threat assigned to a group is invisible for all the users that do not belongs to this group. (vi) A threat that is not assigned to any particular group is visible for all users.

4.5 Threats Encapsulation Result

In the described use case, several threats have been defined in various locations. Four of these threats are located in the area of Warszawa: one from the Gdańsk police center, one from the Kraków one and two from Warszawa. These threats are represented on the map by labels with numbers respectively 1, 2, 3. In Figure 7, the views of the map for different users are depicted. Only the user *PolicemanWarszawa* can see all the threats related to investigations. The users *PolicemanKrakow* and *PolicemanGdansk* can see only the threats from one investigation. Additionally, ordinary user cannot see any of the threats related to investigations.

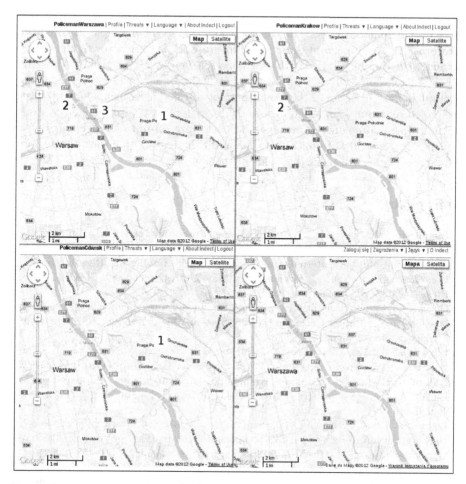

Fig. 7. Map view for various users (*PolicemanWarszawa, PolicemanKrakow, PolicemanGdansk* and a guest user)

5   Conclusions and Future Work

One of the tasks within the INDECT Project is a development of a Web-based system for knowledge acquisition and management. Selected existing solutions of applying intelligent techniques to threat information systems with GIS component have been investigated [6]. Once the main assumptions and requirements were defined, the development has been done iteratively. The prototyping process revealed some usability issues and helped to improve the user experience. Social features proved to be the engaging asset of the system, as important as the simple interface ensuring *instant gratification* through the visualization on the interactive map. The approach to semantics representation now used in the STM will be extended in the future. Currently, only basic semantics is represented with use of tags and categories. They are closer to the model of folksonomies, where users provide custom tags, that can be a foundation for a simple hierarchy of categories [1]. An important direction of future work is to refactor the hierarchy currently existing in STM with the used of a selected OWL 2.0 profile. All of the important relations will be identified and formalized. This will allow for having a complete formal model of the threat ontology. Rule-based technologies [7] will be applied to support on-line decisions of the system. It is also planned to work on the rule-based engine to manage and customize output channels, including reports. Moreover, it is important to maintain the testing to ensure the system will be useful and will meet the needs of the end-users.

References

1. Baumeister, J., Nalepa, G.J.: Verification of distributed knowledge in semantic knowledge wikis. In: Lane, H.C., Guesgen, H.W. (eds.) FLAIRS-22: Proceedings of the Twenty-Second International Florida Artificial Intelligence Research Society Conference, Sanibel Island, Florida, USA, May 19-21, pp. 384–389. FLAIRS, AAAI Press, Menlo Park, California (2009)
2. Ciężkowski, P.: Functionality Analysis and Design and Implementation of User Interface for Threats Enregistration in Internet System. Master's thesis, AGH University of Science and Technology (2011)
3. Ligęza, A., Adrian, W.T., Ernst, S., Nalepa, G.J., Szpyrka, M., Czapko, M., Grzesiak, P., Krzych, M.: Prototypes of a Web System for Citizen Provided Information, Automatic Knowledge Extraction, Knowledge Management and GIS Integration. In: Dziech, A., Czyżewski, A. (eds.) MCSS 2011. CCIS, vol. 149, pp. 268–276. Springer, Heidelberg (2011)
4. Ligęza, A., Ernst, S., Nalepa, G.J., Szpyrka, M.: A conceptual model for web knowledge acquisition system with GIS component. Automatyka: półrocznik Akademii Górniczo-Hutniczej im. Stanisława Staszica w Krakowie 13(2), 421–428 (2009)
5. Ligęza, A., Ernst, S., Nowaczyk, S., Nalepa, G.J., Furmańska, W.T., Czapko, M., Grzesiak, P., Kałuża, M., Krzych, M.: Towards enregistration of threats in urban environments: practical consideration for a GIS-enabled web knowledge acquisition system. In: Dańda, J., Jan Derkacz, A.G. (eds.) MCSS 2010: Multimedia Communications, Services and Security: IEEE International Conference, Kraków, May 6-7, pp. 152–158 (2010)
6. Nalepa, G.J., Furmańska, W.T.: Review of semantic web technologies for GIS. Automatyka: półrocznik Akademii Górniczo-Hutniczej im. Stanisława Staszica w Krakowie 13(2), 485–492 (2009)
7. Nalepa, G.J., Ligęza, A.: HeKatE methodology, hybrid engineering of intelligent systems. International Journal of Applied Mathematics and Computer Science 20(1), 35–53 (2010)
8. Waliszko, J., Adrian, W.T., Ligęza, A.: Traffic Danger Ontology for Citizen Safety Web System. In: Dziech, A., Czyżewski, A. (eds.) MCSS 2011. CCIS, vol. 149, pp. 165–173. Springer, Heidelberg (2011)

Unsupervised Feature Selection for Spherical Data Modeling: Application to Image-Based Spam Filtering

Ola Amayri and Nizar Bouguila

Faculty of Engineering and Computer Science, Concordia University,
Montreal, Qc, Canada, H3G 2W1
o_amayri@ece.concordia.ca, bouguila@ciise.concordia.ca

Abstract. Understanding the relevance of extracted features in domain-specific sense is a matter at the heart of image classification. In this paper, we propose a feature selection framework that allows more compactness of the statistical model while holding good generalization to unseen data. Both feature selection and clustering are based on well-established statistical models that provide natural choice when the data to model are spherical. Moreover, we develop a probabilistic kernel based on Fisher score and mixture of von Mises model (moVM) to feed Support Vector Machines (SVM). The selection process evaluates the relevance of features through a principled feature saliency approach. The unsupervised learning is approached using Expectation Maximization (EM) for parameter estimation along with Minimum Message Length (MML) to determine the optimal number of mixture components. We argue that the proposed framework is well-justified and can be adjusted to different problems. Experimental results involving the challenging problem of image-based spam filtering show the merits of the proposed approach.

Keywords: Von Mises mixture, feature selection, minimum message length, Support Vector Machines, Fisher score, image-based spam.

1 Introduction

The manifestation of digital images has simultaneously offered data analysts broad features of diverse characteristics to explore and analyze. Classification of these features poses economic and utility challenges, including a large number of feature vectors per dataset and extremely high-dimensional vectors. Over the last decades, researchers have paid great deal of attention to develop appropriate image features and accurate models that are robust and invariant under certain transformations. In particular, several local descriptors, such as scale-invariant feature transform (SIFT) [6], have been introduced. These local descriptors are associated to detected keypoints that are stable over different possible ranges of image invariance and orientation [15,18]. Accordingly, image classification based on local descriptors is characterized by vast number of feature vectors which are relatively sparse and high-dimensional. The features in these vectors do not

A. Dziech and A. Czyżewski (Eds.): MCSS 2012, CCIS 287, pp. 13–23, 2012.

contribute equally to classification. Thus, selecting the most relevant features to reduce the dimensionality of feature vectors space is an important step in several applications and generally leads to better modeling and generalization capabilities [16,22,14,17,26]. Nonetheless selecting an appropriate model that solves all aspects of application at hand is a major challenge as distinct approaches needed for particular aspects of an application often depend on different representational choices. For instance, although classification based on Gaussian mixtures (GMM) has provided good performance in some applications, recent works have shown that GMM is sensitive to noise and irresistible to outliers when dealing with high-dimensional data[1] [12]. In [12], authors showed that a good alternative is to use spherical distributions such as moVM, which is based on cosine distance. Numerous researchers employed feature selection for image classification but to the best of our knowledge none of them has considered the case where these feature vectors are spherical so far.

Our proposed work builds on two insights. First, the general trends in state of the art methods for feature selection approaches are either too general to describe the data or incapable to scale for high dimensional representation of the data at hand. By contrast, these restrictions have been alleviated in [4,5] where authors proved that using data-oriented models yields to surprisingly improved performance in diverse applications. In this paper, we propose a novel approach which infer the spherical nature (i.e. L_2-normalized) of visual feature vectors via a spherical distribution represented by moVM to categorize images. In addition, as hybrid generative discriminative frameworks have shown improved performance as compared to their generative or discriminative counterparts, we propose the combination of moVM and SVM upon integrating feature selection. We are mainly motivated by the fact that hybrid models are suitable choice for high-dimensional classification problems especially image categorization which is highly dependent on the nature of examples distribution in hand. Indeed, a given image is better represented by not only one vector but by a bag of vectors which can be modeled themselves by a mixture of distributions. Thus, we develop a hybrid framework that models image descriptors, in an unsupervised way, using moVM from which Fisher kernel is generated for SVMs. The unsupervised learning of movM is approached using an EM algorithm for parameter estimation along with MML criterion to determine the appropriate number of mixture components. The second insight is: in spite of the substantial results achieved using feature selection approach this usually occurs in cost of computational complexity. In this paper, however, the computational cost is based on the EM estimation framework cost where both E- and M-steps have a complexity of $O(NMD)$ which is the same complexity associated with standard EM-based learning approaches. Despite this modest cost, the proposed approach is experimentally shown to be both tractable and effective. Our empirical experiments involve the challenging problem of image-based spam filtering.

---

[1] As it basically relies on the quadratic distance (usually Euclidian or Mahalanobis) between features and their mean to convey the similarity.

The rest of this paper is organized as follows. In Section 2 we present our unsupervised feature selection model. A detailed approach to learn the proposed model is proposed in Section 3 where we develop a Fisher kernel, also. Experimental results are presented in Section 4. Finally, we conclude the paper in Section 5.

2 Proposed Model

In the following, we start by illustrating the procedure of constructing feature vectors based on local descriptors for a given image. Next we present our model for simultaneous clustering and feature selection. An approach to estimate the different parameters of this model is presented in the next section where we develop also a Fisher kernel to tackle the problem of spherical data sequences classification using SVM. See Figure 1 for a summary of the proposed framework.

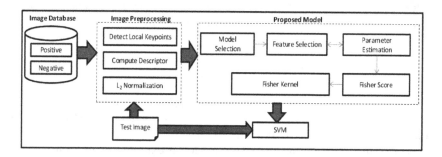

Fig. 1. Proposed classification framework

2.1 Image Representation

Let $\mathcal{I} = \{(I_n, C_n)|n = 1, \ldots, N, \forall C_n \in \{1, \ldots, M\}\}$ be a dataset composed of N images, where M is the number of model classes. For each training image I_n, we start by extracting patches around detected interest points. Then, we compute a local descriptor on each detected key point. Thus, each image I_n is described by a set of D-dimensional feature vectors representing the local descriptors: $\mathcal{X}^n = \{\boldsymbol{X}_i^n = (X_{i1}^n, \ldots, X_{iD}^n)\}$. Thereby, we propose in our framework to normalize these vectors using L_2 normalization[2] such that $(\boldsymbol{X}_i^n)^T \boldsymbol{X}_i^n = 1$ which has shown to increase the robustness to various changes, such as illumination changes in images [24]. If we consider L_2-normalized vectors, then each image can be modeled accurately using movM distributions as we shall see in the following.

---

[2] It is noteworthy that once the feature vectors are L_2 normalized they can be visualized as points on a hypersphere, which can be naturally modeled using spherical distributions.

2.2 Mixture Model with Feature Saliency

vM distribution[3] [10] is a probability distribution dedicated to represent data concentrated on the circumference of a unit circle. In our case, we shall suppose that the elements in each X_i^n are independent and that each element follows a vM distribution which gives the following:

$$p(X_i^n|\mu, \kappa) = \prod_{d=1}^{D} \frac{1}{2\pi I_0(\kappa_d)} \exp\{\kappa_d \mu_d^T Y_{id}^n\} \tag{1}$$

where I_0 is the modified Bessel function of the first kind and order zero [20], $\mu = (\mu_1, \ldots, \mu_D)$, $\mu_d = (\mu_{d1}, \mu_{d2})$ is the mean direction, $\kappa = (\kappa_1, \ldots, \kappa_D)$, κ_d is the concentration parameter and $Y_{id}^n = (Y_{id1}^n, Y_{id2}^n)$ such that $Y_{id1}^n = X_{id}^n$ and $(Y_{id}^n)^T Y_{id}^n = 1$. Generally, a set of vectors comprised of examples that vary in their characteristics and represent dissimilar information and hence belong to many clusters and can be modeled by a finite mixture of distributions. Thus, let $p(X_i^n|\Theta_M)$ be a mixture of M distributions represented by Eq. 1. The probability density function of a M-components movM is given by

$$p(X_i^n|\Theta_M) = \sum_{j=1}^{M} p(X_i^n|\theta_j) p_j \tag{2}$$

where $\Theta_M = \{P = (p_1, \ldots, p_M), \theta = (\theta_1, \ldots, \theta_M)\}$ denotes all the parameters of the mixture model such as $\theta_j = (\mu_j, \kappa_j)$ and P represents the vector of mixing parameters which are positive and sum to one.

The clustering based on finite mixture models is explored by grouping similar documents, where this similarity depends basically on the features that represent each document. Indeed, It has proven over the years the fallacy assumption that the more features representing the document the better discrimination capability the classifier has [16,22,14,17,26]. This can be due to the presence of noisy and non-informative (i.e. irrelevant) features that generally highly drop the performance. To this aim, we adopt the approach proposed in [16], in the case of the Gaussian mixture, that assigns smaller weights to irrelevant feature by defining feature saliency as the probability that a feature d is relevant $\rho_{jd} = 1$, following the assumption that a given feature is irrelevant if it follows a common density across clusters while maintaining the independency of class label [16,5,4]. Thus, our model, to take feature selection into account, can be written as:

$$p(X_i^n|\Theta) = \sum_{j=1}^{M} p_j \prod_{d=1}^{D} (\rho_{jd} p(Y_{id}^n|\theta_{jd}) + (1 - \rho_{jd}) p(Y_{id}^n|\lambda_{jd})) \tag{3}$$

where $\Theta = \{\Theta_M, \{\rho_{jd}\}, \{\lambda_{jd}\}\}$, $\lambda_{jd} = (\mu_{jd|\lambda}, \kappa_{jd|\lambda})$ is the vM from which the irrelevant feature is drawn, and ρ_{jd} denotes the weight of the dth feature on cluster j.

---

[3] Also known as the circular normal distribution [10] and maximum entropy distribution in [7].

3   Model Learning

The model selection problem can be viewed as one that helps finding the optimal trade-off between the complexity of the model and goodness of fit. Over the years, many approaches have been proposed [2,1,27,25]. In this paper, we propose the consideration of MML criterion to find the optimal number of mixture components [2]:

$$MessLen(M) \simeq -\log h(\Theta) + \frac{1}{2}\log|F(\Theta)| + \frac{N_p}{2}(1 + \log\frac{1}{12}) - \log p(\mathcal{X}^n|\Theta) \quad (4)$$

where $h(\Theta)$ is the prior probability, $p(\mathcal{X}^n|\Theta)$ is the likelihood, $F(\Theta)$ is the expected Fisher information matrix which is generally approximated by complete-data Fisher information matrix in the case of finite mixture models, $|F(\Theta)|$ is its determinant, and $N_p = M(1 + 5D) - 1$ is the number of free parameters to be estimated in case of moVM.

In order to define $MessLen$ for moVM, in what follow, we assume the independence of the different groups of parameters, which facilitate the factorization of both prior $h(\Theta)$ (which can be modeled by Jeffrey's prior for each group of parameters [5]) and Fisher information matrix $F(\Theta)$. Thus,

$$MessLen(M) = \frac{1}{2}(M + 5MD)\log N + D\sum_{j=1}^{M}\log p_j + \sum_{j=1}^{M}\sum_{d=1}^{D}\log\rho_{jd} \quad (5)$$

$$+ \sum_{j=1}^{M}\sum_{d=1}^{D}\log(1 - \rho_{jd}) + \frac{1}{2}\sum_{j=1}^{M}\sum_{d=1}^{D}\log(|F_1(\theta_{jd})|) + \frac{1}{2}\sum_{j=1}^{M}\sum_{d=1}^{D}\log(|F_1(\lambda_{jd})|)$$

$$+ \frac{N_p}{2}(1 + \log\frac{1}{12}) - \log p(\mathcal{X}^n|\Theta) - \sum_{d=1}^{D}\sum_{j=1}^{M}[\log(1 + \kappa_{jd}^2) + \log(1 + \kappa_{jd|\lambda}^2)]$$

where $F_1(\theta_{jd}) = N_{jd}^2\kappa_j\frac{I_1(\kappa_{jd})}{I_0(\kappa_{jd})}[1 - (\frac{I_1(\kappa_{jd})}{I_0(\kappa_{jd})})^2 - \frac{I_1(\kappa_{jd})}{I_0(\kappa_{jd})}]$ is the Fisher matrix for one-dimension [7] such that N_{jd} is the number of one-dimensional observations of the dth relevant feature assigned to cluster j. Now, we develop the equations that allow to learn our mixture model while considering simultaneously the relevancy of features. To achieve this goal, we adopt common EM [28] approach which generates a sequence of models with non-decreasing log-likelihood on the data. Thus, the new objective function is given by:

$$S(\Theta, \mathcal{X}^n) = -MessLen(M) + \xi\left(1 - \sum_{j=1}^{M}p_j\right) + \sum_{j=1}^{M}\sum_{d=1}^{D}\nu_{jd}\left(1 - \rho_{jd}\right) \quad (6)$$

where ξ and ν_{jd} are Lagrange multipliers to satisfy the constraints $\sum_{j=1}^{M}p_j = 1$ and $\sum_{j=1}^{M}\sum_{d=1}^{D}\rho_{jd} = 1$ for $j = 1,\ldots,M$, respectively. Thus, straightforward manipulations allow us to obtain the following by maximizing Eq. 6:

$$p(j|\mathbf{X}_i^n) = \frac{p_j\prod_{d=1}^{D}\left(\rho_{jd}p(\mathbf{Y_{id}}^n|\theta_{jd}) + (1 - \rho_{jd})p(\mathbf{Y_{id}}^n|\lambda_{jd})\right)}{\sum_{j=1}^{M}p_j\prod_{d=1}^{D}\left(\rho_{jd}p(\mathbf{Y_{id}}^n|\theta_{jd}) + (1 - \rho_{jd})p(\mathbf{Y_{id}}^n|\lambda_{jd})\right)} \quad (7)$$

$$p_j = \frac{\sum_{i=1}^{N} p(j|\boldsymbol{X}_i^n) - 1}{N - 1} \tag{8}$$

$$\rho_{jd} = \frac{\sum_{i=1}^{N} p(j|\boldsymbol{X}_i^n) \frac{\rho_{jd} p(\boldsymbol{Y_{id}}^n|\theta_{jd})}{\rho_{jd} p(\boldsymbol{Y_{id}}^n|\theta_{jd}) + (1-\rho_{jd}) p(\boldsymbol{Y_{id}}^n|\lambda_{jd})}}{\sum_{i=1}^{N} p(j|\boldsymbol{X}_i^n)} \tag{9}$$

$$\boldsymbol{\mu}_{jd} = \frac{\sum_{i=1}^{N} p(j|\boldsymbol{X}_i^n) \frac{\rho_{jd} p(\boldsymbol{Y_{id}}^n|\theta_{jd})}{\rho_{jd} p(\boldsymbol{Y_{id}}^n|\theta_{jd}) + (1-\rho_{jd}) p(\boldsymbol{Y_{id}}^n|\lambda_{jd})} \boldsymbol{Y}_{id}^n}{\sum_{i=1}^{N} p(j|\boldsymbol{X}_i^n) \frac{\rho_{jd} p(\boldsymbol{Y_{id}}^n|\theta_{jd})}{\rho_{jd} p(\boldsymbol{Y_{id}}^n|\theta_{jd}) + (1-\rho_{jd}) p(\boldsymbol{Y_{id}}^n|\lambda_{jd})}} \tag{10}$$

$$\boldsymbol{\mu}_{jd|\lambda} = \frac{\sum_{i=1}^{N} p(j|\boldsymbol{X}_i^n) \frac{(1-\rho_{jd}) p(\boldsymbol{Y_{id}}^n|\lambda_{jd})}{\rho_{jd} p(\boldsymbol{Y_{id}}^n|\theta_{jd}) + (1-\rho_{jd}) p(\boldsymbol{Y_{id}}^n|\lambda_{jd})} \boldsymbol{Y}_{id}^n}{\sum_{i=1}^{N} p(j|\boldsymbol{X}_i^n) \frac{(1-\rho_{jd}) p(\boldsymbol{Y_{id}}^n|\lambda_{jd})}{\rho_{jd} p(\boldsymbol{Y_{id}}^n|\theta_{jd}) + (1-\rho_{jd}) p(\boldsymbol{Y_{id}}^n|\lambda_{jd})}} \tag{11}$$

and

$$\boldsymbol{\mu}_{jd} = \frac{\boldsymbol{\mu}_{jd}}{\|\boldsymbol{\mu}_{jd}\|}, \qquad \boldsymbol{\mu}_{jd|\lambda} = \frac{\boldsymbol{\mu}_{jd|\lambda}}{\|\boldsymbol{\mu}_{jd|\lambda}\|} \tag{12}$$

$$A(\boldsymbol{\kappa}_{jd}) = \frac{-\boldsymbol{\mu}_{jd}^T \sum_{i=1}^{N} p(j|\boldsymbol{X}_i^n) \frac{\rho_{jd} p(\boldsymbol{Y_{id}}^n|\theta_{jd})}{\rho_{jd} p(\boldsymbol{Y_{id}}^n|\theta_{jd}) + (1-\rho_{jd}) p(\boldsymbol{Y_{id}}^n|\lambda_{jd})} \boldsymbol{Y}_{id}^n}{\sum_{i=1}^{N} p(j|\boldsymbol{X}_i^n) \frac{\rho_{jd} p(\boldsymbol{Y_{id}}^n|\theta_{jd})}{\rho_{jd} p(\boldsymbol{Y_{id}}^n|\theta_{jd}) + (1-\rho_{jd}) p(\boldsymbol{Y_{id}}^n|\lambda_{jd})}} \tag{13}$$

Since, we can not find tractable form for $A^{-1}(\kappa)$, we use Newton-Raphson iterations to find κ, where:

$$\kappa_{jd}^{new} = \kappa_{jd}^{old} - \frac{\partial \log p(\mathcal{X}^n|\Theta)}{\partial \kappa_{jd}} \left(\frac{\partial^2 \log p(\mathcal{X}^n|\Theta)}{\partial^2 \kappa_{jd}} \right)^{-1} \tag{14}$$

Similarly we can calculate $\kappa_{jd|\lambda}^{new}$.

3.1   Fisher Kernel

Authors in [13] have shown that a generative model can be used in a discriminative context by extracting Fisher scores from the generative model and converting them into a Gram Kernel usable by SVMs. Each component of Fisher score is the derivative of the log-likelihood of the sequence $\mathcal{X}^n$ with respect to particular parameter. In the following, we shall show the derivations of the Fisher kernel for our model. Through the computation of gradient of the log likelihood with respect to our model parameters: $p_j, \boldsymbol{\kappa}_{jd}, \boldsymbol{\mu}_{jd}, \rho_{jd}, \boldsymbol{\mu}_{jd|\lambda}$ and $\boldsymbol{\kappa}_{jd|\lambda}$ where $j = 1, \ldots, M$, we obtain

$$\frac{\partial \log p(\mathcal{X}^n|\Theta)}{\partial p_j} = \sum_{i=1}^{N} \left[\frac{p(j|\boldsymbol{X}_i^n)}{p_j} - \frac{p(j=1|\boldsymbol{X}_i^n)}{p_1} \right], \quad 1 < j \le M \tag{15}$$

$$\frac{\partial \log p(\mathcal{X}^n|\Theta)}{\partial \mu_{jd}} = \frac{\sum_{i=1}^{N} \boldsymbol{Y}_{id}^n \rho_{jd} p(j|\boldsymbol{X}_i^n)}{\| \sum_{i=1}^{N} \boldsymbol{Y}_{id}^n \rho_{jd} p(j|\boldsymbol{X}_i^n) \|} \tag{16}$$

$$\frac{\partial \log p(\mathcal{X}^n | \Theta)}{\partial \mu_{jd|\lambda}} = \frac{\sum_{i=1}^{N} Y_{id}^n (1 - \rho_{jd}) p(j | X_i^n)}{\| \sum_{i=1}^{N} Y_{id}^n (1 - \rho_{jd}) p(j | X_i^n) \|} \tag{17}$$

$$\frac{\partial \log p(\mathcal{X}^n | \Theta)}{\partial \kappa_{jd}} = \sum_{i=1}^{N} p(j | X_i^n) \rho_{jd} \Big[\boldsymbol{\mu}_{jd}^T \boldsymbol{Y}_{id}^n - NA(\kappa_{jd}) \Big] \tag{18}$$

$$\frac{\partial \log p(\mathcal{X}^n | \Theta)}{\partial \kappa_{jd|\lambda}} = \sum_{i=1}^{N} p(j | X_i^n) (1 - \rho_{jd}) \Big[\boldsymbol{\mu}_{jd|\lambda}^T \boldsymbol{Y}_{id}^n - NA(\kappa_{jd|\lambda}) \Big] \tag{19}$$

$$\frac{\partial \log p(\mathcal{X}^n | \Theta)}{\partial \rho_{jd}} = \sum_{i=1}^{N} p(j | X_i^n) \Big[\frac{p(\boldsymbol{Y_{id}}^n | \theta_{jd}) - p(\boldsymbol{Y_{id}}^n | \lambda_{jd})}{\rho_{jd} p(\boldsymbol{Y_{id}}^n | \theta_{jd}) + (1 - \rho_{jd}) p(\boldsymbol{Y_{id}}^n | \lambda_{jd})} \Big] \tag{20}$$

where $p(j | X_i^n)$ is the posterior we found previously in E-step. It is noteworthy that in Eq. 15, we take into account the fact that the sum of the mixing parameters equals one and thus there are only $M - 1$ free mixing parameters. See Algorithm 1 for complete proposed learning.

Algorithm 1. Initialization and Complete Estimation Algorithm

---

1: Apply spherical K-means [12] to obtain the initial parameters for each component.
2: Iterate the two following steps until convergence:

    1. **E-Step:** Update $p(j | X_i^n)$ using Eq. 7
    2. **M-Step:** Update p_j, ρ_{jd}, $\boldsymbol{\mu}_{jd}$, $\boldsymbol{\mu}_{jd|\lambda}$, κ_{jd} and $\kappa_{jd|\lambda}$ using Eqs. 8,9, 12, 14, respectively.

3: Calculate the associated message length $MessLength(M)$ using Eq 5.
4: Select the optimal model M^* such that $M^* = \arg \min MessLength(M)$.
5: Calculate SVM kernel matrix using Fisher Kernel based on moVM parameters given in 15, 16, 17, 18, 19 and 20, respectively.

---

4 Experimental Results: Image-Based Spam Filtering

In this section, we evaluate the effectiveness of the proposed learning framework on high-dimensional data involving the challenging image-based spam filtering problem. The main goal of our experiments is to investigate the advantages of performing feature selection in hybrid generative discriminative framework by comparing: hybrid learning of moVM with (HVMFS) and without feature selection (HVM), clustering based on moVM (generative models only) with (VMFS) and without feature selection (VM). We compare our results with another state of the art method which is GMM where previous scenarios were applied also, namely, HGMMFS, HGMM, GMMFS, and GMM. Details about the learning of GMM can be found in [16]. The libsvm[4] software was used for SVMs classifier.

---

[4] http://www.csie.ntu.edu.tw/~cjlin/libsvm/

Image-based spam email circumvents easily classic text based spam filters, thus some approaches have been proposed to detect the nature of email from its image content. Most of proposed approaches consider, however, only the textual content of the image and ignore its rich low-level visual content (e.g. color, texture, shape) which can be very helpful as clearly shown for instance in previous works about content-based image indexing and retrieval [21]. Moreover, spam images may not contain text (e.g. the picture of an object without text to advertise a website). Only few papers have considered the low level visual content of spam images as a solution to make filters more robust and smarter [8,19]. Motivated by the recent success of local descriptors in computer vision applications, the authors in [11] proposed the modeling of images using the so called visual keywords (i.e. quantization of local descriptors) which are then classified as spam or ham using SVM. This approach has some merits since the local descriptor used (i.e. SIFT) is robust to several geometric transformations that may be used by spammers, but the quantization step applied can cause the loss of important information about the image content as shown in [23].

For our experimental evaluation, we use Princeton dataset[5] which consists of 1071 emails that spread into 178 categories and Dredze dataset [8] which contains emails from publicly well-known SpamArchive datasets along with many personal emails of the authors. An important step in our application is the extraction of local features which well-describe given images. To this seek, we evaluate two main approaches in what follows. The first one is based on detecting local regions on each image, using difference-of-Gaussian (DoG) detector, which we describe using their SIFT descriptors [6], giving 128 dimensional vector for each local region. Then, we normalize our vectors using L_2-normalization. Note that using this approach each image is represented by a bag of vectors which can be modeled using a mixture of distributions. The second one is the Bag-of-visual-words (BoVW) approach, thereby each image is represented by a single vector of frequencies instead. To this aim, after we compute the local descriptors using approach one, extracted vectors are clustered using the K-Means algorithm providing 900 visual-words vocabulary. Each image in the datasets is then represented by a 900-dimensional vector describing the frequencies of a set of visual words, provided from the constructed visual vocabulary. Having these feature vectors, we have reduced their dimensionality via our proposed feature selection framework. After we prepared our dataset we randomly split dataset 10 times into training and testing sets, then we start spam filtering. Figure 2 shows the number of components found by our proposed algorithm and when adopting moVM and GMM using different scenario. Evaluation results using both set of local descriptor and BoVW approaches after considering different scenarios are shown on table 1. According to the results, it is clear that the presence of irrelevant features affects both the classification accuracy and estimation of model components. Note that the best results were obtained when hybrid framework was applied using feature selection. Moreover, using moVM shows improvement over GMM in all the scenarios. It is noteworthy that the proposed approach,

---

[5] http://www.princeton.edu/cass/spam/spam_bench/

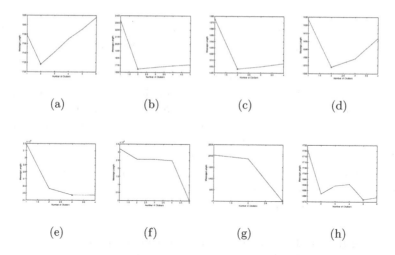

(a)   (b)   (c)   (d)

(e)   (f)   (g)   (h)

Fig. 2. Number of clusters determined to represent spam and legitimate emails using 2(a)HVMFS. 2(b)HVM. 2(c)HGMMFS. 2(d)HGMM. 2(e)VMFS. 2(f)VM. 2(g)GMMFS. 2(h)GMM.

Table 1. Classification Accuracy in (%) For Spam Dataset using different scenarios and image representation choices

	bag of vectors				BoVW			
	HVMFS	HVM	HGMMFS	HGMM	VMFS	VM	GMMFS	GMM
Accuracy	89.01	86.45	85.31	84.78	77.89	77.52	77.95	70.45

unlike many feature selection in literature, does not need labels. However, the independency of features assumption in proposed model yields to certain restrictions because it defines the relevance of particular feature with respect to different mixture components. Therein, the redundancy between features is not considered. Thus, there is no guarantee that the strongest features subset represents the best feature subset. To resolve such issue, we can consider Mutual Information (MI) between features in a given model. Indeed, we find the approximation of class conditional distribution between particular feature vector and given classes with respect to its prior probability. Thus, we can find Shannon entropy MI while another forums of MI entropy is possible like Renyi entropy (which is a general forum of Shannon entropy) and α-degree entropy which have been discussed in [3,9].

5   Conclusion

The identification of informative features is of great interest and fundamental in different data mining and computer vision applications. This paper serves as basis for designing a classifier, based on local features, that are likely to produce improved image classification performance. In particular, our approach is based on the representation of images by bags of vectors for which the moVM has been shown to be an effective model. In particular, we have demonstrated that benefit may be achieved by using the moVM to generate SVM kernels. Experimental results have proved the adequacy of the proposed framework by achieving improved performance in most cases while reducing the number of features. The potential behind these improvements exists because many of nowadays classification often requires incorporation of specific domain knowledge when performing general image classification.

Acknowledgments. The completion of this research was made possible thanks to the Natural Sciences and Engineering Research Council of Canada (NSERC).

References

1. Akaike, H.: A new look at the statistical model identification. IEEE Transaction on Automatic Control 19(6), 716–723 (1974)
2. Baxter, R., Oliver, J.: Finding Overlapping Components with MML. Statistics and Computing 10(1), 5–16 (2000)
3. Ben-Bassat, M.: Use of Distance Measures, Information Measures and Error Bounds in Feature Evaluation, ch. 35, pp. 773–791. Elsevier Science Pub Co.
4. Bouguila, N.: A model-based approach for discrete data clustering and feature weighting using map and stochastic complexity. IEEE Transactions on Knowledge and Data Engineering 21(12), 1649–1664 (2009)
5. Boutemedjet, S., Bouguila, N., Ziou, D.: A hybrid feature extraction selection approach for high-dimensional non-gaussian data clustering. IEEE Transactions on Pattern Analysis and Machine Intelligence 31(8), 1429–1443 (2009)
6. Lowe, D.G.: Distinctive Image Features from Scale-Invariant Keypoints. International Journal of Computer Vision 60(2), 91–110 (2004)
7. Dowe, D.L., Oliver, J.J., Baxter, R.A., Wallace, C.S.: Bayesian estimation of the von mises concentration parameter. In: Proceedings of the Fifteenth International Workshop on Maximum Entropy and Bayesian Methods, pp. 51–59. Kluwer Academic (1995)
8. Dredze, M., Gevaryahu, R., Elias-Bachrach, A.: Learning fast classifiers for image spam. In: CEAS (2007)
9. Erdogmus, D.: Information theoretic learning: Renyi's entropy and its applications to adaptive. Ph.D. thesis, University of Florida (2002)
10. Fisher, N.I.: Statistical analysis of circular data, 1st edn. Cambridge University Press, Cambridge (1993)
11. Hsia, J.H., Chen, M.S.: Language-model-based detection cascade for efficient classification of image-based spam e-mail. In: Proceedings of the IEEE International Conference on Multimedia and Expo, ICME 2009, pp. 1182–1185. IEEE Press, Piscataway (2009)

12. Dhillon, I.S., Modha, D.S.: Concept Decompositions for Large Sparse Text Data Using Clustering. Machine Learning 42(1-2), 143–175 (2001)
13. Jaakkola, T.S., Haussler, D.: Exploiting generative models in discriminative classifiers. In: Proceedings of Advances in Neural Information Systems (NIPS), pp. 487–493. MIT Press (1998)
14. Jain, A., Zongker, D.: Feature selection: evaluation, application, and small sample performance. IEEE Transactions on Pattern Analysis and Machine Intelligence 19(2), 153–158 (1997)
15. Koenderink, J.J.: The structure of images. Biological Cybernetics 50(5), 363–370 (1984), http://dx.doi.org/10.1007/BF00336961
16. Law, M.H.C., Figueiredo, M.A.T., Jain, A.K.: Simultaneous feature selection and clustering using mixture models. IEEE Trans. Pattern Anal. Mach. Intell 26, 1154–1166 (2004)
17. Lemos, A., Caminhas, W., Gomide, F.: Evolving fuzzy linear regression trees with feature selection. In: 2011 IEEE Workshop on Evolving and Adaptive Intelligent Systems (EAIS), pp. 31–38 (April 2011)
18. Lindeberg, T.: Scale-space theory: A basic tool for analysing structures at different scales. Journal of Applied Statistics, 224–270 (1994)
19. Liu, Q., Qin, Z., Cheng, H., Wan, M.: Efficient modeling of spam images. In: Proceedings of the Third International Symposium on Intelligent Information Technology and Security Informatics. IEEE Computer Society, Washington, DC (2010)
20. Mardia, K.V.: Statistics of directional data. Academic Press (1972)
21. Mehta, B., Nangia, S., Gupta, M., Nejdl, W.: Detecting image spam using visual features and near duplicate detection. In: Proceedings of the 17th International Conference on World Wide Web, WWW 2008, pp. 497–506 (2008)
22. Mitra, P., Murthy, C., Pal, S.: Unsupervised feature selection using feature similarity. IEEE Transactions on Pattern Analysis and Machine Intelligence 24(3), 301–312 (2002)
23. Perronnin, F., Dance, C.R.: Fisher kernels on visual vocabularies for image categorization. In: CVPR. IEEE Computer Society (2007)
24. Herbrich, R., Graepel, T.: A PAC-Bayesian Margin Bound for Linear Classifiers: Why SVMs Work. In: Proceedings of Advances in Neural Information Processing Systems (NIPS), pp. 224–230 (2000)
25. Rissanen, J.: Modeling by shortest data discription. Automatica 14, 465–471 (1987)
26. Saeys, Y., Inza, I.N., Larrañaga, P.: A review of feature selection techniques in bioinformatics. Bioinformatics 23(19), 2507–2517 (2007)
27. Schwarz, G.: Estimating dimension of a model. Annals of Statistics 6, 461–464 (1978)
28. Titterington, D., Smith, A., Makov, U.: Statistical Analysis of Finite Mixture Distributions. John Wiley & Sons, Chichester (1985)

Classification of Malware Network Activity

Gilles Berger-Sabbatel and Andrzej Duda

Grenoble Institute of Technology, CNRS Grenoble Informatics Laboratory UMR 5217
681, rue de la Passerelle, BP 72
38402 Saint Martin d'Hères Cedex, France
{Gilles.Berger-Sabbatel,Andrzej.Duda}@imag.fr

Abstract. In the previous work, we have designed and implemented a
platform with tools for capturing malware, running botnets in a
controlled environment, analyzing their interactions with a botmaster,
testing methods and techniques for mitigating botnet nuisance, and eventually disrupting them. We have used the platform to gather a large
number of malware and observe its network activity.

In this paper, we present an approach to malware classification based
on the observation of the malware communication behavior. First, we
show that traditional methods based on antivirus tools are not suitable
for classification. Then, we define the method based on observing the
communication pattern of executing malware. We report on the classification results obtained with the proposed method. Unlike classification
done by existing antivirus tools, the proposed method results in selective
and consistent classification.

1 Introduction

In the previous work, we have developed a platform with tools for capturing
malware (viruses, worms), running botnets in a controlled environment, and analyzing their interactions with a botmaster. A botnet is a network of zombie
computers compromised by some malware. Botnets are coordinated by a botmaster through a Command and Control channel (C&C) to which the malware
connects to get instructions. A botmaster can use botnets to perform malicious
activities. The overall goal of our activities is to test methods and techniques for
mitigating botnet nuisance and eventually disrupting them.

We have recently used the platform to gather a large number of malwares and
observe their network activity. A honeypot captures some malicious code that
we can run on a designated machine to analyze its interaction with the C&C
center for possible communication with a botmaster or detect the attemps of
malware proliferation as well as other malicious activities such as port scans or
Denial of Service attacks (DoS).

We have encountered the problem of *malware classification*: we capture a large
number of malware to analyze, which is a time consuming process run under the
control of an operator, so the analysis should only be done on the most interesting
malware and needs to avoid different instances of the same code. So, before the

A. Dziech and A. Czyżewski (Eds.): MCSS 2012, CCIS 287, pp. 24–35, 2012.

analysis phase, we need to classify malware and pass to the analysis phase only for well chosen, specific malware.

In this paper, we present an approach for malware classification based on the observation of the communication behavior. First, we show that traditional methods based on antivirus tools are not suitable for classification. Then, we define the method based on observing the communication pattern of executing malware. We report on the classification results obtained with the proposed method.

2   Classification of Botnet-Related Malware

When a honeypot captures malware, it computes the MD5 hash on the content of the executable files. The hash indicates that two different files have different content, but the difference does not mean that the files contain different malware. Most of the proliferating malware uses a technique called *polymorphism* to try to escape from the detection by antivirus tools: when a virus proliferate, it modifies its code in a way that does not alter its functionalities. Even without polymorphism, executable code may embed specific data that render it different from other instances of the same malware. Hence, two files with different MD5 hash values may actually contain the same malware. Whenever we capture a executable file, we need to decide whether it is new malware requiring analysis or a new instance of an already known executable file. Moreover, we need to know if the captured malware is interesting from the research point of view. So, we need a method to classify captured malware into different classes.

2.1   Classification Criteria

To be useful, such a classification should have two properties:

Selectivity: malware in different classes should belong to different botnets.
Consistency: malware in a same class should belong to the same botnet.

We can tolerate exceptions to the first property, because it would only result in a few unnecessary analyses, which is not so important. However, the second property is more critical: if some classes contain malware belonging to several botnets, it means that we need to check enough malware in the classes to make sure they belong to the same botnet. Hence, we do not meet our objective of performing only one analysis per class.

2.2   Classification Based on Antivirus Tools

A straightforward classification method consists in using an antivirus tool. It detects and identifies malware by searching some patterns in the code (a signature). However, a signature may match several different malware[1]. Furthermore, the same malware code is often used by several botnets. Hence, classification based on the detection by an antivirus tool may lead to inconsistent results.

---

[1] This can be a feature if it allows to reduce the size of the signature data base.

To experiment with this approach and evaluate the quality of the classification, we have first used ClamAv, a free antivirus tool running under Linux. The problem with this approach is that ClamAv only identifies about 60% of the captured malware. Furthermore, the classification has resulted in 421 classes. We have analyzed a few samples of each class and found that this classification is neither selective, nor consistent.

We have also used another tool—VirusTotal[2] to verify if the quality of the results depends on the tool used for the classification. VirusTotal makes use of several other sophisticated antivirus tools.

We have submitted 2342 malware samples to VirusTotal. Table 1 shows the results of twelve tools that have detected the largest number of files[3].

Table 1. Results obtained from VirusTotal

Anti virus	Number of identified files	Percentage of identified files	Number of signatures	Files/signature
Symantec	2266	96.8	81	28
Microsoft	2272	97.0	158	14.4
Avast	2277	97.2	217	10.5
BitDefender	2287	97.7	1118	2.0
McAfee	2299	98.2	623	3.7
Panda	2300	98.2	187	12.3
NOD32	2300	98.2	359	6.4
GData	2307	98.5	1113	2.1
AVG	2314	98.8	1005	2.3
AntiVir	2320	99.1	698	3.3
Ikarus	2336	99.7	168	13.9

The results show that only a few percent of malware escape from detection for all tools. However, the tools perform differently, if we consider the number of signatures: with only 2 files per signature, we can expect that a classification based on BitDefender would not be selective (the number of signatures corresponds to the number of different viruses identified in the sample files by a given antivirus tool). We would prefer Ikarus (the highest detection rate and a rather small number of signatures) or Symantec (a smaller number of signatures and an adequate detection rate) as long as their results are consistent.

We have also submitted the captured malware to online sandbox analyzers (Norman[4], Anubis[5], and MWanalysis [6]) that identified some malware connecting to botnets and found that the classification based on Ikarus is neither consistent, nor selective. We have not attempted to repeat the same operation with other

---

[2] https://www.virustotal.com/
[3] The results should not be considered as a comparison of antivirus performance, because our set of malware represents only a part of malware encountered on the Internet.
[4] http://onlineanalyzer.norman.com
[5] http://anubis.iseclab.org/
[6] http://mwanalysis.org/

antivirus tools, but this characteristic of the detection process is confirmed by other studies [1].

Polymorphism leads to the need of the identification of the same malware with different signatures. Moreover, a botnet can even use several different versions of code, so that classification based on antivirus will not be selective either.

3   Classification Based on Malware Behavior

We consider that the most important activity of most (if not all) bots consist in trying to contact their C&C center. When malware is installed on a victim computer, we can observe network activity including DNS requests and connection attempts to some hosts on the Internet that are characteristic of a given botnet. Hence, a different approach to malware classification can be to consider their *behavior*, i.e. a sequence of particular communication operations they perform or attempt to perform.

A possible approach to examine the malware behavior is to use tracing or emulation that would allow to detect not only Internet communication attempts, but also system activities including file access, process creation, listening to sockets [1,5,6]. However, there are versions of malware able to detect emulation or apply countermeasures to render tracing ineffective. So, the solutions based on tracing or emulation often fail to detect the network activity of malware. This fact was confirmed by the results we obtained from online sandbox analyzers based on these principles. Another drawback of the approach based on emulation or tracing is that the systems need to be instrumented and consume much resources.

Another solution that we propose is to let the infected computer run "in the wild", and intercept, monitor, and analyze its communications. Such a way of operation can be achieved with the support of a smart gateway able to observe and register interesting events, and prevent malicious activities. But even with these preventive measures, an infected computer should not be allowed to communicate directly with the Internet without the presence of an human operator able to detect and immediately mitigate unexpected dangerous behavior. Furthermore, malware usually has multiple solutions to contact its C&C and it often tries them in a random order. Hence, when the malware connects to the C&C, we cannot obtain the information on the other solutions it may use and we fail to relate it to other instances of the same malware that contacted their C&C in another way.

Hence, we propose an approach providing the bot with a DNS service, so that it can try to connect to the C&C, but no real Internet connectivity. In this way, connection attempts will fail, and the malware will try all solutions for connecting to the C&C. As a result, we obtain the required information: the resolved DNS names, the replies (including all failed requests), and attempted connections. We can then process this information to classify malware. Before describing the details of the classification and its results, we present the platform that supports the proposed method.

4   Platform Supporting Behavior Based Classification

We have designed and implemented a platform with tools for capturing malware, running botnets in a controlled environment, analyzing their interactions with a botmaster, testing methods and techniques for mitigating botnet nuisance, and eventually disrupting them [3,2]. The platform is composed of honeypots, an online analysis environment, a Command and Control monitoring software, and an offline analysis environment.

The essential part of the platform is MWNA (Malware Network Analyzer), a software designed to analyze communications between a computer infected by malware (the victim) and the Internet. MWNA runs on a gateway between the victim and the Internet.

Figure 1 presents the structure of MWNA. It is composed of two programs communicating through TCP sockets: the *filter* and the *reporter*. We also provide a wrapper to start the two others programs in a single command, the *launcher* (not shown in the figure).

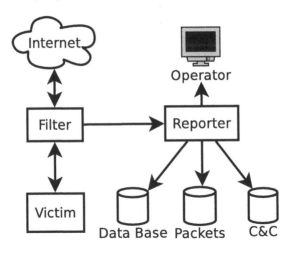

Fig. 1. Structure of MWNA

The *filter* intercepts messages passing through the network stack in the kernel and analyzes them. The *netfilter_queue* mechanism of Linux is used to intercepts packets. The *filter* analyzes them up to the application level when necessary and sends interesting events to the *reporter*. It also detects and report TCP connection establishment and termination.

The *filter* processes the following application protocols:

- DNS: replies of A type are processed and a list of known hosts is maintained. CNAME, TXT, and MX replies are also parsed and reported.
- IRC: the setup of the connection is analyzed until a channel is joined. Interesting data are collected and reported (nickname, user name, channel, passwords etc.),

- HTTP: GET, POST, PUT, and HEAD methods are identified. Replies are parsed. Accessed URL and content types are extracted and reported.
- SMTP: the protocol is processed until a DATA message is intercepted (the final step before actually sending mail). Then, the destination address and the forwarding host are reported, and further packets are dropped, so that no email is actually sent (the mail is usually a spam).

The *filter* monitors malicious activities:

- Interaction with C&C: in the current state of the program, C&C can mainly be detected for an IRC-based protocol, but we are also able to detect non standard textual or binary protocols and HTTP with a few false positives. When an IRC communication with C&C has been detected, all communications between the victim and the C&C are recorded.
- ICMP and TCP scans are detected and blocked. The detection of TCP scans is based on the observation of unanswered connection establishment attempts to different unknown hosts (for which no DNS query has been done). When the number of connections per unit of time exceeds some threshold, the destination port is blacklisted and packets sent to this port are dropped. The number of hosts and the unit of time are configurable. ICMP scans are detected in a similar way. The detection works fairly well—scans are detected after only a few attempts with no false positives.
- Denial of Service attacks (DoS): TCP SYN flooding DoS attempts are detected by monitoring connection establishment attempts to the same host. As for scans, the detection is based on threshold values of the number of connection attempts per unit of time. However, we observe some false positive due to the repetition of connection attempts to a host that does not answer. UDP flooding is processed in a similar way. Application level DoS attacks such as HTTP flooding are currently not yet processed.

As the *filter* is interfaced with the kernel, it must be executed with administrator privileges. Once it is started, it stops after the following events:

- operator interrupts the program,
- after a defined interval of time (optional).
- after a period of inactivity (optional).
- as soon as the communication with C&C has been detected (optional).

The *filter* can also work in an offline mode in which the victim cannot send any packet to the Internet. In this mode, the victim can still communicate with the gateway and use its services. In this case, we can observe DNS queries and replies, and connection attempts.

The *reporter* receives messages from the *filter* when it detects interesting events, or when packet transmissions need to be recorded. It is in charge of displaying messages for the operator, recording events in a Sqlite database, storing communication traces in a file, and recording C&C interactions.

The *reporter* does not need to execute with root privileges, which allows to use high level tools (data base, graphic user interface) without security risks.

Files are owned by the user running the *reporter*, so that their processing does not require administrator privileges.

On modern multi-core processors, the separation in two processes also results in better performance and allows to perform the disk I/O activity outside of the critical path of packets flowing through the kernel.

5   Algorithm for Malware Classification

In this section, we present the details of the classification algorithm. It uses the information gathered during the observation of the malware behavior on the platform.

5.1   Observing the Malware Behavior

Once the honeypot has captured malware, we install it on a virtual machine running Windows XP SP3.

The observation of the malware behavior takes the following steps:

1. The malware is copied to the start-up directory of the Windows disk image.
2. MWNA is started for a maximum time of 10 minutes and a maximum idle time of 2 minutes.
3. The virtual machine is started.
4. When MWNA stops, the virtual machine is stopped.
5. A new clean disk image is copied.

Thes operations are performed by the way of shell scripts running on the host machine.

MWNA stores the results of the analysis in the database: DNS replies, connection attempts, and attempted scans. The whole process takes from 3 to 12 minutes depending on the network activity generated by the malware.

5.2   Classification Algorithm

Currently, the classification is based on DNS replies. We define a class C of malware as the set of malware using DNS names from a given set $\{Names(C)\}$.

In fact, many malware versions only use a subset of the DNS names of their class. This may lead us to classify two malware in different classes if they use different names. When this happen, we may later encounter malware that uses DNS names from both classes.

For every malware M, the classification proceeds as follows:

1. Fetch the malware network activity from the database:
   (a) Let $\{Names(M)\}$ be the set of names queried by M.
2. If M performs no network activity (no connection attempts, no DNS queries, no scans), then classify it as INACTIVE.
3. If M performs only scans, then classify it as SCANONLY[7].

---

[7] Connection attempts from scans are recorded in the database only if they timeout before the scan is detected.

4. If M performs no DNS queries, then classify it as CONNECTIONSNODNS.
5. If M performs only MX type DNS queries, classify it as MXONLY.
6. If every DNS request done by M fails, classify it as DEAD. We consider that a DNS request fails when it returns no result or NXDOMAIN, or when it returns addresses such as 127.0.0.1 or 1.1.1.1, which have probably been sinkholed [7].
7. If there exists a single class C such that $\{Names(M)\} \cap \{Names(C)\} \neq \emptyset$, then:
   (a) Classify M in class C,
   (b) $\{Names(C)\} \leftarrow \{Names(C)\} \cup \{Names(M)\}$.
8. Else, if there exists no class C such that $\{Names(M)\} \cap \{Names(C)\} \neq \emptyset$, then:
   (a) Create a new class C with $\{Names(C)\} \leftarrow \{Names(M)\}$.
   (b) Classify M as a member of class C,
9. Else, if there exists several classes C_i such that $\{Names(M)\} \cap \{Names(C_i)\} \neq \emptyset$, then: let M unclassified.

6   Classification Results

During near 2 years, our honeypot have captured 3608 different files. We have analyzed and classified 1467 malwares captured since january 2011. This gives the following results:

- 248 files are INACTIVE—they do not perform any network activity. We discuss this class below in Section 6.1.
- 211 files are DEAD (the DNS request they perform fails).
- 75 malwares perform SCANONLY. Such malware probably represents autonomous code that does not participate in a botnet.
- 15 malwares are in the CONNECTIONSNODNS class. They attempt connections, but do not perform any DNS queries. The destinations of the connections are currently unreachable or refuse connections, so that we could not perform further analysis.
- 5 malwares only perform MX DNS queries on gmail.com (MXONLY class), and scan port 445. They do not try to connect to gmail.com.
- 5 malwares are left unclassified, as they use names belonging to two classes.
- Other malware files are classified into 43 classes[8]. We discuss these classes in Section 6.3.

6.1   Inactive Malware

Malware classified as INACTIVE raises the following problem: a malware transmitted through the Internet is supposed to have network activity at least to propagate itself. Some of the files are not even a valid Windows executable file, which would mean that our honeypot failed to download them correctly.

For the other files, we can only conjecture the following:

---

[8] When clarity require it, we name classes after a part of the DNS name they use.

1. The malware detects that it has been captured by a honeypot and refuses to operate. We have used a low interaction honeypot that emulates the behavior of a vulnerable system to capture malware. A high interaction honeypot would be harder to detect, but is also harder to setup and manage in a secure and efficient way.
2. The malware delays its activity, so that it may stay undetected during our analysis.
3. The malware needs some features or resources unavailable in the system to operate correctly.
4. The malware detects that it runs on a virtual machine and refuses to function.

We have setup a platform allowing to test automatically malware on a real machine. It is composed of two machines:

- A dual boot Linux/Windows victim machine, setup to reboot under Linux unless instructed otherwise. Windows is setup to automaticaly reboot after 6 minutes. Under Linux, a script is executed to restore a clean state of the Windows partition, and install on its startup directory the malware to test. Then the machine is rebooted under Windows.
- A gateway, on which a script synchronized with the victim is in charge of lauching MWNA for a limited time when the victim is ready to reboot under Windows.

With this setup, we re-analyzed inactive malwares, and did not find network activity. However, the same test including older malwares allowed to detect activity on about half of all malwares. So, we cannot draw clear conclusions about the influence of virtual machines, but it should be marginal.

Other tests with an older version of Windows XP gave significantly less network activity, so that we can tell that the OS version matters. So, several tests should be done as long as previous tests fails to detect network activity, including tests on real machines, tests on different OS versions, and tests with a significantly longer time, including one reboot after the infection.

Only if all these tests fail, we could possibly conclude that the detected malware captured by a honeypot refuses to operate as it should.

6.2   Unclassified Malware

Malware is left unclassified when it uses DNS names belonging to several classes. In this case, we could decide to merge the classes, but this could lead to inconsistent classes. Practically, when we encounter this kind of situation, we should have already analyzed at least one file in the classes and determined their C&C. If the C&C are found the same, then we should merge the classes.

Hence, if malware belongs really to two classes, the classes should have already been merged, else we need to analyze the malware before taking any decision.

6.3   Other Classes of Malware

In 43 classes that exhibit network activity, 18 contain only one malware, 4 classes contain more than 50 malwares. The average number is 17.6 malwares per class.

Let us now consider the number of DNS names per class:

- 12 classes use only one name.
- Only 2 classes use more than 5 names.

For every class, we have determined the total number of names, the fraction of names common to all malware, the minimum, maximum, and the average number of names per malware, and the standard deviation. Table 2 presents the statistics for classes with more that one malware and one name.

Table 2. Statistics for classes

Class name	Cardinality	Names	Common	Min	Max	Avg	StdDev
facebookvideocentral	12	2	2	2	2	2	0
ahrampress	3	3	3	3	3	3	0
bravefinder	2	3	3	3	3	3	0
compunass	29	4	4	4	4	4	0
bitcity	30	4	1	2	3	2.3	0.4
trenz	2	5	2	3	4	3.5	0.5

In most classes, malware uses exactly the same set of names, so we expect these classes to be consistent. However, in classes *bitcity* and *trenz*, malwares have only a subset of names in common, so that the question of the consistency of these classes is open.

The inspection of these classes and a more complete analysis of the malwares they include shows that they are actualy consistent: all the malwares try a random subset of names.

Some malwares in the class *trenz* use more than 100 DNS names (most of them result in a "non existent domain" reply, hence, they are not accounted in the statistics above). Most of these names seem to be a random sequence of 6 letters followed by ".com". We conjecture that this malware tried to use some kind of the fast-flux method [8,4] combined with a more traditionnal C&C discovery protocol.

A more complete analysis of samples of each classes showed that several classes which use different names connect actualy to a same IRC channel : they use a different host, but use the same port number, channel name, etc... This should result from an evolution of malwares.

Hence, we can tell that our method meets our objectives :

- Even if is not perfectly selective, it allows to avoid most unuseful analysis.
- It appears to be consistent : we did not find malwares with significantly different behaviors in the same classes.

However, a classification only based on DNS names should not be the only criterion to decide to do further investigations on malware. Any significant change in the behavior should be taken into consideration including a significant change in the number of names, connection attempts to adresses without DNS resolution, or scans, etc...

Port numbers may also be used in classification, to further minimize the risk on inconsistency. Bu we cannot realy use the couple IP name/port number or IP address/port number : IP addresses may change quite often in the life of a malware, and the use of aliases may render impossible to relate the IP adress used by a connection attempt with the IP name actualy used by the malware.

The next step in malware processing should then be to let an analysis run until the C&C has been detected, which means that we need a reliable and conservative C&C detection if we intend to do this in an unattended mode.

In our experimentation, we have classified a set a malware including malwares captured several months before their analysis, which means that we have possibly analyzed obsolete malware and loose some information. Our objective is rather to analyze and classify malware right after their captures, so that we can detect new malware and analyze more deeply their activities. Hence, when we classify malware, already existing classes should have already been analyzed and we should have more information about them such as the port numbers they use, the C&C protocol, and possibly the details of these protocols (IRC nicknames or channels, HTTP URL, etc.).

7   Conclusions

In this paper, we have presented the problem of the classification of malware with respect to their botnet activity. We have found that the classification based on antivirus tools would not have required properties of consistency and selectivity.

We have proposed a classification method based on the passive observation of malware network activity without actually letting them communicate with the Internet. We have found that much malware would need several analysis on different platforms, including real machines for some of them.

We have given the preliminary results of our classification method showing that if we can expect to get consistent and relatively selective classes, the detection of other significant changes in the network behavior should also lead us to perform further analysis of the malware at least until the detection of the C&C.

With an efficient method to classify malware, we expect to be able to concentrate on new threats and find adequate ways for their confinement.

Acknowledgments. This work was partially supported by the EC FP7 project INDECT under contract 218086.

References

1. Bailey, M., Oberheide, J., Andersen, J., Mao, Z.M., Jahanian, F., Nazario, J.: Automated Classification and Analysis of Internet Malware. In: Kruegel, C., Lippmann, R., Clark, A. (eds.) RAID 2007. LNCS, vol. 4637, pp. 178–197. Springer, Heidelberg (2007)
2. Berger-Sabbatel, G., Duda, A.: Analysis of Malware Network Activity. In: Dziech, A., Czyżewski, A. (eds.) MCSS 2011. CCIS, vol. 149, pp. 207–215. Springer, Heidelberg (2011)
3. Berger-Sabbatel, G., Korczyński, M., Duda, A.: Architecture of a Platform for Malware Analysis and Confinement. In: Proc. MCSS 2010: Multimedia Communications, Services and Security, Cracow (2010)
4. Caglayan, A., Toothaker, M., Drapaeau, D., Burke, D., Eaton, G.: Behavioral analysis of fast flux service networks. In: Proceedings of the 5th Annual Workshop on Cyber Security and Information Intelligence Research: Cyber Security and Information Intelligence Challenges and Strategies, CSIIRW 2009, pp. 48:1–48:4. ACM, New York (2009)
5. Carsten, W., Holz, T., Freiling, F.: Toward automated dynamic malware analysis using cwsandbox. IEEE Security and Privacy 5, 32–39 (2007)
6. Fedynyshyn, G., Chuah, M.C., Tan, G.: Detection and Classification of Different Botnet C&C Channels. In: Calero, J.M.A., Yang, L.T., Mármol, F.G., García Villalba, L.J., Li, A.X., Wang, Y. (eds.) ATC 2011. LNCS, vol. 6906, pp. 228–242. Springer, Heidelberg (2011)
7. Leder, F., Werner, T., Martini, P.: Proactive botnet countermeasures - an offensive approach. Technical report, Institute of Computer Science IV, University of Bonn, Germany (2009)
8. Nazario, J., Holz, T.: As the net churns: Fast-flux botnet observations. In: 3rd International Conference on Malicious and Unwanted Software, Fairfax, pp. 24–31 (October 2008)

Analysis of Impact of Lossy Audio Compression on the Robustness of Watermark Embedded in the DWT Domain for Non-blind Copyright Protection

Piotr Czyżyk, Janusz Cichowski, Andrzej Czyżewski, and Bożena Kostek

Gdansk University of Technology, Multimedia Systems Department,
Narutowicza 11/12, 80 -233, Gdansk, Poland
ksm@sound.eti.pg.gda.pl

Abstract. A methodology of non-blind watermarking of the audio content is proposed. The outline of audio copyright problem and motivation for practical applications are discussed. The algorithmic theory pertaining watermarking techniques is briefly introduced. The system architecture together with employed workflows for embedding and extracting the watermarks are described. The implemented approach is described and obtained results are reported. The possible attacks on the embedded watermark are described and the procedure of simulating the attacks is explained. The research is focused on the influence of lossy compression on the embedded watermark degradation. The peak signal to noise ratio and bit error rate are analyzed and compared. Advantages and disadvantages of the proposed approach are discussed. Future work and some possible improvements to the introduced methodology are explained.

Keywords: non-blind audio watermarking, discrete wavelet transform.

1   Introduction

Recent developments in digital technology such as high-speed broadband networks, decreasing cost of disk storage space or increasing capacity of portable memory changed the way of production, distribution and protection of multimedia content. Simultaneously, the easiness of copying digital data causes a big threat of piracy. It also implies the need of protecting copyrighted multimedia data such as music recordings, movies, software and books. The problem of preserving the copyrights had not existed in such scale until the development of Internet together with FTP (file transfer protocol) and peer-to-peer software. The historic methods of copyright protection were designed for the analog domain and are useless for digital data protection. Copyright holders are in need for a novel approach to protect their property in a way that would be designed especially for the content.

The most popular approach for protection both digital audio and video data involves watermarking. The idea of digital watermarking assumes the addition of imperceptible changes related to the watermark binary data to the digital signal. Knowledge of watermarking algorithms and watermark interpretations allows

A. Dziech and A. Czyżewski (Eds.): MCSS 2012, CCIS 287, pp. 36–46, 2012.

protecting multimedia data in two different ways described in the literature [10]: active (i.e. through a dedicated multimedia player, which plays music/video only if algorithm finds a watermark) or passive (i.e. piracy detection using data embedded as fingerprint allowing identification of the suspected user ID, email or IP address).

Many digital watermarking methods were developed and tested over the years. A few of those methods were designed to secure audio data. The spread spectrum watermarking proposed by Bloom and Cox [6] is the most common one. Another popular method is echo hiding designed by Ciarkowski and Czyżewski [7]. Dutta, Gupta and Pathak proposed a watermarking scheme based on DCT (Discrete Cosine Transform) in their paper [9]. The newest approach to audio watermarking is based on DWT (Discrete Wavelet Transform) domain. There are also novel approaches employing DWT together with SVD (Singular Value Decomposition) [11][16] or DCT-SVD [4] for efficient digital audio watermarking. An example of a watermarking system based on DWT can be found in the literature [13]. The watermarking solution proposed in this paper was developed especially for copyright protection of the audio content available via multimedia web service for automatic sound restoration [8]. The main objectives of the developed algorithm was fast processing and high robustness against lossy audio compression.

The remainder of the paper is organized as follows. An audio watermarking operating scheme in the wavelet transform domain is described in Section 2. Also processing workflows are shown with a brief description of the watermark embedding process. The principle of the non-blind extraction employing DWT and motivation for downgrading procedure application are presented in Section 3. The measurement methodology, comparison of different formats of audio lossy compression and results explanation are included in Section 4. The summarizing Section contains observations and conclusions focused on future system implementations to reduce possible drawbacks.

2   Watermarking Scheme

A text information, of the length of 10 ASCII [15] characters (8 bits per character), is generated and translated into a binary sequence, providing the input for the watermark embedding module. The watermark length is 96 bits in total (80 bits of information plus 8 bits of start sequence and 8 bits of end sequence). A simplified procedure of embedding the watermark is shown in Fig. 1. Watermarking embedding takes place in the DWT domain.

As presented in Fig. 1, the original file is divided into non-overlapping frames containing 1024 samples each. The frame can also have a width of 256, 512 or 2048 samples. The designed application accepts files with the sample rate 44.1 kHz. The wider the frame, the less artifacts are generated during the watermark embedding, but at a cost of smaller watermark bitrate. Among audio signal samples each frame contains one bit of watermark. The watermark is embedded in the second level low frequency DWT transformation of the frame. The values of samples in frames are modified according to the proposed relation (Eq. (1)):

$$bit = 1 \Rightarrow s^2_{LP_{wat}}[n] = s^2_{LP_{org}}[n] + \alpha \cdot \overline{s_{org}}$$

$$bit = 0 \Rightarrow s^2_{LP_{wat}}[n] = s^2_{LP_{org}}[n] - \alpha \cdot \overline{s_{org}}$$

(1)

where:

$\overline{s_{org}}$ — average sample value of the entire file

$s^2_{LP_{wat}}[n]$ — output sample value of the DWT second level low-pass component

$s^2_{LP_{org}}[n]$ — original sample value from the current DWT second level low-pass component

α — watermarking strength parameter

The α parameter represents the watermarking strength and is inversely proportional to the fidelity of the watermarked file (understood in terms of identity of the content of the watermarked file in comparison to the source file).

There are two embedding modes in stereo signals and one in mono signals. The first mode, called Mono/Stereo 1, embeds the same watermark signal in both channels of a stereo signal, which entails redundancy. The second mode, Stereo 2, uses two

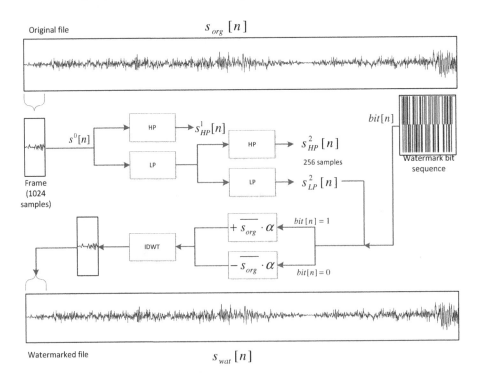

Fig. 1. The process of file watermarking

channels as separate media for watermarking, which provides a larger watermark bitrate. In this study the Stereo 1 mode is used. The dependence of the frame width and watermark bitrate is shown in Tab. 1. The watermark bitrate is not strictly related to audio bitrate, because the frame width influence on data capacity able to be watermarked. The two bitrates should not be confused.

After the embedding process a frame is transformed back into the time domain and added into the output stream.

Table 1. The dependence of watermarking mode, frame width and watermark bitrate

Frame width [bits]	256	512	1024	2048
- Mono mode - Stereo 1 mode [bits/sec]	172	86	42	22
- Stereo 2 mode [bits/sec]	344	172	86	42

3   Non-blind Extraction

As important as watermark embedding is its errorless extraction. The non-blind extraction mechanism [6] requires availability of the original and the watermarked signals to extract the information hidden in the watermark. The idea of extraction procedure is presented in Fig. 2.

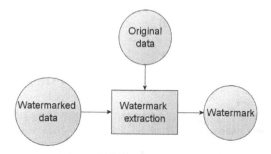

Fig. 2. Non-blind watermark extraction scheme

The disadvantage of providing the original signal for watermark extraction is compensated by the high imperceptibility and high capacity of the watermark, both being desirable in audio applications. Moreover, the watermarking strength can be maintained adequate, so it does not introduce any audio quality degradation.

The watermark extraction is similar to the embedding process. The extraction scheme is shown in Fig. 3. For the extraction of the watermark two files are required: the original uncompressed one ('*file_org*') and the watermarked file ('*file_wat*'). If the watermarked file is compressed into different format then some preprocessing must be made. Moreover, if files have different quality the downgrading procedure is required for quality rate compensation. The '*file_org*' must be compressed to the exact format of '*file_wat*'.

The difference between low frequency parts of the second level DWT of two files ('*sig_diff*') represents the watermark signal. The '*sig_diff*' is searched for in order to find the expected beginning sequence of the watermark and afterwards it is divided into non-overlapping frames containing 256 (because of 2nd level DWT decomposition) samples each. From each frame one watermark bit is extracted, if the mean frame value of the '*sig_diff*' is greater than zero the extracted bit equals 1, otherwise the extracted bit equals 0. Finally, the binary sequence is translated into text and can be presented in an easily readable form.

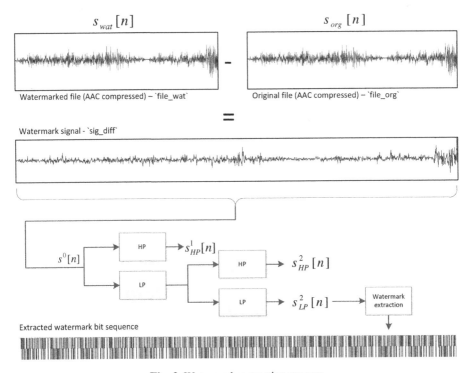

Fig. 3. Watermark extraction process

4   Simulation of Attacks

On the user's side many attacks that affect the robustness of the watermark can be performed, since the audio file may be processed in many different ways like re-sampling, filtration, time warping, cropping, lossy compression etc. All types of unexpected processing could pose a threat to the extraction of a bit sequence hidden in the embedded watermark. Simple modifications of audio signal are treated as potential watermarks attacks described in the literature [12][14]. The possible processing is not always aimed at watermark deletion or corruption, however either intentional or unintentional modifications are sufficiently dangerous and they can lead to a water-mark degradation.

It is infeasible to embed a watermark being robust against all types of attacks. The lossy audio compression is the most possible attack type in the digital data flow. The commonly used lossy compression formats such as: AAC (Advanced Audio Coding), MP3 (MPEG-1/MPEG-2 Audio Layer 3) and OGG (Xiph.org - Ogg Vorbis) allow the users to modify data size at the cost of data quality. Encoding a file to another format influences the watermark as is shown in Fig. 4. The waveform in Fig. 4(a) presents the extracted watermark signal, which represents the difference between original and watermarked file in the second level low frequency DWT domain while using the text watermark of the length of 10 ASCII characters [15]. The same procedures of embed-ding and extracting the watermark were done for the signal shown in Fig. 4(b), after applying the MP3 lossy compression with CBR (constant bitrate) equal to 320 kbps. The audio sample containing a classical music was used in the experiment.

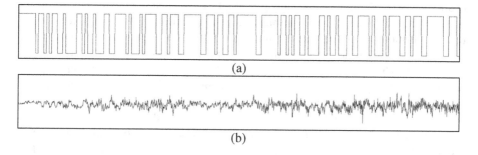

Fig. 4. The watermark signal extracted from: a) raw audio data, b) compressed audio data

The waveforms presented in Fig. 4 are significantly different, but both contain the same watermark. The implemented extraction algorithm is able to extract the binary watermark sequence without errors from (a) and (b) waveforms. However, the algo-rithms had to be tested for different music genres, lossy compression codecs and for different compression levels to ensure the correctness of the watermark extraction.

To simulate realistic situations a large test set is required. The set employed for si-mulations was generated automatically considering variations of four music genres (classical, jazz, pop, rock), three types of lossy compression formats (AAC, MP3, OGG) and six quality levels for each codec. The recompression procedure was ex-ecuted using the Audacity [3] open source software. Unfortunately, the chosen codecs do not have a unified compression rate which would ensure a similar audio quality for each level of compression. The different quality rates for codecs are compensated using the ways described in the Audacity software manual [3]. The reference parame-ter is MP3 CBR bitrate, the minimum value '128 kbps' in this codec should provide similar audio quality as for the value '5' in the OGG quality scale and the value '250' in the AAC quality scale. The maximum reference value '320 kbps' in the MP3 codec should provide a similar audio quality as the value '10' in the OGG quality scale and the value '500' in the AAC quality scale. The six quality values were chosen with the linear interval for each scale. Moreover, each music file was watermarked with three different values of the α parameter as in Eq. (1). Firstly, the value of α was heuristically adopted as 10 (base value), because this value provided inaudible changes in audio signals as well as errorless watermark detection which is illustrated in Fig. 4. Two other values were chosen as the half of the base value ($\alpha = 5$) and as a double of the base value equal to $\alpha = 20$.

Finally, the set of 288 audio files including 72 reference audio files (original files, without watermark compressed with specific codec and quality) and 216 watermarked audio files were generated. The large amount of generated files enabled a simulation of potential attacks as is described in the following section.

5   Experiments and Results

Typically the BER (bit error rate) is used for the analysis of the watermark extraction quality [2][5]. BER shows how many bits of all were erroneously extracted. Basically it shows the efficiency of the watermarking system and its ability to recover embedded information. In the performed simulations BER was equal to zero in each test case regardless of music genre, compression type, compression quality or watermarking strength. The remaining part of this paper focuses on another important property of watermarking systems – the fidelity of the watermarked audio file. The watermark in an audio file shouldn't degrade original file in a perceptible way because this kind of system would be useless to both producer and user of audio content.

In simulation experiments the reference watermark signal obtained from uncompressed files, as shown in Fig. 3, was compared to the watermark signal extracted in the same way from files attacked with a lossy compression. Measurements of the differences between obtained watermark signals are shown using PSNR (peak signal to noise ratio) given by Eq. (2) [1]:

$$PSNR = 10\log_{10}\left(\frac{MAX^2}{MSE}\right)$$

(2)

$$MSE = \frac{1}{l}\sum_{n=0}^{l}\left(x[n] - y[n]\right)^2$$

where:
$PSNR$	- peak signal to noise ratio
MSE	- mean square error
MAX	- maximum value of used data type
l	- total signal length in sample
$x[n]$	- current sample value in the reference signal
$y[n]$	- current sample value in the analyzed signal

The simplest definition of PSNR starts out from MSE which measures the absolute difference between two signals x and y. When MSE equals 0 then signals are identical and PSNR is not computed, otherwise the MSE is compared to the MAX value (in case of float precision data with range <-1.0, 1.0>, MAX equals 1.0). The PSNR allows presenting the results in the dB scale. Smaller values of PSNR signifies bigger differences in signals, thereby higher values are better in the context of the executed simulations and measures.

The results obtained during simulations are reported in Fig. 5. The x axes are related to each other with normalized quality rates separately for each codec described in the previous section.

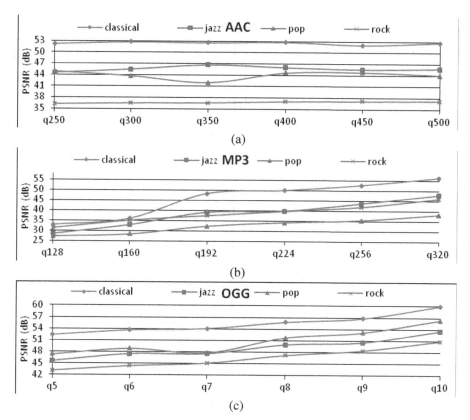

Fig. 5. Results of compression attack simulations for $\alpha = 10$: a) AAC, b) MP3, c) OGG

The first iteration of measurements (with the base value of $\alpha = 10$) was very promising because no errors occurred. Moreover, the BER (bit error rate) in each test case was equal to zero which proved an errorless watermark extraction. A very positive results of the watermark extraction are due to employing the identical compression parameters and software for preparing the reference and watermarked files. The non-blind extraction is very efficient if the analyzed audio files have the same quality rates and compression artifacts. The charts in Fig. 5 present the compression influence on the watermark together with differences caused by a variety of music types. The PSNR level directly depends on the music genre, the highest PSNR value is obtained for classical music, the lowest for rock and pop. The increase of the PSNR is observed for MP3 and OGG opposite to the AAC encoder which provides a constant PSNR level independently from quality level. The further measurement series assumed a modification of the α parameter for two other values $\alpha = 5$ and $\alpha = 20$ (see Eq. 1). Results are presented in Fig. 6.

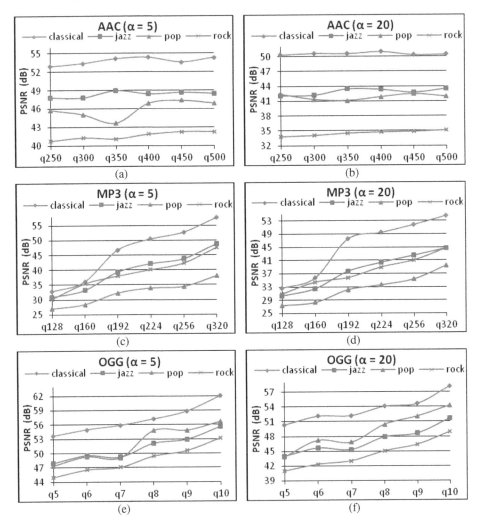

Fig. 6. Results of compression attack simulations for: a) AAC, $\alpha = 5$; b) AAC, $\alpha = 20$; c) MP3, $\alpha = 5$; d) MP3, $\alpha = 20$; e) OGG, $\alpha = 5$; f) OGG, $\alpha = 20$

The results are surprising, because no errors were observed. The PSNR values vary ca. ± 3 dB compared to results obtained for $\alpha = 10$. The watermarks are properly decrypted for each test case, although the encoder and music genres influence the PSNR.

The watermarking strength parameter α was tested with some different values of this parameter. Regardless of the chosen value of α, no errors in the extracted binary watermark were detected. Nevertheless, an increase of the parameter α caused a significant reduction of PSNR values for each measurement.

6 Conclusions

The proposed watermarking system is still in early stages of development, but preliminary results are very promising. The extraction of watermark from files that were attacked with lossy compression were successful – the obtained BER was equal to 0.

Further tests employing other files representing different music genres should be performed. Also the auditory tests should be carried out to ensure the imperceptiveness of the inserted watermark and to verify the perceived quality of the modified audio files. Additional measurements of watermark robustness are required in conjunction with other types of attacks such as: stereo downmix to mono, multiple recompression and D/A $\rightarrow$ A/D conversion.

Acknowledgements. The research was founded within the project No. SP/I/1/77065/10 entitled: "Creation of universal, open, repository platform for hosting and communication of networked resources of knowledge for science, education and open society of knowledge", being a part of Strategic Research Program "Interdisciplinary system of interactive scientific and technical information" supported by The National Centre for Research and Development (NCBiR, Poland).

References

1. Al-Haj, A.M.: Advanced Techniques in Multimedia Watermarking: Image, Video and Audio Applications, New York (2010)
2. Al-Haj, A., Mohammad, A.: Digital Audio Watermarking Based on the Discrete Wavelets Transform and Singular Value Decomposition. European Journal of Scientific Research 39(1), 6–21 (2010)
3. Audacity Manual website,
   http://manual.audacityteam.org/man/Main_Page
4. Lei, B.Y., Soon, I.Y., Li, Z.: Blind and Robust Audio Watermarking Scheme Based on SVD–DCT. Signal Processing 91(8), 1973–1984 (2011)
5. Bhat, K.V., Sengupta, I., Das, A.: A New Audio Watermarking Scheme Based on Singular Value Decomposition and Quantization. Multimedia Tools and Applications 52(2-3), 369–383 (2011)
6. Bloom, J.A., Cox, I.J., Fridrich, J., Kalker, T., Miller, M.L.: Digital Watermarking and Steganography, Boston (2008)
7. Ciarkowski, A., Czyżewski, A.: Performance of Watermarking-Based DTD Algorithm under Time-Varying Echo Path Conditions. Intelligent Interactive Multimedia Systems and Services 6, 69–78 (2010)

8. Czyżewski, A., Kostek, B., Kupryjanow, A.: Automatic Sound Restoration System - Concepts and Design. In: International Conference on Signal Processing and Multimedia Applications, pp. 1–5 (July 2011)

9. Dutta, M.K., Gupta, P., Pathak, V.K.: Perceptible Audio Watermarking for Digital Rights Management Control. In: 7th International Conference on Information, Communications and Signal Processing, vol. 1, pp. 55–59 (2009)

10. Furht, B., Kirovski, D.: Multimedia Encryption and Authentication Techniques and Applications, Florida (2006)

11. Lalitha, N.V., Suresh, G., Sailaja, V.: Improved Audio Watermarking Using DWT-SVD. International Journal of Scientific and Engineering Research 2(6), 1–7 (2011)

12. Lang, A., Dittmann, J., Spring, R., Vielhauer, C.: Audio Watermark Attacks: from single to profile attacks. In: Proceedings of the 7th Workshop on Multimedia and Security, New York (2005)

13. Maha, C., Maher, E., Chokri, B.A.: A blind audio watermarking scheme based on Neural Network and Psychoacoustic Model with Error correcting code in Wavelet Domain. In: 3rd Int. Symp. on Communications, Control and Signal Processing, pp. 1138–1143 (2008)

14. Petitcolas, F.A.P., Anderson, R.J., Kuhn, M.G.: Attacks on Copyright Marking Systems. In: Aucsmith, D. (ed.) IH 1998. LNCS, vol. 1525, pp. 218–238. Springer, Heidelberg (1998)

15. RFC 20: ASCII format for Network Interchange, ANSI X3.4 (October 1969)

16. Vongpraphip, S., Ketcham, M.: An Intelligence Audio Watermarking Based on DWT-SVD Using ATS. In: Global Congress on Intelligent Systems, GCIS 2009, pp. 150–154 (May 2009)

Platon Scientific HD TV Platform in PIONIER Network

Mirosław Czyrnek, Jędrzej Jajor, Ewa Kusmierek, Cezary Mazurek,
Maciej Stroiński, and Jan Weglarz

Poznan Supercomputing and Networking Center, Poznan, Poland
{majrek,jedrzej,kusmiere,mazurek,stroins,weglarz}@man.poznan.pl

Abstract. Platon TV is a country-wide platform that supports HD content production, processing and delivery for research and educational purposes. The
scope of operation and a wide area of scientific subjects covered, in combination with the technological requirements of high resolution, high quality content
makes design of such a platform a challenging task. In this paper we present
how the intended functionality is implemented by Platon TV for various stages
of content lifecycle. We describe the platform architecture and mechanisms that
ensure required security and reliability.

Keywords: multimedia, HD TV, CDN.

1 Introduction

Steady growth of the networking infrastructure capacities provided along with high
quality of service and on-demand network provisioning systems [1], paves the way
for the development of the network demanding applications, such as HD videoconferencing [2,3] and media delivery [4,5], virtual and augmented reality [6] or 4K content
distribution [7] to mention just a few. Platon HD TV is a platform for network demanding applications developed to make it possible to use high quality multimedia
for knowledge sharing, education and science popularisation.

High resolution is the key content characteristics for applications that deal with
events whose representation requires high level of details, such as the course and the
results of a scientific experiment or a medical surgery. HD is required also for events
that are transmitted and projected onto a large screen in a conference or lecture hall,
for example. Hence, there is a need for high resolution on a micro and macro scale.
There is a long list of potential applications of HD content such as research results
popularization, education, knowledge sharing and dissemination. The group of end
users includes, but is not limited to, universities, research laboratories, medical clinics
and schools.

Platon HD TV is a project whose goal is to provide a country–wide platform that
hosts applications supporting the entire lifecycle of HD content, starting with its production to the delivery to end users. The platform comprises a variety of components
for content processing at its different stages and connects them to provide a complete
processing path. The design of such a system is challenging due to the scope of operation, content characteristics and related technological requirements. Content available

A. Dziech and A. Czyżewski (Eds.): MCSS 2012, CCIS 287, pp. 47–57, 2012.

through Platon TV covers various scientific subjects in liberal arts and technology. Such a universal range of topics creates new opportunities for applications whose development is enabled with a set of tools provided by Platon TV but also constitutes additional challenge. In this paper we present solution that addresses all these issues.

Platon HD TV is one of the services offered to the research community by the Platon project [8] through deployment of dedicated infrastructure and applications in the PIONIER [9] network. Platon service container includes, beside scientific HD TV, also HD videoconferencing, eduroam [10], campus computations and archiving service. All these services will become available in the mid-2012, as a result of joint effort of the PIONIER consortium members, funded by the European Regional Development Fund through Innovative Economy Program (Action: Scientific Infrastructure Development). In this paper we present the HD multimedia production and delivery infrastructure. We start this presentation with the description of Platon TV content production and delivery model in Section 2. We focus on the assumptions that guided system design process. Then, we present the platform architecture, its components and their relationships in Section 3. Section 4 introduces mechanisms that ensure security and reliability. We conclude the paper in Section 5.

2   Content Production and Delivery Model

The goal of Platon HD TV is to support the entire HD content lifecycle from its production to delivery of the final result to end users. In order to achieve this goal an integrated service platform was designed and implemented to provide all the component services required to manage and process content at the various stages of its lifecycle. In this section, first we present the assumptions that guided platform design process. Then, we introduce the functional components of the platform and describe how they satisfy the requirements. We complete the model presentation with the description of a typical content processing path.

2.1   Design Assumptions

Platon TV platform was designed to operate under the following assumptions. There are a number of content production studios that are a part of Platon TV as well as external content providers. Both groups constitute a primary source of content for the system. The providers can operate independently from one another in different geographical regions and in different scientific areas. There is one logically centralized content repository which enables content sharing among various platform users. There are a number of end user (content recipient) groups which differ from one another in terms of the area of interest, the purpose of using the platform, experience, age, and for which different content should be offered. As a consequence of this assumption, there should be services and applications available, that make it possible to prepare content offer for a specific user group and to generate new content by processing the primary source content, e.g., by grouping, merging, annotating, linking and so on. Given the scope of operation, the platform should exhibit a high degree of scalability and reliability of all its components. Platon TV content is primarily HD multimedia with all the consequences for bandwidth, storage and computational power requirements. However, content could also be accessed on a mobile device for example, which may not be capable of rendering HD content. The platform offers access to content available on demand as well as content transmitted live.

2.2   Functional Components

Based on the assumptions listed in the previous subsection, the following functional components of the platform are identified (Fig. 1). *Content production system* consists of a number of TV studios equipped for production of HD content. *Content repository* receives content provided by Platon TV studios and external providers, and makes it available for delivery and further processing. *Live broadcast system* is used for managing virtual TV channels that comprise VoD content from the repository as well as feeds from production studio live encoders. *Content delivery system* is a distributed, highly scalable CDN for delivery of VoD and live content to end users. *Content processing system* provides tools for content editing and presentation. A number of *portals* provide end users with access to content and platform services.

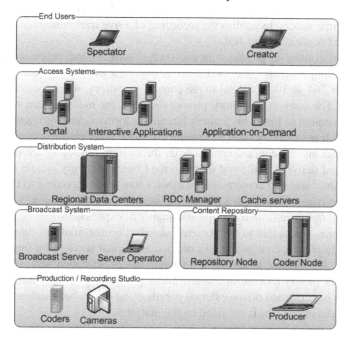

Fig. 1. Platon TV Functional Components

All these components can be divided into three groups according to content lifecycle stage in which they are involved. The media production group includes production studios, live broadcast system and content processing system. The media delivery group consists of content repository and content delivery system. All portals belong to the media access group.

2.3   Content Lifecycle

Let us now describe a typical content lifecycle and indicate which platform components are involved at its various stages. We distinguish two types of content, live and on-demand, which go through different initial processing steps and are described separately.

Live full HD content originates from the studio facilities where it is produced with broadcast HD-SDI standard compatible equipment, encoded with H.264 or VC-1 codecs and provided as a media stream over IP to other platform components. Smooth Streaming technology, which requires up to 10 Mbps per full HD stream, is used for delivery to end users. The live streams (feeds) from encoders are managed through the live broadcast system which allows one to define live broadcast schedule in the form of a virtual channel. The virtual channel can comprise live content and on-demand media available from the content repository. Moreover, with the live broadcast system it is possible to record the live transmission for archival purposes, perform simple post-processing and export results to the repository. Such approach shortens the time needed to publish the recording of a live transmission as on-demand asset after the broadcast is finished.

On-demand content is produced in the studio facilities and processed using the non-linear editing systems. The final asset is exported using high quality encoding profile (e.g. H.264 @ 10-35 Mbps) to the content repository, where content metadata is generated. The content repository provides storage for media assets, mechanisms for recoding content to lower bit Smooth Streaming formats, and for managing distribution policy. It is a primary source of media assets for other platform components.

Live streams and on-demand content from the content repository are provided for distribution and delivery to the end-users, to the Content delivery system (CDS). The CDS caches the requested content and distributes live feeds according to the end-user requests at the edge nodes. With a monitoring system and performance evaluation the CDS is able to optimize resource usage and quality of experience.

The end-users get access to the content published on the platform through portals which not only present content but also provide additional services for interactive use of the assets. The portal tools enable instantiation of so called sub-portals which present content in way tailored to a specific user group interests and needs. The content may also be presented on the external portals using the embed mechanism and the dedicated media player that is provided by the platform.

3   Platon TV Platform Architecture

Platon TV platform architecture has been designed to integrate functional components responsible for designated content flow (lifecycle) stages. In this section we introduce Platon TV architecture, describe its elements and explain how they are connected.

The architecture is designed for HD content which means that its elements must be able to satisfy network bandwidth, storage and computational power requirements typical for high resolution media. Given the scope of platform operation its components also should exhibit high degree of scalability and reliability as well as be easily extensible, and they are designed to follow these guidelines. Platon TV architecture contains all functional elements described in Section 2.2. Now we concentrate on the platform structure and the relationships between its elements that are described in the following subsections.

3.1   Media Production Studios

The Platon TV media production infrastructure consists of three types of studios: production studios, recording studios with somewhat limited functionality compared to a production studio, and a mobile studio – an Outside Broadcasting VAN (OB-VAN). Each studio can operate independently from other studios and support the entire content production process, although, a recording studio offers limited functionality as far as live production is concerned. The mobile studio (OB-VAN) is designed to support live broadcast and recording of events in places which are out of the reach of the stationary studios. The media production infrastructure is distributed geographically and deployed in Metropolitan Area Networks and HPC Centres.

Such a structure of the production system enables a number of content providers to operate the production studios independently from one another in different geographical regions and in different scientific areas.

3.2   Live Broadcast System

Live Broadcast System consists of a number of servers that function as intermediaries between production infrastructure and Content Delivery System. The servers are placed in Regional Content Centres. Thus, the Live Broadcast System structure is tightly coupled with the Content Delivery System.

The Live Broadcast System is designed to support live transmission directly from a studio and a virtual channel broadcast. A virtual channel is equivalent to a live transmission from the end users point of view but for the Live Broadcast System it is transmission of a set of content available on-demand, according to a predefined schedule. Virtual channel schedule also may include live feed from a studio for a more flexible solution. Such an approach enables broadcast of a large number of virtual channels, even though the number of production infrastructure elements and the corresponding live feeds is limited.

3.3   Content Repository

Content Repository is the primary source of content for other platform components. It provides storage, content format processing and management services. The repository is conceptually centralized but implemented as a distributed, centrally managed system. It consists of a number of storage nodes and recoding nodes distributed geographically. Content replication and archiving services are used to ensure high content availability. The recoding nodes process content to provide various formats which are accessible to end users. Although, repository stores primarily HD content, also lower quality formats are generated and made available for users.

The repository receives content from the studios, external content providers and from content processing system. The repository provides content to Content Delivery System, Live Broadcast System and content processing system. Hence, the repository enables content sharing among various Platon TV users, content producers as well as content recipients.

3.4   Content Delivery System

Content Delivery System has two-level hierarchical architecture based on concept of content distribution deployed in PIONIER network. The first level consists of Regional Content Centres (RCC). Each RCC contains a number of proxy/cache servers responsible for media services delivery to the end-users. The architecture of the CDS is based on the concept of content distribution deployed in PIONIER network and used by Polish Public Television [11,12], for example.

3.5   Content Processing System

Since Platon TV users may vary in terms of the area of interest, the purpose of using the platform, experience, age and so, a set of tools is needed to prepare content for presentation for different user groups. For example, the same scientific experiment will be presented in a different way to a researcher working in a given area and to students of a given subject. It will also need a different supplemental material in both cases. In order to satisfy this requirement Content Processing System provides Content Management Systems and a number of interactive applications which we collectively refer to as Application-on-Demand (AoD). The former is designed to create content offer, i.e., to manage the content and the way it is presented to end users by access portals. The latter contains a set of tools for adding supplemental content, such as presentations, graphical elements, audio and text, for generating new content from the source material available in the repository, and for managing live content presentation with its supplemental content.

3.6   Access Portals

A set of interactive applications provide access to the available services and content. An interactive portal is the primary application providing access to content, content browsing and searching tools. In addition, the Portal provides communication services for its users and between users and content produces. Registered users have an opportunity to comment on the particular video he/she was watching and to provide its evaluation.

4   Security and Reliability Mechanisms of PLATON TV Services and Applications

Platon TV offers a complete end-to-end distribution platform solution for both live event broadcasting and on demand multimedia content. The content delivery system guarantees good performance and reliability, while also providing high level of security. In this Section we described security and reliability mechanisms deployed in Platon TV. We start with the CDN, its server placement and content distribution strategies. Next we address the reliability and security issues with respect to the distribution network. Hardware and service redundancy is described in the following two subsections.

4.1   CDN Mechanisms

Several design patterns and architecture characteristics are common practices often used in CDNs [13]. One of them is the concept of geographically distributed caches, positioned in peering points or inside major ISP's networks, which deliver the content directly to end users from the nearest cache server inside the ISP's network. This approach is used to avoid congestion and latency issues in the Internet and on peering points between the ISP's and CDN networks and is also a very good reliability design strategy. In the Platon TV project, physical servers are collocated in 22 datacenters of the PIONIER consortium partners. Each of the datacenters has connections with several Internet Service Providers, building a mesh of connections, allowing the CDN to redirect the end user to the nearest available server. The internal topology of the CDN allows groups of servers, as well as individual servers within each group, to communicate with one another, giving multiple possible paths for content distribution. The end devices, called the Edge Servers, store the multimedia content in their internal caches, when the request to provision the content is made. At any time, the cached content can be removed according to the cache replacement strategy freeing disk space needed to cache new content. This is a safe operation, as the content can be always downloaded from the repository. In this architecture the repository itself is responsible for keeping a persistent copy of each material and for providing reliable access to the content, thus it was necessary to install two instances of the repository, which replicate the contents with one another.

4.2   Network Security and Reliability

CDNs often use separate network isolated from the public Internet for internal content distribution and for content provisioning to end users. Platon TV also uses this approach with a Level-3 Virtual Private Network, which is one of the network-layer services provided by the Polish Optical Internet Network backbone MPLS. The virtual routed network, isolated from other users of the backbone, connects all 22 datacenters housing Platon TV equipment. While keeping the traffic isolated from other clients, this architecture also provides automatic best path selection and path failover in case of an underlying link failure. Because the Polish Optical Internet is built on the multi-10 gigabit DWDM technology (multiple 10 gigabit lambdas on network links), it provides very fast transfer medium.

Aside from content distribution, the internal network is also used to provide safe administrative access to all devices of the system – ranging from the switch consoles, server's hardware remote management modules, to systems' remote desktops and remote shells, as well as to provision new devices from the central deployment server.

4.3   Data Protection Mechanisms

In order to achieve the highest possible end-to-end security within the Platon TV, several measures are employed for data protection, starting with physical security, through network security to application layer security.

All devices that comprise the CDN – servers, switches and disk arrays – are split between 22 partners of the PIONIER consortium and are collocated in secure datacenters, where direct access is strictly controlled and only authorized administrators can gain physical access. The CDN provides public services, and therefore, the network access cannot be restricted to specific IP addresses. User authentication and authorization is however used to provide controlled access to administrative applications – the content repository, broadcast server controller and public portal administration. The network traffic separation mentioned in the previous Section provides data protection at the network level.

The application layer security is used for content access control since there are a number of legal restrictions regarding the possibilities to share, stream and broadcast the digital content which is available through the CDN. Each individual digital material has several properties - Access Channels - that define the accessibility of the content to specific user groups. This mechanism is very flexible, and makes it possible to restrict access based on geographical location of the end user, the user's Internet Service Provider, the user group or type of the user's end device. User's geographical location is obtained from the external GeoIP database, while the association between the user's IP, ISP name and network is kept in the internal database and updated with data from public Whois servers, as well as from BGP sessions. Of course these control mechanisms also require a mechanism for URL authentication to detect and prevent the incidents of URL theft. To satisfy this requirement, the multimedia content URLs use special attribute – the ticket – a unique time-based value, generated each time the URL is requested by the end-user. The ticket identifies a particular user and request, and cannot be reused by other malicious users, because the URL expires within minutes. The tickets are encrypted, to prevent users from tampering with the ticket data. Even if an external website were to embed the link with such URL, the link would not be functional, preventing others from publishing the content without the necessary permission.

4.4   Hardware Redundancy

A Content Delivery Network is a specific kind of a computer system in terms of reliability. Because of the video content nature, and the streaming protocols requirements of, it is difficult to limit the influence of the interruptions in network and system accessibility on the video quality. Infrastructure problems – ranging from congested links, high packet loss, and packet or frame reordering to heavily loaded servers, machine failures, and application issues, affect the overall users' experience, giving ragged video, freezing or missing frames or even stream playback interruptions. Therefore it is essential to design a redundant architecture.

All Platon TV servers use redundant physical components - power supplies from two different power lines, hard disks with RAID-1 or RAID-50 mirroring and multiple aggregated network adapters. These measures limit the negative effect of the most common failures which happen in datacenters. The network infrastructure consists of redundant switches in each of the datacenters, with multiple uplinks to the core network. Multiple network connections use the 802.3ad Dynamic Link Aggregation (LACP) protocol to provide both bandwidth scalability and link redundancy. The best path and failover link selection between the Platon routers and the core network is done by the Open Shortest Path First (OSPF) dynamic routing protocol. The core network uses the Multiprotocol Label Switching technique to provide fast alternative path selection.

4.5   Service Redundancy

End users interact with the CDN in two points, one of them is an Edge Server responsible for provisioning the content, and the other one is the access service, which redirects the user request for a particular multimedia content to the designated Edge Server. Naturally, the access service is a critical component in such architecture. The Platon TV CDN has multiple access services, deployed in 5 regional datacenters, supported by load-balancing and high-availability clusters of three servers in each of these locations. This configuration gives both high reliability and high transaction throughput of this service. Last but not least, there are several software components called distribution system and repository managers, each running in a different physical location, which communicate and cooperate with one another to monitor and control all servers in the system, and manage the distribution process. The manager is responsible for node monitoring, and automatically redirecting new requests to less loaded servers, to dynamically balance system load.

4.6   Other Mechanisms

The system was built with the security in mind and all deployed systems are hardened from the moment of installation. Nevertheless, this is not sufficient, as many new threats arise each day. To keep the system safe, mandatory security audits and periodical patch installations are performed on all components of the system. The access to the system for its administrators is performed over an IPSec VPN through a firewall, which is a gateway between the internal private network and the Internet.

The above summary of features is only a brief description of the Platon TV architecture, designed for scalability and security. But not only these technical features and multiple levels of redundancy ensure maximum availability make the system so reliable. A very important role plays the dedicated Network Operations Center which manages the incidents and supervises the Content Delivery Network 24 hours a day, 7 days a week, providing a single point of contact for all issues regarding the system.

5   Conclusions

Platon TV is a platform that offers a complete solution for HD content production and delivery. Beside the standard elements necessary to provide such functionality, Platon TV also offers a number of tolls that allow one to shape content offer and tailor it to a specific user group needs as required for a large variety of target audiences and universal range of scientific topics. The platform architecture encompasses all these solutions and integrates them into a complete content processing path for live and on-demand content. The system design addresses needs for high reliability and security which must be considered in combination with the technological requirements imposed by HD content characteristics.

References

1. Ciulli, N., et al.: Architectural approaches for the integration of the service plane and control plane in optical networks. A Computer Networks Journal, Optical Switching and Networking 5(2+3), 94–106 (2008) ISSN 1573-4277
2. Gharai, L., Lehman, T., Saurin, A., Perkins, C.: Experiences with High Definition Interactive Video Conferencing. In: IEEE International Conference on Multimedia & Expo. (ICME), Toronto, Canada (July 2006)
3. Alcober, J., et al.: High Definition videoconferencing. The future of collaboration in healthcare and education. In: eChallenges e-2009 Conference (2009) (in press)
4. Kuśmierek, E., Czyrnek, M., Mazurek, C., Stroinski, M.: iTVP: Large-scale content distribution for live and on-demand video services. In: Zimmermann, R., Griwodz, C. (eds.) Multimedia Computing and Networking SPIE-IS&T Electronic Imaging. SPIE, vol. 6504, Article CID 6504-8 (2007)
5. Perkins, C., Gharai, L., Lehman, T., Mankin, A.: Experiments with Delivery of HDTV over IP Networks. In: Proceedings of the 12th International Packet Video Workshop, Pittsburgh (April 2002)
6. Matyska, L., Hladka, E., Holub, P., Liska, M.: High Quality Large Scale Virtual Classroom. In: Proceedings of the 14th International Conference of European University Information Systems (EUNIS 2008), Arhus, Denmark (2008)
7. Binczewski, A., Glowiak, M., Idzikowski, B., Stroinski, M., Strozyk, M.: New generation media in research and entertainment. Presented at TERENA Networking Conference 2010, Vilnius, Lithuania, May 31-June 3 (2010)
8. PLATON website,
   http://www.platon.pionier.net.pl/online/?lang=en

9. Binczewski, A., Meyer, N., Nabrzyski, J., Starzak, S., Stroiński, M., Węglarz, J.: First experiences with the Polish Optical Internet. Computer Networks 37(6), 747–760 (2001)
10. Eduroam website, http://www.eduroam.pl/
11. Czyrnek, M., Mazurek, C., Stroiński, M.: Two-level content delivery system in PIONIER Optical Network for interactive TV services. In: Bartkowiak, M., et al. (eds.) Proceedings of 11th International Workshop on Systems, Signals and Image Processing, IWSSIP 2004, Ambient multimedia, Poznań, PTETiS, September 13-15, pp. 457–460 (2004)
12. Czyrnek, M., Mazurek, C., Stroiński, M.: Management and publishing of digital content on the iTVP Platform. In: Bartkowiak, M., et al. (eds.) Proceedings of 11th International Workshop on Systems, Signals and Image Processing, IWSSIP 2004, Ambient multimedia, Poznań, PTETiS, September 13-15, pp. 481–484 (2004)
13. Czyrnek, M., Kuśmierek, E., Mazurek, C., Stroiński, M., Węglarz, J.: CDN for Live and On-Demand Video Services over IP. In: Buyya, R., Pathan, M., Vakali, A. (eds.) Content Delivery Networks. LNEE, vol. 9, ch.13, pp. 317–342 (2008) ISBN: 978-3-540-77886-8

Multi-camera Vehicle Tracking Using Local Image Features and Neural Networks

Piotr Dalka

Multimedia Systems Department,
Gdansk University of Technology,
Gdansk, Poland
`piotr.dalka@sound.eti.pg.gda.pl`

Abstract. A method for tracking moving objects crossing fields of view of multiple cameras is presented. The algorithm utilizes Artificial Neural Networks (ANNs). Each ANN is trained to recognize images of one moving object acquired by a single camera. Local image features calculated in the vicinity of automatically detected interest points are used as object image parameters. Next, ANNs are employed to identify the same objects captured by other cameras. Object tracking is supplemented by spatial and temporal constraints defining possible transitions between cameras' fields of view. Experiments carried out were focused on identification of the same vehicles in different cameras. The results achieved prove that the algorithm is sufficiently effective for multi-camera object tracking provided that the cameras' orientations with respect to moving objects and to the ground are similar.

Keywords: multi-camera object tracking, moving object segmentation, local image features, SURF; object identification.

1 Introduction

Tracking objects that move in large public areas and cross fields of view of multiple cameras is a required feature of advanced, distributed surveillance systems. Such knowledge is necessary for automatic route reconstruction and therefore – for behavior analysis and automatic detection of various events. The task becomes very complicated in case of large gaps between cameras and many possible routes between particular observation points.

Many approaches related to multi-camera tracking are presented in the literature. Most of them require a calibrated camera field of view and assume that a site model is known [1]. Therefore matching can be performed on a limited set of candidates based on spatial constraint introduced by camera layout. In order to track moving objects locally in each camera and globally in the ground plane, particle filters [2] or Kalman filters [3] are often employed.

Various features and comparing techniques are used to match the same object images captures by different cameras. In [4] geometric and intensity features are modeled as multivariate Gaussian and match with Mahalanobis distance measure.

A. Dziech and A. Czyżewski (Eds.): MCSS 2012, CCIS 287, pp. 58–67, 2012.
© Springer-Verlag Berlin Heidelberg 2012

Gaussians of mixtures are also used to represent object appearance in [5]. Appearance models are associated witch each other by searching a function maximizing model dissimilarity measure.

The solution described in the paper does not require calibrated fields of view of cameras. However, some knowledge regarding site topography needs to be provided. The algorithm utilizes local features of objects and employ neural networks to learn each object appearance and then to recognize it in another camera. This approach seems to be novel as the majority of solutions found in the literature employ various matric-based matching algorithms. ANNs were chosen because of their great generalization capabilities, fast learning and small footprint of the trained ANN, comparing to other identification solutions (e.g. SVM); the latter two properties are especially important in the light of practical implementation of the method in the real-time video processing system. Experiments were carried out with vehicle images only; however the solution is universal and does not rely on vehicle-specific characteristics. Section 2 of the paper describes low-level image routines leading to moving object segmentation in each camera. The next section presents multi-camera tracking algorithm, including definition of parameters used for object appearance description and application of artificial neural networks for object identification. Section 4 describes experiments carried out and their results, and section 5 concludes the paper.

2   Moving Object Segmentation

The first step in multi-camera object tracking involves independent object segmentation in video streams from all cameras. This process consists of two steps. First, moving objects are detected in each video frame, independently. The algorithm based on background modeling method utilizing Gaussian Mixtures is used for this purpose as it proved to be effective in earlier experiments [6]. The results of background modeling are processed by detecting and removing shadow pixels.

Next, movements of the detected objects (blobs) are tracked in successive image frames using a method based on Kalman filters that allow predicting object position in the current frame based on observations in previous frames. By comparing results of background subtraction in the current frame with predicted object positions it is possible to correlate each tracker (Kalman filter representing one real object) with the detected moving object (including partial occlusions), so the movement of each object is tracked continuously. In case of conflicts, when the assignment of detected moving objects to trackers is ambiguous (overlapping, covered, splitting and merging objects, etc.), the correct association is provided by the algorithm that iteratively analyses blob-tracker clusters and assigns blobs to trackers based on their visual similarity [7].

Only segmented images of moving objects are used for further processing. Furthermore, there is no need to analyze all images of every object because their variations on frame by frame basis are often insignificant. Therefore an object image is analyzed only if the defined amount of time t elapsed since the last image acquisition and only if the object moved a predetermined distance d. For the purpose of experiments values for t and d are calculated according to the following equations:

$$t = \frac{3}{f} \qquad\qquad d = 0.05 \cdot \frac{w+h}{2} \qquad\qquad (1)$$

where f denotes number of frames acquired by a camera per second, w and h represent video frame width and height.

This approach helps to reduce amount of data for analysis 3 to 10 times, depending on object movement characteristics.

3   Multi-camera Tracking

Multi-camera object tracking algorithm is presented in Fig. 1. Its task is to identify the same object in video frames from different cameras. In order to facilitate the task, spatial and temporal constraints regarding possible transitions between cameras are implemented to limit number of possible candidates for matching.

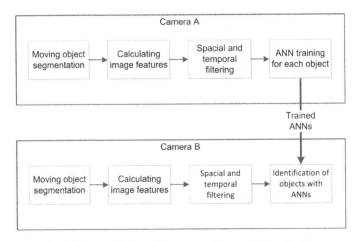

Fig. 1. Scheme of the multi-camera object tracking algorithm

Objects appearing in the current camera are matched with feasible objects from other cameras based on their image features. Matching is made with Artificial Neural Networks (ANNs). Each ANN is trained to recognize images of one particular object in the given camera and then it is employed to recognize the same object captured by a different camera.

3.1   Spatial and Temporal Filtering

Identification of objects moving between cameras' fields of view by comparing their visual features only is a difficult and error prone task in busy and complex scenes. Therefore, spatial and temporal rules are employed. Spatial rules involve hot areas in each camera field of view. For an object leaving a video frame in one of hot areas, its data are sent only to the nearest cameras, depending on the object's direction of movement. New objects in a video frame are treated as candidates for comparing if they appeared in a hot area only. Furthermore, the time duration of object transition between cameras is limited. Objects sent from neighboring cameras are compared with new objects appearing in appropriate hot areas only.

In a video frame of each camera, hot areas are selected manually. They represent entry/leave regions in a video frame and are defined as polygons placed on roads and other areas suitable for objects of interest movement (Fig. 2). Only objects entering the camera field of view from the outside are analyzed. This condition is met if an object appeared in the vicinity of a hot area, entered the hot area, left the hot area and it moves in the direction of the frame center. Similarly, the object is considered leaving the camera field of view if it entered a hot area, left the hot area and left the camera field of view, moving in the direction opposite to the direction of the frame center.

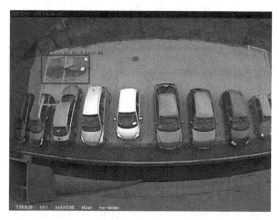

Fig. 2. Example of a vehicle (red rectangle and green, transparent mask) entering camera field of view in the left hot area (orange polygon)

3.2   Object Image Features

Object image features chosen for object identification in different cameras' fields of view should be robust against object pose changes and different optical settings of a camera (i.e. white balance). Therefore Speed Up Robust Features have been selected for this task.

Speeded Up Robust Features (SURF) [8] is a scale- and rotation-invariant local image descriptor around a selected interest point. SURF parameters are based on gray-level images and Haar wavelet responses in horizontal and vertical directions therefore they are less sensitive to color differences than color-based features, e.g. histograms. One SURF descriptor is a 64-element vector describing interest point vicinity. The wavelet responses are invariant to illumination offset. Invariance to contrast is achieved by turning the descriptor into a unit vector. Additionally, usage of many local features around several interest points instead of one global feature vector for the whole image facilitates object identification despite the change in its pose, as long as there is a sufficient overlap of object fragments visible in both images.

Interest points (their location, orientation and size) are chosen automatically with Fast-Hessian detector. Because the number of interest points may vary significantly for different images, an approach has been implemented to reduce fluctuations in the number of interest points. Detection of interest points is performed iteratively. Whenever the number of interest points K exceeds the target range $[D_K - 10\%, D_K + 10\%]$, where $D_K = 25$, the Fast-Hessian detector threshold is incremented or decremented and the detection is repeated. Iterations are terminated when K is in the defined range or after I iterations, where I is set to 30 based on initial experiments.

3.3   Object Identification

An Artificial Neural Network (ANN) is employed to recognize the same object registered by different cameras. ANN is trained on images of the single object captured by one camera and then it is used to recognize the same object in images acquired by another camera. Positive training samples are formed by the object features described in the previous section. Negative samples are formed by features of other objects that passed the field of view of the camera during the last T minutes. Therefore the ANN is trained to distinguish one particular object from other, possible similar objects that could be found in the same area and time.

For the purpose of object identification, a feed-forward ANN with one hidden layer is used. Because the number of image features related to one image of an object varies and depends on the number of interest points detected, it is not possible to feed ANN with all features of one image at one time. Therefore ANN is fed with SURF features calculated for each interest point, independently. Thus the number of inputs corresponds with the length of a SURF vector and is equal 64. Based on initial experiments, the number of neurons in the hidden layer is set to the half of ANN inputs. Bipolar sigmoid transfer functions are used in all neurons.

There are two outputs from the network. An expected ANN output is equal $(1, -1)$ for a SURF vector belonging to the valid object and $(-1, 1)$ if it belongs to other objects. During recognition, two ANN outputs (instead of one) make possible to get information both on the SURF vector similarity and the ANN response reliability.

In order to identify the object S seen in camera C_1 in video frames acquired from camera C_2, it is necessary to find the most similar object out of all candidates for matching. Let O_i, $i = 1...NO$ denotes ith objects for matching from among NO objects, I_{ij}, $j = 1...NI_i$ represents jth image of the ith object out of NI_i images of O_i and V_{ijk}, $k = 1...NV_{ij}$ defines kth SURF vector of the image I_{ij} from among NV_{ij} vectors. During the identification, all SURF vectors V_{ijk} of every image of all objects found in C_2 are classified with ANN that represents the object S received from C_1.

The ANN response r_{ijk} for the input vector V_{ijk} is calculated as follows:

$$r_{ijk} = 0.5 \cdot (o_1 - o_2 + 1), \quad r_{ijk} \in [0,1] \tag{2}$$

where o_i denotes ith element of the output vector scaled from $[-1, 1]$ range to $[0, 1]$ range. The response r_{ijk} represents the similarity of vector V_{ijk} to local features of the object S.

Additionally, each response r_{ijk} is assigned a weight w_{ijk} that represents ANN response reliability:

$$w_{ijk} = \frac{\max(o_1, o_2)^2}{o_1 + o_2} \tag{3}$$

ANN responses for all vectors of the image I_{ij} are aggregated as the weighted mean of the responses R_{ij}, according to the equation:

$$R_{ij} = \sum_{k=0}^{NV_{ij}} w_{ijk} \cdot r_{ijk} \bigg/ \sum_{k=0}^{NV_{ij}} w_{ijk} \tag{4}$$

Object S is matched with object O_i if there is an image I_{ij} of object O_i having the highest aggregated response R_{ij} from among all responses of other objects' images, according to the condition:

$$\exists_j \forall_{m \neq i} \forall_n R_{ij} \geq R_{mn} \tag{5}$$

If there are more objects satisfying condition (5), images with the same aggregated response R are discarded and the condition (5) is evaluated again amongst the most similar objects only. If there are more ANNs that match to the same object, ANN is assigned to the object with the highest value of the aggregated response.

4   Experiments and Results

For the purpose of experiments, a prototype video surveillance installation covering a part of the parking lot around an office building has been established (Fig. 3). The installation contains 8 fixed cameras with different orientations in relation to the ground (Fig. 4). The gaps between cameras' fields of view vary from a meter to a few dozen meters (e.g. a right angle turn between cameras no. 11 and 8). There is only one route connecting the parking entrance and the exit from the monitored area; it crosses fields of view of all cameras. Polygons denoting hot areas were labeled manually.

Approximately 10 hours of video streams from each camera, covering one weekday, have been used for experiments. Only vehicle images were analyzed. Vehicles (often several of them in the same time) were tracked during their movements in the monitored area with spatial and temporal rules only (section 3.1) in order to acquire sample images of the same vehicle in different cameras automatically. The results have been verified manually.

The database contains 12274 images of 246 different vehicles moving in the monitored area. Each vehicle transited from one camera to another at least once. The total number of transitions between cameras' fields of view is equal to 1901. Sample vehicle images from different cameras are presented in Fig. 5.

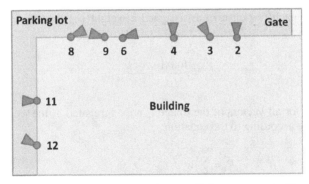

Fig. 3. Placements of cameras used in the experiments; digits correspond with camera numbers

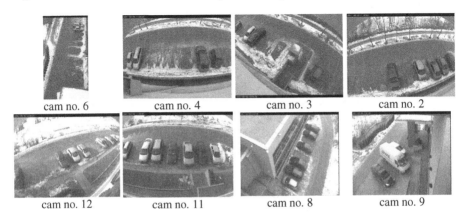

cam no. 6   cam no. 4   cam no. 3   cam no. 2

cam no. 12   cam no. 11   cam no. 8   cam no. 9

Fig. 4. Sample frames from all video cameras used in experiments; entering/leaving hot areas are marked with orange polygons

2   3   4   6   9   8   11   12

Fig. 5. Sample images of a vehicle entering (upper row) and leaving (lower row) a parking lot captured by each camera on its route; camera number at the bottom

For the purpose of experiments regarding vehicle identification based on its visual features, ANNs were trained with images of each vehicle in one camera and that the vehicle was recognized in all other cameras that it appeared in, not only in the nearest one according to the direction of movement. The positive training samples were formed by images of the vehicle S in a camera C_1. All images of other vehicles that appeared in the camera C_1 during the last $T = 10$ minutes were used as negative training samples. The positive testing samples were formed by images of the vehicle S in camera C_2, where $C_1 \neq C_2$. In order to create negative testing samples, images of all vehicles that appeared in the field of view of C_2 during next $T = 10$ minutes were used. If there were no vehicles found for negative training or testing samples, the calculations for the object S and cameras C_1 and C_2 were omitted. Additionally it was assured that negative training and testing samples does not contain images of the same vehicles, therefore during identification negative testing samples belonged to vehicles whose images were not used for ANN training.

The total number of identification performed is 5805. Because of insufficient negative training or testing samples, 409 identifications were impossible to perform. There were approx. 7 vehicles in negative training and testing sets on average. Each vehicle was represented by 10 images on average. It means that ANNs were trained with more negative than positive samples and their task was to find one valid vehicle in a set of approx. 8 vehicles. This is a very challenging task that is unlikely to happen in the real scenarios with the application of spatial and temporal constraints.

Table 1 contains summary results of vehicle identification for a specific camera as the source and as the destination of each transition analyzed. The worst results (below 50% of correctly identified vehicles) were obtained for camera no. 6, 8, 9 and 12 as the source and 6, 8 and 9 as the destination. Based on Fig. 4 and 5 it may be noticed that cameras no. 6, 8 and 9 have completely different orientations in relation to the road, comparing to other cameras. The majority of cameras observe moving vehicles from the side. On the other hand, the cameras no. 6, 8 and 9 are oriented more or less parallel to the road axis. Furthermore, in case of the same moving vehicle, cameras no. 6 and 9 are able to see it from the front, while camera no. 8 – from the back, and vice-versa, depending on the direction of movement. And lastly, the angle between the camera number 9 and the ground is very low while the same angle in case of camera no. 8 is almost perpendicular. These setup differences cause that images of vehicles in these cameras, comparing to images from other ones, are completely different. The results presented in table 1 show that the solution presented in the paper cannot cope with such drastic changes in object appearances. Therefore cameras no. 6, 8 and 9 were excluded from further analysis.

Table 2 presents results of vehicle identification depending on the pair of the cameras. It may be noticed that the algorithm can handle rotation of the camera no. 3, comparing to cameras no. 2 and 4, providing classification effectiveness over 85% for

Table 1. Summary results of vehicle identification for a specific camera as the source and as the destination of transitions

No. of camera	Camera as the source			Camera as the destination		
	Identifica-tions	Correct	%	Identifica-tions	Correct	%
2	912	681	74.7%	911	582	63.9%
3	937	635	67.8%	939	520	55.4%
4	923	644	69.8%	924	563	60.9%
6	868	340	39.2%	875	345	39.4%
8	533	166	31.1%	535	225	42.1%
9	704	275	39.1%	709	315	44.4%
11	454	245	54.0%	447	298	66.7%
12	376	134	35.6%	367	252	68.7%

each pair of cameras in this group. The worst results were obtained in case of transitions originating from camera no. 12 because its field of view differs the most from other remaining cameras.

Table 2. Results of vehicle identification for each pair of cameras

Transistion be-tween cameras	Number of identifications	Correct	%	Transistion be-tween cameras	Number of identifications	Correct	%
4 → 2	195	190	97.44	2 → 3	218	187	85.78
2 → 4	194	188	96.91	4 → 11	65	54	83.08
2 → 11	63	58	92.06	12 → 11	57	45	78.95
2 → 12	50	46	92.00	3 → 12	52	41	78.85
11 → 12	55	50	90.91	11 → 4	66	43	65.15
3 → 2	217	196	90.32	11 → 2	63	39	61.90
4 → 12	52	46	88.46	11 → 3	66	34	51.52
4 → 3	203	177	87.19	12 → 4	53	22	41.51
3 → 11	65	56	86.15	12 → 2	51	19	37.25
3 → 4	202	174	86.14	12 → 3	53	14	26.42
All	**2040**	**1679**	**82.30**				

All experiments presented in this section assume that the vehicle for identification is present in the testing set. The task of finding missing objects (not present in the destination camera field of view during identification with ANNs) is not cover in this paper. However, it was noticed that aggregated responses R for all images of positive testing vehicles are significantly higher that the responses of negative testing vehicle (Table 3). This property could be exploited to establish a minimum required value of the aggregated response R for the best-matched image. If the threshold is not exceeded than it may be assumed that there are no images of the searched object present in the testing set.

Total accuracy of identification of vehicles moving in different cameras' field of view varies depending on the setup of source and destination cameras. In case of similar camera orientations, the algorithm presented in the paper is sufficient for real applications. However in case of highly different poses of objects, a new approach needs to be implemented. In case of vehicles, their color data should be exploited provided that a mechanism for compensating differences in the color intensities between cameras is developed.

Table 3. Statistics of aggregated responses R of ANNs for all images of positive (valid) and negative (invalid) vehicles

Camera pairs	Positive vehicles		Negative vehicles	
	Mean of R	Standard deviation of R	Mean of R	Standard deviation of R
All	0.3061	0.1223	0.1670	0.06600
Without no. 6, 8, 9	0.4268	0.1065	0.1654	0.0796

5   Conclusions

The paper presents a solution for tracking moving objects in multiple cameras' fields of view. The experiments carried out were focused on identification of vehicles in different cameras. Artificial Neural Networks were employed for this task; each ANN learns one object appearance based on its images in one camera and then it recognizes its images captured by another camera. The results achieved prove that feature vector containing local image features (SURF) is sufficient to identify vehicles provided that the orientations of cameras are similar. In case of significant differences (e.g. perpendicular axes of view) the method described in the paper fails because of too large dissimilarities between vehicle images.

Future work will be focused on improving the method in order to allow more variations in camera orientations. The accuracy of the method in case of person tracking will be validated and an attempt to exploit objects' color will be made. Additionally, other object image descriptors will be studied and ANN training scheme will be altered in order to include images of an object from all previous cameras.

Acknowledgements. Research is subsidized by the European Commission within FP7 project "INDECT" (Grant Agreement No. 218086). The author wishes to thank the Gdansk Science and Technology Park for their help in establishing the test bed for the experiments described in the paper.

References

1. Javed, O., Safique, K., Rasheed, Z., Shah, M.: Modeling inter camera space-time and apperacnce relationships for tracking across non-overlapping views. Computer Vision and Image Understanding 109, 146–162 (2008)
2. Du, W., Piater, J.: Multi-camera People Tracking by Collaborative Particle Filters and Principal Axis-Based Integration. In: Yagi, Y., Kang, S.B., Kweon, I.S., Zha, H. (eds.) ACCV 2007, Part I. LNCS, vol. 4843, pp. 365–374. Springer, Heidelberg (2007)
3. Chilgunde, A., Kumar, P., Ranganath, S., WeiMin, H.: Multi-Camera Target Tracking in Blind Regions of Cameras with Non-overlapping Fields of View. In: British Machine Vision Conference BMVC, Kingston, September 7-9 (2004)
4. Cai, Q., Aggarwal, J.K.: Tracking human motion in structured environments using a distributed camera system. IEEE Trans. Pattern Anal. Mach. Intell. 2(11), 1241–1247 (1999)
5. Jeong, K., Jaynes, C.: Object matching in disjoint cameras using a color transfer approach. Machine Vision and Application 19, 443–455 (2008)
6. Czyzewski, A., Dalka, P.: Moving object detection and tracking for the purpose of multimodal surveillance system in urban areas. In: Tsihrintzis, G.A., et al. (eds.) New Directions in Intelligent Interactive Multimedia. SCI, vol. 142, pp. 75–84. Springer, Heidelberg (2008)
7. Czyzewski, A., Dalka, P.: Examining Kalman filters applied to tracking objects in motion. In: 9th Int. Workshop on Image Analysis for Mult. Interact. Services, pp. 175–178 (2008)
8. Bay, H., Tuytelaars, T., Van Gool, L.: SURF: Speeded Up Robust Features. In: Leonardis, A., Bischof, H., Pinz, A. (eds.) ECCV 2006, Part I. LNCS, vol. 3951, pp. 404–417. Springer, Heidelberg (2006)

Polish Speech Dictation System
as an Application of Voice Interfaces

Grażyna Demenko[1,2], Robert Cecko[1], Marcin Szymański[1], Mariusz Owsianny[1],
Piotr Francuzik[1], and Marek Lange[1]

[1] Poznań Supercomputing and Networking Center,
Polish Academy of Sciences, Poznań, Poland
[2] The Institute of Linguistics, Adam Mickiewicz University, Poznań, Poland
`{grazyna.demenko,marcin.szymanski,mariusz.owsianny,`
`piotr.francuzik,marek.lange}@speechlabs.pl, cecko@man.poznan.pl`

Abstract. This paper presents the results of the project realized at PSNC and supported by The Polish Ministry of Science and Higher Education – "Integrated system of automatic speech-to-text conversion based on linguistic modeling designed in the environment of the analysis and legal documentation workflow for the needs of homeland security", aiming at developing a Polish speech dictation (or Large Vocabulary Continuous Speech Recognition, LVCSR) system designed with the use of a phonetically controlled large vocabulary speech corpus and a large text corpora. The functions of the resulting system are outlined, the software architecture is presented briefly, then the example applications are demonstrated and the recognition results are discussed.

Keywords: speech recognition, dictation, voice interfaces.

1 Introduction

The objective of the speech-to-text project is the implementation of the speech recognition technology based on a very large corpora, especially for dedicated end-users in Poland – for the System of Justice, for the Police, the Border Guard and other services responsible for public security. The solution can be applied in making suit documentation, writing notes, protocols, recording inspections/post-mortem examinations, formal texts, legal documents, police reports, in dictating verdicts and verdict justifications, preparing stenographic records from meetings, sessions and materials from police operations, making summaries and in voice-controlled technical systems. The project is strictly correlated with the activity of the Polish Platform for Homeland Security within the area of creating innovative technology and computer tools to support law and public security institutions.

The structure of the remaining parts of the paper is as follows: Section 2 outlines the architecture of the system; Section 3 introduces the standalone software as well as example integration with a third-party application; in Section 4 the system accuracy is evaluated; the paper is concluded in Section 5, where possible future directions are also signaled.

A. Dziech and A. Czyżewski (Eds.): MCSS 2012, CCIS 287, pp. 68–76, 2012.

2   Features and Architecture

The Section 2.1 briefly lists the features and requirements of the dictation system. In Section 2.2 the overview of the system architecture is given.

2.1   Environment and Main Features

The present dictation system for Polish is designed to run on Microsoft Windows operating systems (XP or later). It was entirely written in C# and developed under the Microsoft .NET Framework 4.0 platform, with the use of Task Parallel Library (TPL). The system can work in two modes: off-line – the speech signal is read from a file, or on-line – the speech signal is acquired directly from an audio device (microphone). Depending on the quality/speed preset used (cf. Section 4), it is possible to obtain real-time recognition, which is understood as the behavior in which recognition is not slower than the playback of an utterance (the resulting sentence is presented with a small delay).

The recommended hardware resources for the software are: Intel Core i5 processor running Microsoft Windows 7 Pro 64-bit with 4GB of RAM. It is also possible to use the dictation system under 32-bit system with 2GB of RAM, at the expense of the lower quality of language modeling. Because of the intensive use of parallelism as well as the size of acoustic and linguistic models, the present system tends to use a maximum of hardware resources, with respect to both CPU and memory.

The dictation system can be used through one of two general interfaces: (1) as a standalone application (editor) that was designed to offer a maximum of features while remaining user-friendly and reasonably simple, thus hiding many internal details, or (2) through operating system integration, which allows to enter the recognized text directly into an active edit control of any third-party software. Section 3 gives more details.

In addition to its main feature which is, naturally, the textual presentation of a recorded sentence, the software also performs rudimentary text formatting which includes dates and numbers as well as capital letters and punctuation on sentence boundaries.

Since it is not possible for any speech recognition solution to contain all possible words that any user can utter, the system offers a possibility to add custom words.

It is well known that speech decoding is performed faster and more accurately once an original, speaker-independent acoustic model is tailored to a specific voice. For this reason, the presented dictation system also offers the speaker adaptation procedure which demands a target user to record ca. 250 fixed sentences.

2.2   System Architecture

The overview of the architecture of the Polish dictation system is presented in Fig. 1. It may be seen as one following general client-server architecture (although both sides can, and usually do, run on the same physical machine). The client is responsible only for delivering the audio signal and then presenting the recognition result. The server detects the speech, performs the recognition and formatting. The following subsections describe the acoustic and linguistic parts of the server.

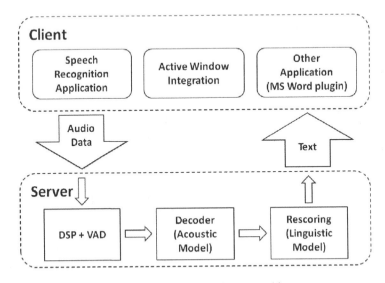

Fig. 1. Overview of Dictation System architecture

Voice Detections and Parameterization

The digital signal processing (DSP) and Voice Activity Detection (VAD) stage is presented in Fig. 2. The raw PCM audio signal from the client is passed to the DSP module, which divides it into separate observations (25 ms window with 10 ms stepping) and calculates the Mel-frequency cepstral coefficients (MFCC). The special LDA (Linear Discriminant Analysis) transformation is computed over those parameters in order to obtain VAD Parameters which are subsequently compared to statistical speech/silence/noise distributions held in VAD Acoustic Model. Once the Voice Activity Detection module determines the utterance boundaries, the list of the former parameters (i.e. MFCC) are passed to the next stage.

Decoding and Rescoring

The actual recognition stage, presented in Fig. 3, consists of two main operations: decoding and rescoring.

The Decoder is built upon Viterbi algorithm [1] (with some modifications, mainly to allow more effective parallel processing) and works over a Recognition Network being a word-loop of ca. 320-thousand dictionary entries with imposed unigram log-probabilities. As a result, it produces recognition hypotheses in form of the Word

Lattice. The decoder is also the module that uses the Acoustic Model, which is the collection of statistical distributions of Polish triphones' acoustic features. The model was trained over the material selected from the Jurisdict database designed specifically for the present dictation system whose target users are judges, lawyers, policemen and other public officers. The aforementioned database is a phonetically controlled large vocabulary corpus and contains recordings of speech delivered in quiet office environments by over 2000 speakers (a total of over 1155 hours of speech) from 16 regions of Poland.

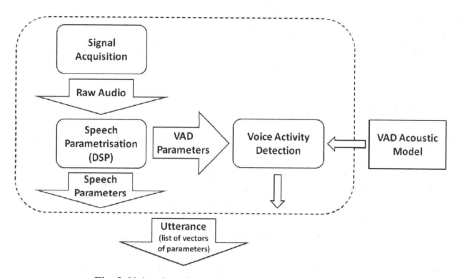

Fig. 2. Voice detection and speech parameterization stage

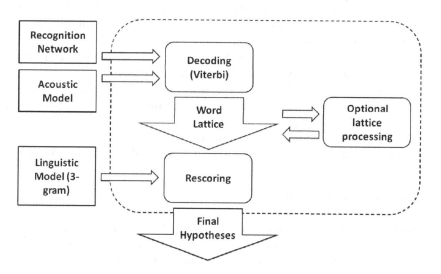

Fig. 3. Speech decoding and hypotheses rescoring

The Acoustic Model was estimated using HTK [2], with the standard training procedure for Triphone Continuous Density Hidden Markov Model. See [3-5] for more details on database design and [4,5] for details on Acoustic Model training.

The Word Lattice elements are attributed with appropriate acoustic probabilities. All lattice hypotheses are then evaluated using N-Gram Linguistic Model in the Rescoring module (see eg. [6]). The hypothesis with the best probability is returned as the recognized text. The Linguistic Model is currently an interpolation of two back-off trigram models (i.e. models that assign a word probability based a word in question and two preceding words) built with SRILM toolkit [7], estimated on over 4GB of automatically normalized text, including newspaper articles and legal-domain texts (judgments, law acts, briefs, contracts etc.).

Optional lattice processing module, presented in the Fig. 3, stands for the operation in which non-linguistic events are removed from the Word Lattice.

3   Applications

As already stated, the dictation system can be used through one of two general interfaces.

The standalone application, presented in Fig. 4, is software resembling a simple editor. It allows user to: select the input device (a microphone) for the on-line or disk audio file for the off-line recognition, to record a sentence (or a longer section of speech) and to play it back, to recognize an utterance on one of the 7 possible quality/speed presets and to manually correct the output sentence. The manual correction can be performed by, naturally, typing actually-spoken words, or, more importantly, by selecting one of alternative recognition hypotheses which are available within the context menu for most of output tokens. The recognition speed is presented in the status bar (both nominal and in terms of a real-time ratio); also, for testing purposes, it is possible to load a reference file (i.e. a text file with the proper orthographic transcription), that allows to calculate the recognition accuracy.

The Fig. 5 demonstrates the other interface, which is the operating system integration. This grants the possibility to enter the recognized text directly into an active edit control of any third-party software. The actual application presented in Fig. 5 is e-Posterunek (Polish for "e-PoliceStation"). The dictation system integration widget can be seen in the bottom-right corner.

However, the latter case does not allow to use alternative hypotheses unless a special plug-in is implemented (such a plug-in is being prepared for Microsoft Word).

4   Recognition Results

The speech recognition evaluation is commonly based on a recognition accuracy that is the ratio of the number of correctly recognized words minus the number of inserted words to the total number of words in the reference (perfect recognition result).

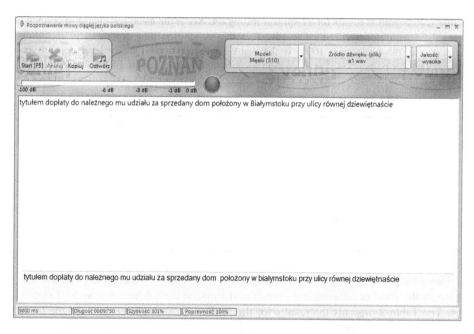

Fig. 4. Standalone speech dictation application. See Sec. 3 for description.

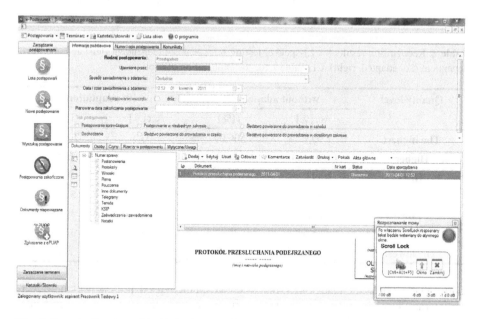

Fig. 5. Speech dictation with third-party software, e-Posterunek (e-PoliceStation), as an example of active window integration. Speech dictation widget is visible in the lower-right corner.

Table 1. Recognition results for 97-speaker testing set. **Acc** is the mean percentage of correctly recognized words minus inserted words, **T** is the recognition time compared to recording duration (ie. playback time; 100% is the real-time recognition). Quality level depends on internal decoder parameters. The evaluation was run on Intel® Core Duo E8400 3.00GHz, 4GB RAM.

Quality level	Acc[%]	T[%]
The highest	88,7	745,14
Higher	88,1	370,42
High	87,3	197,5
Mean	**86**	**117,1**
Low	84,1	74,95
Lower	82,2	56,9
The lowest	79,6	45,42

Here, we show the recognition accuracy for two distinct test corpora. The Table 1 presents the speaker-independent system performance for a 6-hour set with 97 speakers. The Table 2 presents the results for the 3-hour set with 13 speakers. In the latter case, however, the subjects separately recorded the adaptation sets, which allowed the evaluation of the speaker-dependent scenario. The accuracies and the recognition times are presented separately for each of the 7 quality/speed presets. The accuracy figures have been calculated using the Sclite tool [8].

Table 2. Recognition results for 13-speaker testing set, speaker-independent (left) and speaker-dependent (i.e. adapted, right). Cf. Table 1 for the explanation of captions.

Quality level	without adaptation		with adaptation	
	Acc[%]	T[%]	Acc[%]	T[%]
The highest	90,7	500,09	93	297,08
Higher	90,4	254,89	92,6	161,06
High	89,7	145,51	92,2	97,6
Mean	**87,9**	**82,83**	**90,8**	**52,72**
Low	85,6	56,95	89,2	39,51
Lower	83,5	45,6	88	33,17
The lowest	81,2	37,66	85	28,2

It can be observed that, under the selected quality/speed presets which reflect the internal decoder parameters (heuristic pruning deltas), the system accuracy tends to saturate at ca. 91% in case of non-adapted acoustic models and at ca. 93% in case of adapted-speaker models with ever-increasing recognition time. Conversely, the recognition time tends to saturate at the lower presets with ever-decreasing accuracy. The real-time recognition can be obtained with an accuracy of ca. 85%-88.5%

in case of speaker-independent and over 92% in case of speaker-dependent models. The speaker adaptation yields up to 30% of the reduction of error (adaptation method implemented within the software is the MLLR[9]).

5   Discussion and Future Development

The recognition accuracy and the general software behavior are very promising, making it an already serious candidate for the widespread release. The system is currently under intensive evaluation by a group of Police officers, with a positive overall feedback; also, training and evaluation campaigns are planned within selected lawyers' office courts, Border Guard and other services responsible for public security.

Moore[10] states that a 1000-hour database allows for building a system with a word error rate of ca. 12% when language modeling is applied; he also estimates that at least 100 000 hours of speech is needed to train an ASR system with accuracy comparable to that of a human listener. The system presented in this paper is already showing accuracy close to the former theoretical figure. However, the authors feel that some improvements can still be done, especially in terms of the recognition accuracy (that, ideally, should be improved without sacrificing speed, which is already acceptable in most presets). Some ideas include: tuning of some heuristics used by the decoder, or, possibly, re-allocation of the decoder (or some parts of it) to GPU.

From the functional point of view, the client-server architecture makes it possible to develop a speech recognition service, which would allow a remote (Internet) access, handling many recognition sessions at the same time, both live (on-line) and batch (off-line), possibly with many different acoustic models at the same time.

Acknowledgements. Work supported by grant "Integrated system of automatic speech-to-text conversion based on linguistic modeling designed in the environment of the analysis and legal documentation workflow for the needs of homeland security" (OR 00006707). The authors are currently supported by grant "Collecting and processing of the verbal information in military systems for crime and terrorism prevention and control." (OR 00017012).

References

1. Rabiner, L.R.: A tutorial on Hidden Markov Models and selected applications in speech recognition. Proc. of the IEEE 77(2), 257–286 (1989)
2. Young, S., et al.: The HTK Book (for HTK Version 3.2), Cambridge University Engineering Department (2002)
3. Klessa, K., Demenko, G.: Structure and Annotation of Polish LVCSR Speech Database. In: Proc. of Interspeech, Brighton UK, pp. 1815–1818 (2009)
4. Szymański, M., Klessa, K., Lange, M., Rapp, B., Grocholewski, S., Demenko, G.: Development of acoustic models for the needs of a speech recognition system using large lexical databases. Best Practices - Nauka w obliczu społeczeństwa Cyfrowego, Poznań (2010)

5. Demenko, G., Szymański, M., Cecko, R., Lange, M., Klessa, K., Owsianny, M.: Development of Large Vocabulary Continuous Speech Recognition using phonetically structured speech corpus. In: Proc. Intl. Congress of Phonetic Sciences, Hong Kong (2011)
6. Kneser, R., Ney, H.: Improved backing-off for M-gram language modeling. In: Proc. ICASSP, Detroit, vol. 1, pp. 181–184 (1995)
7. Stolcke, A.: SRILM - An Extensible Language Modeling Toolkit. In: Proc. Intl. Conf. Spoken Language Processing, Denver (2001)
8. Sclite tool kit on-line documentation,
   http://www.itl.nist.gov/iad/mig/tools/
9. Leggetter, C.J., Woodland, P.: Speaker Adaptation of Continuous Density HMMs Using Multivariate Linear Regression. In: Proceedings of ICSLP 1994, Yokohama, Japan (1994)
10. Moore, R.K.: A comparison of the data requirements of automatic speech recognition systems and human listeners. In: Proc. Eurospeech, Geneva (2003)

Latent Semantic Analysis Evaluation of Conceptual Dependency Driven Focused Crawling

Krzysztof Dorosz[1] and Michał Korzycki[2]

[1] Jagiellonian University
Department of Computational Linguistics
ul. Gołębia 24, Kraków, Poland
`krzysztof.dorosz@uj.edu.pl`
[2] AGH University of Science and Technology
Department of Computer Science
Al. Mickiewicza 30, Kraków, Poland
`korzycki@agh.edu.pl`

Abstract. In this paper we study a focused crawler driven by deep semantic analysis provided by the Conceptual Dependency (CD) theory. We test in practice the application of CD scripts as an approach of defining topics (queries) in a focused crawler and its robustness in evaluating real text structures extracted from HTML documents. In order to benchmark its efficiency in comparison to classical approaches, apart from human evaluation we also provide an evaluation of the result set based on its internal similarity using Latent Semantic Analysis (LSA). The performed measurement brings us to the conclusion that the CD theory is well suited for evaluating the similarity of HTML documents provided a specific query, as it achieves a high precision measured through human evaluation. At the same time we observe the drawbacks of LSA used in the same context.

Keywords: focused crawling, topic crawling, conceptual dependency, LSA.

1 Introduction

With the growing size and complexity of available information on the Web, new challenges appear in the field of Information Retrieval. The classical approach so far have been to download and store large amounts of information, index it in a way that will help to access this information, and allow to query this index. With the growing amount and complexity of available data and the shortening time window of its relevance, the mentioned approach seems to have reached the limits of its applicability. A query system based on accessing preindexed information must suffer from the lack of timeliness and low precision due to the amount of information retrieved.

Vertical search engines, distinct from the general-purpose Web search engines, are one of the ways of providing a solution for the appearing issues. If used in

A. Dziech and A. Czyżewski (Eds.): MCSS 2012, CCIS 287, pp. 77–84, 2012.
© Springer-Verlag Berlin Heidelberg 2012

a web context, they typically use a focused crawling engine that attempts to access and index only those Web pages that are relevant to a specified search topic. This approach leads much faster to more precise and concise search results than accessing a general-purpose index of pre-downloaded content.

In this paper we present and evaluate the application of a deep semantic analysis method through focused web crawling for retrieving text documents from a large document corpus such as the World Wide Web. The evaluation is based on comparing our approach with classical unsupervised methods for semantic text analysis, that will show clearly the advantages of the presented method.

2   Related Work

Focused crawling was introduced by Chakrabarti et. al. in [1]. A vast literature on the topic describes the usage of different approaches for query representation and recognition in the context of focused crawling, including boolean models with bag of words and keywords [2] or statistical techniques with naïve Bayesian methods [12] and LSI [6]. There are however no examples of the use of Conceptual Dependency for focused crawlers. To represent a query for a focused crawler we use the Conceptual Dependency theory introduced by Schank [11].

Latent Semantic Analysis (also referred to as Latent Semantic Indexing) is a highly efficient technique for analyzing relationships between documents and querying them [8]. It was introduced in the seminal work of Landauer et al. [3]. In this method, a matrix is constructed with rows corresponding to specific words, and columns corresponding to specific documents. We use some specific ways of populating this matrix, as suggested by Dumais [5], with some additional modifications related to the specifics of the language of the discussed corpus [7].

3   Focused Crawling

Focused crawling was introduced by [1] as a new approach for implementing crawling systems. Unlike a generic crawler which tries to cover an as large as possible subgraph of the Web to index a large scope of documents, a focused crawler is strictly driven by a single query. Basically it means that this kind of crawler tries to minimize the number of downloaded documents and at the same time tries to maximize the number of retrieved documents relevant to the query. Focused crawlers are commonly called also topic, topical, or vertical crawlers as they are built for retrieving very specific information from the Web. In order to focus the crawling we need to define a similarity function which returns a numerical value between 0 and 1 describing the similarity P of a document to a topic (query). Having this measure of page importance (a greater similarity means, that this document is more relevant) a rank of found URI anchors in the processed documents can be computed. This crawler part is especially important, because it gives the crawler the ability to intuitively drive its operation to those unknown URIs that most probably point to relevant documents and in the same time it avoids searching vast amount of probably irrelevant web addresses. Considering the practical time limitation for a single topic crawling and the scale of the Web, an improved crawling strategy performance is crucial even for finding any results at all. Despite the fact that focused crawling reduces significantly the scope of the browsed Web subgraph, it still requires to use a dedicated URI database capable of handling a very large crawling frontier [4]. The crawling frontier is the queue of all founded URI adresses prepared to be downloaded by the crawler in the next step.

The implementation of the similarity function P is a crucial part of such a focused crawler. The ordering of URIs in the crawling frontier is depends on the value of P. For example, the crawler can promote links found in the context of documents that have been already classified as relevant. A crawler can also promote links containing text that is already similar in any way to the driven query. There are many different approaches that were examined [9,10] in the literature. Those approaches differ in methods for topic representation and recognition as well as in measures for crawler performance evaluation.

In contrast to others our work is devoted to testing a focused crawler driven by conceptual dependency theory based deep semantic analysis, as described in the following chapters. We believe that this method can bring noticeable advantage in high precision search, as it is well suited for dealing with the textual content of web pages.

4 Conceptual Dependency Scripts

The Conceptual Dependency theory (CD) was introduced by [11]. The idea behind Schank's theory is an event oriented semantic structure that characterize every textual information. Basically the CD theory depends on the concept of roles with sentence like structures. The typical roles described by Schank are: actor, object, action, direction. We will demonstrate it by taking the following sentence as an example:

Joe gives Mary a book.

Here, the following roles can be matched:

*((**actor**: "Joe") (**action**: "give") (**object**: "a book") (**direction**: "Mary"))*

This single script sentence representation in an example of a synthetic event description. More complex events also can be formed in a CD structure using a sequence of simple events, what will be referred as an analytical event description . As a further example, the event describing the typical action of purchasing an item can be easily represented as a sequence of activities that need to occur in order to classify this event as a "purchase":

Somebody comes to a shopping area.
Somebody is choosing an item.

Somebody is paying for an item.
A shop is transferring the possession of an object to somebody.
Somebody lives a shopping area.

Each sentence describes a simple event that needs to occur. In fact we can consider more complex scenarios of events, as they could have variants, and in fact be represented as trees or, in general, as directed graphs. Considering this simple example the following script will represent one of possible information structure given in analytic way:

(((actor: ?) (action: "go", "come") (direction: "shop", "bakery", "e-commerce site"))
((actor: ?) (action: "choose", "browse") (object: ?))
((actor: ?) (action: "pay") (direction: "shop", "cashier"))
((actor: "shop", "cashier") (action: "give", "send") (object: ?) (direction: ?))
((actor: ?) (action: "leave", "logout") (direction: "shop", "bakery", "e-commerce site"))
)

Matching a script is the process of filling roles with appropriate text tokens found in a single sentence. It is possible to have only a partial match, as the text often does not contain the full information about the event. After a successful match this method not only gives us the fact that the text matches a pattern but also extracts additional information, i.e. it returns the set of filled roles that are represented as text tokens (words).

5   LSA Based Similarity Measurement

Latent Semantic Analysis is a widely used technique in vectorial semantics for analyzing relationships between documents. In this method, a matrix is constructed with rows corresponding to specific words, and columns corresponding to specific documents. In its simplest scheme - the matrix contains the count of words occurring in each document. This large matrix is then reduced to a low-rank approximation through Singular Value Decomposition, leading to a much smaller "concept" space. The documents are then compared using cosine similarity in this reduced concept space with 1 representing very similar documents and -1 very dissimilar ones. As has been shown [3], this low-rank representation of the knowledge actually improves over the original full matrix in many information retrieval applications.

As the system retrieves documents from the web, a corpus is built and expanded. This corpus is used to constantly train a Latent Semantic Analysis model, updating it with each retrieved entry. We use as the training data the whole set of retrieved documents to this specific point of time. The model is built based on a term/document matrix populated through a classical [5] term-frequency/inverted document frequency scheme. As the analyzed texts are in a highly inflectional language, lemmatisation, a common [7] preprocessing technique is used. The term space is reduced beforehand to base forms, using a dictionary based approach where possible, and a dictionary trained stemmer where a dictionary is insufficient. That approach permits us to find much sooner (i.e. in a smaller corpus) matching terms, without the need to have a corpus covering most of the inflexional forms.

For similarity measurements, the documents matching the Conceptual Dependency script are used. The matching documents retrieved so far are converted to the trained N-dimensional LSA space (for $N = 200$) as described above. Each of them is then used in turn as a query (as described in the seminal work of Deerwester, Dumais et al. [3]) and its cosine similarity is measured toward each of the rest of the matching documents. For a specific set of CD script matching documents, a set of similarities (represented by numbers ranging from -1 to 1) between each other is gathered. From this set we will consider its maximal, median and average values.

6   Testbed and Results

Our test was performed live on the site *pajeczyna.pl* using our own crawling software. The analyzed site is a service in Polish with classified advertisements, similar to *craigslist.com*. Our search topic was for adverts related to human organ traffic. We prepared the following CD script for determining if a text contain an offer related to human organ traffic:

*((**actor**: "zdrowy", "nałóg", "palący", "młody", "młoda", "pośrednik", "dawca", "mężczyzna", "kobieta") (action: "odstąpić", "kupić", "sprzedać", "sprzedanie", "potrzebować", "potrzebny", "dać", "oddać", "sprzedaż") (**object**: "nerka", "szpik", "wątroba", "śledziona", "narząd") (**direction**: "przeszczep", "przeszczepić", "transplantacja", transplantować", "pośrednictwo", "pomoc", "pomóc") (**circumstance**: "tanio", "niedrogo", "pilny", "RH+", "RH-", "ARH+", "ABRH+", "BRH+", "0RH+", "ARH-", "ABRH-", "BRH-", "0RH-"))*

Basically we searched for actions of trading objects like kidney, bone, liver, spleen or organ. The occurance of additional roles was also promoted - reason (transplantation, support) or the circumstance (cheap, urgently, blood type).

The search was started from one URI containing a posting of organ traffic retrieved using a generic search engine, and was limited strictly to the *pajeczyna.pl* domain and its subdomains. The crawler downloaded in total 2275 HTML documents, out of which 9892 text documents were retrieved through a HTML segmentation process and saved as a flat text corpus. From the corpus, the CD extractor retrieved 16 documents matching them to the search query. Human evaluation of the given topic gave us the observation that out of 16 retrieved documents 1 document was incorrectly classified as matching. Thus, the precision

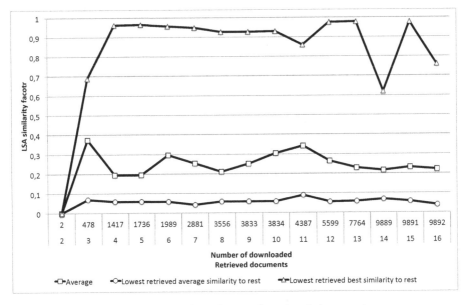

Fig. 1. LSA based similarity of retrieved documents

in this case is 93.75%. The whole set of downloaded documents was also subjected to human evaluation, with a result of 1 document found that was relevant but not retrieved by the crawler. The recall in this case would also have a value 93.75%. High values of precision and recall prove that the CD method provides a very good topic representation for this case. Having the dataset annotated with the relevance of particular documents to the submitted topic, we subjected this result set to LSA similarity measurement, to test if the retrieved documents are similar in the LSA sense.

Figure 1 presents the results of evaluating the retrieved documents through LSA. Those figures were obtained by building the LSA model online on the corpus of downloaded documents while they are being retrieved. The represented values on the graph come from taking the set of all cosine similarities between each of the documents that have been matched by the conceptual dependency script. The maximum, minimum, average and median values of this set are represented on the graph above.

As can be observed, the average similarity of retrieved documents stabilizes quickly on a certain level, while the maximum and minimum tend to diverge. What is interesting, is the fact, that the minimal, average and median similarities stays at most on the same level (if not diminish slightly) with the increasing size of the retrieved corpus, meaning that the CD script tends to retrieve documents that are more and more dissimilar (in the LSA sense) .

We analyzed further the retrieved content. Taking the retrieved documents and taking the distances from each of them to each of the non-matching documents in the corpus, we can see in figure 2 that the average and median distance

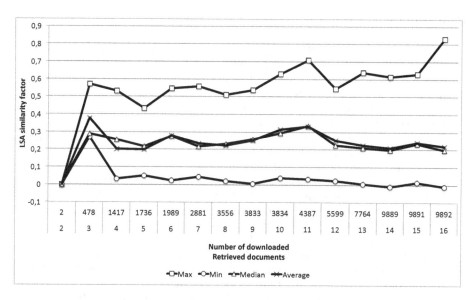

Fig. 2. LSA similarity of retrieved documents to non-matching documents

between the retrieved documents and the rest of the corpus is similar to the median and average distance between matching documents. We can also tell by looking at the respective maxima and minima, that the retrieved documents have much closely corresponding documents in the remaining non-matching corpus than in the matching one.

7   Conclusions

As we have shown above, a Conceptual Dependency script approach for information retrieval permits to create a highly discerning and precise classification scheme for documents.

In the described experiment, we obtained a precision and recall both of the value of 93.75% (1 document unmatched and 1 document matched erroneously). We also clearly demonstrated, that documents that are highly relevant semantically to the submitted query can be retrieved by a Conceptual Dependency script, where a LSA driven approach would clearly fail to discern between matching and non-matching documents, giving a strong argument for the preference of the former method over the latter for a highly discerning mechanism for deep semantic based information retrieval.

Acknowledgements. This work is supported by the European Commission via the EU FP7 INDECT project, Grant No.218086, Research area: SEC- 2007-1.2-01 Intelligent Urban Environment Observation System.

References

1. Chakrabarti, S., van den Berg, M., Dom, B.: Focused crawling: a new approach to topic-specific web resource discovery (1999)
2. Cho, J., Garcia-Molina, H., Page, L.: Efficient crawling through url ordering. Computer Networks and ISDN Systems 30(1-7), 161–172 (1998); Proceedings of the Seventh International World Wide Web Conference
3. Deerwester, S.C., Dumais, S.T., Landauer, T.K., Furnas, G.W., Harshman, R.A.: Indexing by Latent Semantic Analysis. Journal of the American Society of Information Science 41(6), 391–407 (1990)
4. Dorosz, K.: Usage of dedicated data structures for url databases in a large-scale crawling. Computer Science: rocznik Akademii Górniczo-Hutniczej imienia Stanisława Staszica w Krakowie 10, 7–17 (2009)
5. Dumais, S.: Enhancing Performance in Latent Semantic Indexing. Technical report, TM-ARH-017527 Technical Report, Bellcore (1990)
6. Hao, H.-W., Mu, C.-X., Yin, X.-C., Li, S., Wang, Z.-B.: An improved topic relevance algorithm for focused crawling. In: SMC, pp. 850–855 (2011)
7. Kuta, M., Kitowski, J.: Clustering Polish Texts with Latent Semantic Analysis. In: Rutkowski, L., Scherer, R., Tadeusiewicz, R., Zadeh, L.A., Zurada, J.M. (eds.) ICAISC 2010. LNCS, vol. 6114, pp. 532–539. Springer, Heidelberg (2010)
8. Landauer, T.K., Dumais, S.T.: A solution to plato's problem: The latent semantic analysis theory of acquisition, induction, and representation of knowledge. Psychological Review 104(2), 211–240 (1997)
9. Menczer, F., Pant, G., Srinivasan, P., Ruiz, M.E.: Evaluating topic-driven web crawlers (2001)
10. Passerini, A., Frasconi, P., Soda, G.: Evaluation Methods for Focused Crawling. In: Esposito, F. (ed.) AI*IA 2001. LNCS (LNAI), vol. 2175, pp. 33–39. Springer, Heidelberg (2001)
11. Schank, R.C., Tesler, L.: A conceptual dependency parser for natural language. In: Proceedings of the 1969 Conference on Computational Linguistics, COLING 1969, pp. 1–3. Association for Computational Linguistics, Stroudsburg (1969)
12. Zhang, H., Lu, J.: A fuzzy approach to ranking hyperlinks. In: Proceedings of the Fourth International Conference on Fuzzy Systems and Knowledge Discovery, vol. 03, pp. 406–410. IEEE Computer Society, Washington, DC (2007)

Speaker Recognition Based on Multilevel Speech Signal Analysis on Polish Corpus

Szymon Drgas and Adam Dabrowski

Chair of Control and Systems Engineering, Poznan University of Technology,
Piotrowo 3A, Poznan, Poland
{szymon.drgas,adam.dabrowski}@put.poznan.pl

Abstract. This article deals with a new approach to the text-independent speaker verification task. It is namely proposed to combine spectral and the so-called high-level features (prosodic, articulatory, and lexical) in order increase accuracy of speaker verification. The presented experiments were performed using a Polish language corpus called PUEPS. It contains semi-spontaneous telephone conversations (acted emergency telephone notifications) recorded in laboratory conditions. As the Polish language is under resourced and the PUEPS corpus is relatively small, another approach is needed than these known from the well known NIST evaluations. The authors proposed to use the fast scoring instead of more complex classifiers and the AdaBoost algorithm for features combination. Combination of features resulted in equal error rate (EER) reduction for various SNR conditions.

Keywords: Speaker recognition, high-level features, kernel combination, boosting.

1 Introduction

In this article the text-independent speaker verification task is considered in context of emergency telephone conversations. An approach that was chosen was to combine spectral and the so-called high-level features in order increase accuracy [13]. Experiments were performed using Polish language corpus called PUEPS. It contains semi-spontaneous telephone conversations recorded in laboratory conditions. This gives a possibility to control degradation of the speech signal. As the Polish language is under resourced and the PUEPS corpus is relatively small, another approach is needed than these known from NIST evaluations. The authors proposed to use the fast scoring instead of more complex classifiers and the AdaBoost algorithm for features combination.

The article is structured as follows: in Section 2 multi-level features approach to speaker recognition is described. Next, in Section 3 scoring and feature combination methods are presented. In Section 4 an experimental method and the results are showed and discussed. Finally, conclusions are given in Section 5.

A. Dziech and A. Czyżewski (Eds.): MCSS 2012, CCIS 287, pp. 85–94, 2012.
© Springer-Verlag Berlin Heidelberg 2012

2   Multi-level Speaker Recognition

In multi-level speaker recognition systems several types of features are extracted, next they are combined and finally a classifier is used to discriminate speakers. So-called higher level features provide complementary information to classic spectral features and they make system more robust [8,12,13,2,4]. In this work four types of features were used: spectral, prosodic, articulatory, and lexical. Spectral features convey information about timbre which is related to shape and size of the vocal tract. They are based on computation of MFCC (mel-frequency cepstral coefficients) features. Prosodic features carry information about intonation, accent and rhythm. These features are obtained from F0 and intensity contours. Articulatory features are related to the characteristic pronunciation of the speaker. In order to extract these features, neural networks trained to discriminate between articulatory classes were employed. Finally, lexical features correspond to characteristic words or phrases used by the speakers. They were obtained from transcriptions of the utterances in the PUEPS corpus.

3   Fast Scoring and Feature Combination

In the presented speaker recognition system after the feature extraction phase, each conversation side is represented as K vectors. As in this work spectral, prosodic, articulatory, lexical features were used $K = 4$, these vectors are stored in K matrices $\mathbf{X}_1, \ldots, \mathbf{X}_K$ where:

$$\mathbf{X}_k = [\mathbf{x}_{k1} \quad \cdots \quad \mathbf{x}_{kN}] \tag{1}$$

where $\mathbf{x}_{ij}$ is the feature vector of i'th type extracted for j'th conversation side. N denotes the number of conversation sides available in the corpus. In the next phase, from matrices $\mathbf{X}_i$, kernel matrices were computed. In the performed experiments cosine kernel was used. The (j, k) element of the matrix $\mathbf{K}_i$ can be calculated as follows:

$$\mathbf{K}_i(j, k) = \frac{\mathbf{x}_{ij}^T \mathbf{x}_{ik}}{||\mathbf{x}_{ij}|| ||\mathbf{x}_{ik}||} \tag{2}$$

There is no SVM classifier used, but values of the kernel matrix elements are used as decision scores. This can cause a problem when there are more than one training sides per speaker model. It is not obvious how to combine information from the available training sides in order to achieve higher speaker recognition accuracy. In this article, this problem was not addressed. It is because durations of conversation sides collected in the PUEPS corpus are not normalized. This is caused by a type of conversations. Indeed, it is natural that the emergency telephone calls differ in duration. The experiments with varying number of training sides could lead to uninterpretable results, when the times of these sides differ. Another argument for this approach is that, when there is no large dataset available to model background speakers, there is no any benefit from using SVM in terms of verification error. Thus, its rejection simplifies the system and reduces computational complexity. Using kernel matrix element values directly as scores is known in the literature as fast scoring [7].

It is important to use proper kernel in such systems. Each kernel matrix element should reflect similarity between the corresponding sides. For example it is possible to show datasets, for which linear kernel values would not reflect similarity. Instead, the cosine kernel or spherical normalization can be used.

Using fast scoring makes possibility to use the boosting algorithms to combine information from different features. However, values of elements of the kernel matrix should be centered at zero. Thus, the score matrix $\mathbf{S}$ was introduced:

$$\mathbf{S}_i = \mathbf{K}_i - \theta_i^{\mathrm{EER}} \mathbf{11}^T \tag{3}$$

where $\mathbf{1}$ is a vector filled with ones, while θ_i^{EER} is the threshold, for which probability of false alarms is equal to the probability of the miss error for i'th kernel matrix. Each kernel matrix needs to be shifted by this value. After this operation the semi-definite property of the kernel matrices can be lost.

Kernel boosting is a method of the kernel matrix combination, which is similar to the MKL algorithm and the combination coefficients are optimized.

$$\mathbf{K} = \sum_i \alpha_i \mathbf{K}_i \tag{4}$$

where α coefficients determine contribution of features in the resulting kernel matrix. Boosting is based on the AdaBoost algorithm [6]. In the present work score matrices $\mathbf{S}_i$ instead of the kernel matrices were used.

The proposed algorithm can be presented in the following steps:

1. **Input:** Kernel matrices set and corresponding labels: $\{(\mathbf{S}_i, \mathbf{D}_i)\}_{i=1}^k$. Where:

$$(\mathbf{D}_i)_{jl} = \begin{cases} 1 & \text{if } \mathbf{x}_j \text{ i } \mathbf{x}_l \text{ belong to the same class} \\ -1 & \text{otherwise} \end{cases}$$

   $\mathbf{S} \in \mathbb{R}^{N \times N}$, $\mathbf{D} \in \mathbb{R}^{N \times N}$.
2. **Initialization:** $\mathbf{W} = 1/m$, where $m = \frac{n(n+1)}{2}$, $\hat{\mathbf{S}} = 0$
3. **For** each pair $(\mathbf{S}_i, \mathbf{D}_i)$
   (a) $S^+ = \{(i,j) : (\mathbf{S}_i)_{jl}(\mathbf{D})_{jl} > 0\}$, $S^- = \{(i,j) : (\mathbf{S}_i)_{jl}(\mathbf{D}_i)_{jl} < 0\}$
   (b) $W^+ = \sum_{(j,l) \in S^+} (\mathbf{W})_{jl} |(\mathbf{S}_i)_{jl}|$
      $W^- = \sum_{(j,l) \in S^-} (\mathbf{W})_{jl} |(\mathbf{S}_i)_{jl}|$
   (c) $\alpha = \frac{1}{2} \log \left(\frac{W^+}{W^-} \right)$
   (d) $(\mathbf{W})_{jl} = (\mathbf{W})_{jl} \exp(-\alpha (\mathbf{D})_{jl} (\mathbf{S})_{jl})$

   (e) $\mathbf{W} = \frac{\mathbf{W}}{\mathbf{1}^T \mathbf{W} \mathbf{1}}$
   (f) $\hat{\mathbf{S}} = \hat{\mathbf{S}} + \alpha \mathbf{S}$
4. **Output:** Kernel matrix $\hat{\mathbf{S}}$

Boosting algorithm uses exponential loss function $\exp(-x)$ that bounds empirical error. Using this function the optimal kernel weights are determined. Additionally each element of the added kernel matrix is weighted in order to optimize mixing weight mainly using badly classified data.

4   Experiments

4.1   PUEPS Corpus

PUEPS corpus was recorded in order to provide spontaneous speech samples in Polish. It contains a set of telephone conversations with the emergency telephone service.

Data Recording and System Architecture. There were several objectives while PUEPS corpus [3] was designed. The language of speech had to be Polish. Second, there was a need for spontaneous speech. It has been very important as lexical features have sense when the speaker uses his/her own words. It has also significance in case of prosodic features (tempo, and speaking style are less consciously controlled by a speaker) as well as articulatory (speaker articulates words less carefully). Another requirement of this corpus was a laboratory quality of the recordings. This feature allows free signal degradation; for example using Head and Torso simulator as in [5]. Finally, each speaker needed to record several conversations to make automatic speaker recognition experiments possible. Moreover, the time between conversations of one speaker should be longer than day in order to obtain a within-speaker variability that is close to reality.

The records contain the acted telephone conversations with the emergency telephone service. The task of the speakers was to act as a person that is a victim or observer of a crime or an emergency situation. He/she had to report the event to the emergency telephone operator. Only Polish native speaker took part in the recordings. In order to provide the speaker information about the situation to describe, without suggesting words to use, video movies were prepared and presented to the speakers.

At the beginning of an experiment the speaker had to watch such a short movie with a crime scene. After that he or she was calling to the emergency phone and reporting a crime to a person who played role of the police officer. The telephone call was recorded and prepared for the post-processing.

The caller was located in an anechoic room equipped with a terminal, a telephone and a high quality microphone. After watching the movie the participant called over the PBX to the emergency phone operator, who was located in another laboratory. The call was recorded twofold: with a digital call recorder from the telephone line and with an audio recorder from the observer microphone. The call recorder processed signal of the whole conversation with a typical telephone line quality, while the high quality caller voice was stored with the audio recorder for post-processing.

Calls Database Statistics. 30 speakers participated in the recording sessions of PUEPS corpus. All of them were students aged between 19 and 26. Majority of recordings consist male voices (27 speakers) with the rest done by females (3 speakers). The role of the operator that received emergency notifications was performed by 3 persons.

Each speaker recorded 6 conversations. Maximally two conversations were recorded by one speaker during one session. Minimal time between two sessions in which speaker could take part was one week. The mean conversation time is 111 s, while median is 105 s.

4.2   Feature Extraction

The high-level features were extracted in the same way as in earlier authors' works [9]. The shortened description of extraction is presented below.

Spectral Features. In order to convey spectral information to multilevel recognition system, GMM supervectors were used For each recording the MFCC's were extracted and the GMM model was trained. In order to cope with two distributions the Kullback-Leibler (KL) divergence was used. However, it cannot be applied directly, because in our case the Mercer's condition would not be fulfilled. One of possibilities is to use the function that is an upper bound of the KL divergence.

$$d(\mathbf{m}_1, \mathbf{m}_2) = \frac{1}{2} \sum_{i=1}^{C} \omega_i (\mathbf{m}_i^1 - \mathbf{m}_i^2) \boldsymbol{\Sigma}_i^{-1} (\mathbf{m}_i^1 - \mathbf{m}_i^2) \tag{5}$$

where ω_i denotes weights, $\mathbf{m}_i$ mean, and $\boldsymbol{\Sigma}_i$ is a covariance matrix of i'th component. Weights and covariance matrices are the same for all conversations and are equal to parameters of global GMM model for training data. Kernel function can be expressed in the following way:

$$K(\mathbf{m}_1, \mathbf{m}_2) = \sum_{i=1}^{C} \omega_i \mathbf{m}_i^1 \boldsymbol{\Sigma}_i^{-1} \mathbf{m}_i^2 \tag{6}$$

Prosodic Features. The prosodic features are based on a linear approximation of F0 and intensity contours computed similarly as in [1]. The procedure consists of the following steps:

1. fundamental frequency and intensity contours extraction,
2. determination of voiced speech segments,
3. approximation of F0 contours with lines,
4. approximation of intensity contours with lines,
5. quantization of directional coefficients and interval lengths,
6. combination of quantized sequences into one sequence,
7. calculation of a bag of n-grams statistics and forming the resulting vector.

Articulatory Features. In order to catch speaker characteristics connected to pronunciation, articulatory features were extracted. It was done using neural networks trained on 2000 hours of telephone speech [11]. First, spectral features were extracted using PLP (perceptual linear prediction) method. Next, each frame and its context (4 preceding and 4 subsequent frames), has been classified using multi-layer perceptrons (MLPs) in order to extract articulatory features. Two categories of articulatory features were taken into account: place of articulation and articulatory degree (two MLPs were used). The perceptrons had 3 layers. The number of units of the input layer was 39 (dimension of PLP feature vector), the number of hidden units was set according to amount of available data. It was 1900 for MLP for place of articulation and 1600 for degree of articulation. The outputs of the MLPs correspond to places and degrees of articulation summarized in Table 1. Activations at the outputs of the neural networks were quantized and modeled by bigrams.

Table 1. Outputs of the MLPs for acticulatory features extraction

Category	Outputs
place	alveolar, dental, labial, labio-dental, lateral, post-alveolar, rhotic, velar
degree	approximant, closure, flap, fricative, vowel

Lexical Features. Idiolectal aspects of speakers were also taken into account [8]. They were caught by lexical features obtained from manually made transcriptions of the PUEPS corpus. First, dictionaries were constructed for word bigrams. Then for each side the number of occurrences of each word in the dictionary was counted. These numbers of occurrences were used to construct bi-gram vectors for each side.

4.3   Experimental Setup

The fact that the PUEPS corpus was recorded in laboratory conditions gives possibility to perform experiments, in which a degree of the signal degradation is controlled. The so-called "babble noise" was added to original recordings with the following SNRs: 20, 10, and 0 dB. In case of testing verification for each feature type individually error was using data from all speakers from the PUEPS corpus. However, in case of testing combination of features it was necessary do divide corpus into background and test data. The background data were used to train feature weights (using AdaBoost), while test data were used to determine EER. As the number of speakers in the corpus is relatively small experiments were performed for 20 different divisions of speakers. Finally, EERs from all divisions were averaged.

4.4   Results

The DET curves for SNR 20dB are presented in Figure 1. The lowest EER (9.26%) was achieved for spectral features. The about two times higher EER was brought for prosodic features (21.72%). A slightly worse accuracy was obtained for lexical features (29.14%). The worst performance was observed for articulatory features (EER=33.50%).

For spectral features the EER is about two percent higher than in case of experiments conducted by the authors with the Switchboard corpus [10] for one training side condition. The cause of this discrepancy may be the following factors: first, duration of each conversation side and second presence of the background dataset. For the Switchboard database the side duration it is about 2,5 minute while for the PUEPS corpus the average duration is 1 minute and its dispersion is much higher than this of the Switchboard database. Additionally, for the PUEPS corpus no background dataset is available.

Prosodic features, despite the duration and time differences, give similar accuracies. Prosodic features vector has a dimension of 100. This is a not very large number and it can perform well.

Articulatory features for the PUEPS corpus give a higher error than that in case of the Switchboard database [10] and it is equal to 33.5%. Here, the number of elements of the feature vectors is 3600. In case of short conversations (about 1 minute) this number is similar to the number of frames for many conversations in the PUEPS corpus. This can lead to a high noise level.

Lexical features give error about 50% higher than in case of Switchboard. Here the bigram lexicon contains 424 positions.

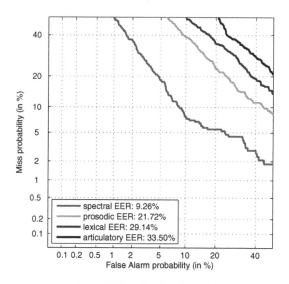

Fig. 1. Results for SNR 20dB

The DET curves for the case, in which SNR is 10 dB, are presented in Figure 2. The increase of the noise level only in a small degree influenced the EER for spectral features. It increased about 0.05%. Prosodic features turned out to be more sensitive to "babble noise" – it increased by about 10.08%. For articulatory features with SNR decrement the EER slightly decreased. It can be caused by masking some unreliable features.

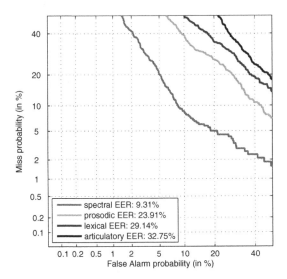

Fig. 2. Results for SNR 10dB

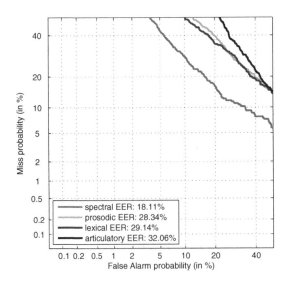

Fig. 3. Results for SNR 0dB

For SNR 0dB (see Figure 3) significant deterioration of accuracy of spectral features was observed – the EER increased 2 times in comparison to the SNR=10dB case. For prosodic features the EER increased by about 18,53%. The EER for articulatory features remained at a similar value, (above 30%). But it is still high. This means that it does not carry much additional speaker-dependent information.

Table 2. Results for features combined with AdaBoost algorithm

SNR	spectral features	combined features
20	9.59	7.54
10	9.92	9.38
0	18.89	16.39

Combined Features. The results of the speaker verification accuracy based on features combination with weights of each feature set determined with the AdaBoost algorithm, are presented in Table 2. In the experiments the spectral, prosodic, and lexical features were combined. The articulatory features were rejected, because of a relatively high EER (more than 30%). Indeed, the preliminary experiments showed that the considered combination together with articulatory features led to some system accuracy deterioration.

For SNR=20% the system based on features combination gives an error of about 21.21% lower than for the system based on spectral features. For SNR=10dB an improvement caused by the features combination is slight - it is about 5%. It is caused by the fact, that the noise influenced spectral features in small degree while accuracy of the prosodic features was decreased. In this case, a combination with high-level features caused the EER decrement by about 13.23%.

5   Conclusions

From the performed experiments the following conclusions can be drawn:

1. Prosodic and lexical features provide a complementary information to spectral features in the speaker verification task
2. Articulatory features do not provide useful speaker-dependent information in multi-level speaker recognition systems
3. AdaBoost and fast scoring methods can effectively be used to combine information from various features.

References

1. Adami, A.G.: Modelling prosodic differences for speaker recognition. Speech Communication 49(4), 277–291 (2007)
2. Baker, B.J.: Speaker verification incorporating high-level linguistic features. PhD thesis, Queensland University of Technology (2008)
3. Balcerek, J., Drgas, S., Dabrowski, A., Konieczka, A.: Prototype multimedia database system for registration of emergency situations. In: SPA Conference (2009)
4. Campbell, W.M., Campbell, J.P., Gleason, T.P., Reynolds, D.A., Shen, W.: Speaker verification using support vector machines and high-level features. IEEE Transactions on Audio, Speech, and Language Processing 15(7), 2085–2094 (2007)

5. Cetnarowicz, D., Drgas, S., Dabrowski, A.: Speaker recognition system and experiments with head / torso simulator and telephone transmission. In: Signal Processing Algorithms, Architectures, Arrangements, and Applications Conference Proceedings (SPA), 2010, pp. 99–103 (September 2010)
6. Crammer, K., Keshet, J., Singer, Y.: Kernel design using boosting. In: NIPS 2002 (2002)
7. Dehak, N., Dehak, R., Kenny, P., Brümmer, N., Ouellet, P., Dumouchel, P.: Support vector machines versus fast scoring in the low-dimensional total variability space for speaker verification. In: Interspeech 2009 (2009)
8. Doddington, G.: Speaker recognition based on idiolectal differences between speakers. In: Eurospeech 2001, pp. 2521–2524 (2001)
9. Drgas, S., Dabrowski, A.: Kernel alignment maximization for speaker recognition based on high-level features. In: Interspeech 2011, pp. 489–492 (2011)
10. Drgas, S., Dabrowski, A.: Kernel matrix size reduction methods for speaker verification. In: 5th Language & Technology Conference (2011)
11. Frankel, J., Magimai-Doss, M., King, S., Livescu, K., Cetin, O.: Articulatory feature classifiers trained on 2000 hours of telephone speech. In: Interspeech 2007 (2007)
12. Reynolds, D., Andrews, W., Campbell, J., Navratil, J., Peskin, B., Adami, A., Jin, Q., Klusacek, D., Abramson, J., Mihaescu, R., Godfrey, J., Jones, D., Xiang, B.: The supersid project: exploiting high-level information for high-accuracy speaker recognition. In: Proc. IEEE International Conference on Acoustics, Speech, and Signal Processing (ICASSP 2003), April 6-10, vol. 4, pp. IV–784–7 (2003)
13. Shriberg, E.: Higher-Level Features in Speaker Recognition. In: Müller, C. (ed.) Speaker Classification I 2007. LNCS (LNAI), vol. 4343, pp. 241–259. Springer, Heidelberg (2007)

Face Detection and Facial Expression Recognition Using a Novel Variational Statistical Framework

Wentao Fan and Nizar Bouguila

Concordia Institute for Information Systems Engineering
Concordia University, QC, Canada
wenta_fa@encs.concordia.ca, bouguila@ciise.concordia.ca

Abstract. In this paper, we propose a statistical Bayesian framework based on finite generalized Dirichlet (GD) mixture models and apply it to two challenging problems namely face detection and facial expression recognition. The proposed Bayesian model is learned through a principled variational approach and allows simultaneous clustering and feature selection. The feature selection process is taken into account via the integration of a background density to handle irrelevant features to which small weights have to be affected. Moreover, a variational form of the Deviance Information Criterion (DIC) is incorporated within the proposed statistical framework for evaluating the correctness of the model complexity (i.e. number of mixture components and number of relevant features). The effectiveness of the proposed model is illustrated through extensive empirical results using challenging real examples.

Keywords: Generalized Dirichlet mixture, variational learning, face detection, facial expression recognition.

1 Introduction

Face detection is a crucial task in computer vision and has been employed in various applications such as video surveillance, image database management and human-computer interfaces. Furthermore, it is a common preprocessing step for a facial recognition system (see, for instance, [12,17,1]). On the other side, facial expression recognition is a type of visual learning process which deals with the classification of facial motion and has been applied in various fields such as image understanding, psychological studies, facial nerve grading in medicine, synthetic face animation and virtual reality. During the last two decades, various technologies have been developed for facial expression recognition (see, for instance [15,21,10] and references therein).

A central step in all the previously proposed approaches for face detection and facial expression recognition is to extract low level features (e.g. color, texture, shape, motion, etc.) from input images. Recently, the bag of visual words model has drawn considerable attention, and has been successfully applied in many applications [9,4,7]. The obtention of these visual words is based on running an

A. Dziech and A. Czyżewski (Eds.): MCSS 2012, CCIS 287, pp. 95–106, 2012.

interest point detector on images. Local descriptors, such as SIFT [20], are then computed on the detected interest points. The obtained descriptors are finally clustered over a set of training images to build a dictionary of visual words. Thus, each object can be viewed as a point in a multi-dimensional space (i.e. a histogram of visual words). The goal of this paper is to propose a novel approach for face detection and facial expression recognition using the notion of visual words by developing a variational framework of finite generalized Dirichlet (GD) mixture models. The motivation of employing the GD mixture is due to its better modeling capabilities in the case of non-Gaussian data, especially those involving proportional data [5,6,7]. Based on the experimental results, the proposed approach is efficient and allows simultaneously the estimation of the parameters and model complexity in terms of both feature selection and determination of the accurate number of mixture components. It is worth mentioning that, to the best of our knowledge the variational learning of GD mixture models with DIC-based model selection has never been considered in the past.

The rest of this paper is organized as follows: Section 2 briefly introduces the finite GD mixture model. In Section 3, we describe our variational algorithm for the proposed model learning. Section 4 is devoted to the experimental results. In Section 5 we present conclusion and directions for future research.

2   Model Specification

Assume that we have a set of N independent and identically distributed vectors $\mathcal{Y} = (\boldsymbol{Y}_1, \ldots, \boldsymbol{Y}_N)$, where each vector $\boldsymbol{Y}_i = (Y_1, \ldots, Y_D)$ is assumed to be sampled from a finite GD mixture model with M components [5]:

$$p(\boldsymbol{Y}_i|\boldsymbol{\pi}, \boldsymbol{\alpha}, \boldsymbol{\beta}) = \sum_{j=1}^{M} \pi_j \mathrm{GD}(\boldsymbol{Y}_i|\boldsymbol{\alpha}_j, \boldsymbol{\beta}_j), \tag{1}$$

where $\sum_{l=1}^{D} Y_{il} < 1$ and $0 < Y_{il} < 1$ for $l = 1, \ldots, D$, $\boldsymbol{\alpha} = (\boldsymbol{\alpha}_1, \ldots, \boldsymbol{\alpha}_M)$ and $\boldsymbol{\beta} = (\boldsymbol{\beta}_1, \ldots, \boldsymbol{\beta}_M)$. $\boldsymbol{\alpha}_j = (\alpha_{j1}, \ldots, \alpha_{jD})$ and $\boldsymbol{\beta}_j = (\beta_{j1}, \ldots, \beta_{jD})$ are the positive parameters of the GD distribution representing component j. $\boldsymbol{\pi} = (\pi_1, \ldots, \pi_M)$ represents the mixing coefficients with the constraints that are positive and sum to one.

According to the mathematical properties of the GD thoroughly discussed in [7], the finite GD mixture model is equivalent to the following

$$p(\boldsymbol{X}_i|\boldsymbol{\pi}, \boldsymbol{\alpha}, \boldsymbol{\beta}) = \sum_{j=1}^{M} \pi_j \prod_{l=1}^{D} \mathrm{Beta}(X_{il}|\alpha_{jl}, \beta_{jl}) \tag{2}$$

where $\boldsymbol{X}_i = (X_{i1}, \ldots, X_{iD})$, $X_{i1} = Y_{i1}$ and $X_{il} = Y_{il}/(1 - \sum_{k=1}^{l-1} Y_{ik})$ for $l > 1$, and $\mathrm{Beta}(X_{il}|\alpha_{jl}, \beta_{jl})$ is a Beta distribution defined with parameters $(\alpha_{jl}, \beta_{jl})$:

$$\mathrm{Beta}(X_{il}|\alpha_{jl}, \beta_{jl}) = \frac{\Gamma(\alpha_{jl} + \beta_{jl})}{\Gamma(\alpha_{jl})\Gamma(\beta_{jl})} X_{il}^{\alpha_{jl}-1}(1 - X_{il})^{\beta_{jl}-1} \tag{3}$$

Indeed, (2) is an important property of the GD mixture, since the independence between the features, in the case of the new data set $\mathcal{X}$, becomes a fact and not an assumption as considered in previous unsupervised feature selection Gaussian mixture-based approaches [16,8].

Next, we assign a binary latent variable $\mathbf{Z}_i = (Z_{i1}, \ldots, Z_{iM})$ to each observation $\mathbf{X}_i$, such that $Z_{ij} \in \{0,1\}$, $\sum_{j=1}^{M} Z_{ij} = 1$, $Z_{ij} = 1$ if $\mathbf{X}_i$ belongs to component j and equal to 0, otherwise. The conditional distribution of latent variables $\mathcal{Z} = (\mathbf{Z}_1, \ldots, \mathbf{Z}_N)$, given the mixing coefficients $\boldsymbol{\pi}$, is defined as

$$p(\mathcal{Z}|\boldsymbol{\pi}) = \prod_{i=1}^{N} \prod_{j=1}^{M} \pi_j^{Z_{ij}}. \tag{4}$$

Notice that, in practice the features $\{X_{il}\}$ are generally not equally important for the clustering task since some features may be "noise" that do not contribute to clustering process. Thus, feature selection is a crucial factor to improve the learning performance. In our work, we adopt the feature selection scheme that has been proposed in [7] by approximating the feature distribution as

$$p(X_{il}) \simeq \left[\mathrm{Beta}(X_{il}|\alpha_{jl}, \beta_{jl}) \right]^{\phi_{il}} \left[\prod_{k=1}^{K} \mathrm{Beta}(X_{il}|\lambda_{kl}, \tau_{kl})^{W_{ikl}} \right]^{1-\phi_{il}} \tag{5}$$

where ϕ_{il} is a binary latent variable, such that $\phi_{il} = 0$ if feature l of component j is irrelevant and follows a mixture of K Beta distributions: $\mathrm{Beta}(X_{il}|\lambda_{kl}, \tau_{kl})$. W_{ikl} is a binary variable such that $\sum_{k=1}^{K} W_{ikl} = 1$. When $W_{ikl} = 1$, it indicates that X_{il} comes from the kth component of the irrelevant beta distribution within component j. Assuming that W_{ikl} represents the elements of $\mathcal{W}$, the marginal distribution of $\mathcal{W}$ is defined as

$$p(\mathcal{W}|\boldsymbol{\eta}) = \prod_{i=1}^{N} \prod_{k=1}^{K} \prod_{l=1}^{D} \eta_{kl}^{W_{ikl}} \tag{6}$$

where η_{kl} represents the prior probability that X_{il} comes from the kth component of the irrelevant Beta distribution, and $\sum_{k=1}^{K} \eta_{kl} = 1$. The prior distribution of ϕ is defined as

$$p(\phi|\epsilon) = \prod_{i=1}^{N} \prod_{l=1}^{D} \epsilon_{l_1}^{\phi_{il}} \epsilon_{l_2}^{1-\phi_{il}} \tag{7}$$

where each ϕ_{il} is a Bernoulli variable such that $p(\phi_{il} = 1) = \epsilon_{l_1}$ and $p(\phi_{il} = 0) = \epsilon_{l_2}$. The vector $\epsilon = (\epsilon_1, \ldots, \epsilon_D)$ represents the features saliencies (i.e. the probabilities that the features are relevant) such that $\epsilon_l = (\epsilon_{l_1}, \epsilon_{l_2})$ and $\epsilon_{l_1} + \epsilon_{l_2} = 1$. For parameters $\boldsymbol{\pi}$, $\boldsymbol{\eta}$ and ϵ, Dirichlet distributions are adopted as the conjugate prior:

$$p(\boldsymbol{\pi}) = \mathrm{Dir}(\boldsymbol{\pi}|\boldsymbol{a}), \qquad p(\boldsymbol{\eta}) = \prod_{l=1}^{D} \mathrm{Dir}(\boldsymbol{\eta}_l|\boldsymbol{c}), \qquad p(\epsilon) = \prod_{l=1}^{D} \mathrm{Dir}(\epsilon_l|\boldsymbol{b}) \tag{8}$$

where the Dirichlet distribution Dir($\cdot$) is defined as

$$\text{Dir}(\boldsymbol{\pi}|\boldsymbol{a}) = \frac{\Gamma(\sum_{j=1}^{M} a_j)}{\prod_{j=1}^{M} \Gamma(a_j)} \prod_{j=1}^{M} \pi_j^{a_j-1} \tag{9}$$

Then, the Gamma distribution is adopted to approximate a conjugate prior over parameters $\boldsymbol{\alpha}, \boldsymbol{\beta}, \boldsymbol{\lambda}$ and $\boldsymbol{\tau}$ as suggested recently in [22], by assuming that the different model's parameters are independent:

$$p(\boldsymbol{\alpha}) = \mathcal{G}(\boldsymbol{\alpha}|\boldsymbol{u},\boldsymbol{v}), \quad p(\boldsymbol{\beta}) = \mathcal{G}(\boldsymbol{\beta}|\boldsymbol{p},\boldsymbol{q}), \quad p(\boldsymbol{\lambda}) = \mathcal{G}(\boldsymbol{\lambda}|\boldsymbol{g},\boldsymbol{h}), \quad p(\boldsymbol{\tau}) = \mathcal{G}(\boldsymbol{\tau}|\boldsymbol{s},\boldsymbol{t}) \tag{10}$$

where $\mathcal{G}(\cdot)$ is the Gamma distribution and is defined as

$$\mathcal{G}(x|a,b) = \frac{b^a}{\Gamma(a)} x^{a-1} e^{-bx}. \tag{11}$$

3   Variational Learning

Variational inference (or variational Bayes) is a deterministic approximation scheme which is used to formulate the computation of a marginal or conditional probability in terms of an optimization problem [2,14,3]. In this section, we develop a variational framework for learning the finite GD mixture model.

To simplify the notation, we define $\Theta = \{\mathcal{Z}, \mathcal{W}, \boldsymbol{\phi}, \boldsymbol{\alpha}, \boldsymbol{\beta}, \boldsymbol{\lambda}, \boldsymbol{\tau}, \boldsymbol{\pi}, \boldsymbol{\eta}, \boldsymbol{\epsilon}\}$. The central idea in variational learning is to find an approximation for the posterior distribution $p(\Theta|\mathcal{X})$ as well as for the model evidence $p(\mathcal{X})$. By applying Jensen's inequality, the lower bound $\mathcal{L}$ of the logarithm of the marginal likelihood $p(\mathcal{X})$ can be found as $\mathcal{L}(Q) = \int Q(\Theta) \ln[p(\mathcal{X}, \Theta)/Q(\Theta)]d\Theta$, where $Q(\Theta)$ is an approximation for $p(\Theta|\mathcal{X})$. Here, we adopt the factorial approximation which is known as *mean field theory* for the variational inference and has been used efficiently by several researchers in the past [24,13]. Thus, $Q(\Theta)$ can be factorized into disjoint tractable distributions, such that

$$Q(\Theta) = Q(\mathcal{Z})Q(\mathcal{W})Q(\boldsymbol{\phi})Q(\boldsymbol{\alpha})Q(\boldsymbol{\beta})Q(\boldsymbol{\lambda})Q(\boldsymbol{\tau})Q(\boldsymbol{\pi})Q(\boldsymbol{\eta})Q(\boldsymbol{\epsilon}) \tag{12}$$

In order to maximize the lower bound $\mathcal{L}(Q)$, we need to make a variational optimization of $\mathcal{L}(Q)$ with respect to each of the factors in turn using the general expression for its optimal solution:

$$Q_s(\Theta_s) = \frac{\exp\langle \ln p(\mathcal{X}, \Theta)\rangle_{\neq s}}{\int \exp\langle \ln p(\mathcal{X}, \Theta)\rangle_{\neq s} d\Theta} \tag{13}$$

where $\langle\cdot\rangle_{\neq s}$ denotes an expectation with respect to all the factor distributions except for s. By applying (13) to each factor of the variational posterior, we obtain the variational solutions as

$$Q(\mathcal{Z}) = \prod_{i=1}^{N} \prod_{j=1}^{M} r_{ij}^{Z_{ij}}, \quad Q(\boldsymbol{\phi}) = \prod_{i=1}^{N} \prod_{l=1}^{D} f_{il}^{\phi_{il}} (1 - f_{il})^{(1-\phi_{il})}, \quad Q(\boldsymbol{\pi}) = \text{Dir}(\boldsymbol{\pi}|\boldsymbol{a}^*)$$

$$\tag{14}$$

$$Q(\mathcal{W}) = \prod_{i=1}^{N}\prod_{k=1}^{K}\prod_{l=1}^{D} m_{ikl}^{W_{ikl}}, \quad Q(\boldsymbol{\eta}) = \prod_{l=1}^{D}\mathrm{Dir}(\boldsymbol{\eta}_l|c^*), \quad Q(\boldsymbol{\epsilon}) = \prod_{l=1}^{D}\mathrm{Dir}(\boldsymbol{\epsilon}|b^*) \tag{15}$$

$$Q(\boldsymbol{\alpha}) = \prod_{j=1}^{M}\prod_{l=1}^{D}\mathcal{G}(\alpha_{jl}|u_{jl}^*, v_{jl}^*), \quad Q(\boldsymbol{\beta}) = \prod_{j=1}^{M}\prod_{l=1}^{D}\mathcal{G}(\beta_{jl}|p_{jl}^*, q_{jl}^*) \tag{16}$$

$$Q(\boldsymbol{\lambda}) = \prod_{l=1}^{D}\prod_{k=1}^{K}\mathcal{G}(\lambda_{kl}|g_{kl}^*, h_{kl}^*), \quad Q(\boldsymbol{\tau}) = \prod_{l=1}^{D}\prod_{k=1}^{K}\mathcal{G}(\tau_{kl}|s_{kl}^*, t_{kl}^*) \tag{17}$$

where $\mathcal{G}$ represents the Gamma distribution and where we define

$$r_{ij} = \frac{\rho_{ij}}{\sum_{d=1}^{M}\rho_{id}}, \quad f_{il} = \frac{\delta_{il}^{(\phi_{il})}}{\delta_{il}^{(\phi_{il})} + \delta_{il}^{(1-\phi_{il})}}, \quad m_{ilk} = \frac{\varphi_{ilk}}{\sum_{d=1}^{K}\varphi_{ild}}$$

$$c_k^* = c_k + \sum_{i=1}^{N}\langle W_{ikl}\rangle, \quad b_1^* = b_1 + \sum_{i=1}^{N}\langle\phi_{il}\rangle, \quad b_2^* = b_1 + \sum_{i=1}^{N}\langle 1-\phi_{il}\rangle$$

$$\rho_{ij} = \exp\left\{\langle\ln\pi_j\rangle + \sum_{l=1}^{D}\langle\phi_{il}\rangle[\widetilde{\mathcal{R}}_{jl} + (\bar{\alpha}_{jl}-1)\ln X_{il} + (\bar{\beta}_{jl}-1)\ln(1-X_{il})]\right\}$$

$$\delta_{il}^{(\phi_{il})} = \exp\left\{\sum_{j=1}^{M}\langle Z_{ij}\rangle[\widetilde{\mathcal{R}}_{jl} + (\bar{\alpha}_{jl}-1)\ln X_{il} + (\bar{\beta}_{jl}-1)\ln(1-X_{il})] + \langle\ln\epsilon_{l_1}\rangle\right\}$$

$$\delta_{il}^{(1-\phi_{il})} = \exp\left\{\sum_{k=1}^{K}\langle W_{ikl}\rangle[\widetilde{\mathcal{F}}_{kl} + (\bar{\lambda}_{kl}-1)\ln X_{il} + (\bar{\tau}_{kl}-1)\ln(1-X_{il})] + \langle\ln\epsilon_{l_2}\rangle\right\}$$

$$\varphi_{ikl} = \exp\left\{\langle 1-\phi_{il}\rangle[\widetilde{\mathcal{F}}_{kl} + (\bar{\lambda}_{kl}-1)\ln X_{il} + (\bar{\tau}_{kl}-1)\ln(1-X_{il})] + \langle\ln\eta_{kl}\rangle\right\}$$

$$u_{jl}^* = u_{jl} + \sum_{i=1}^{N}\langle Z_{ij}\rangle\langle\phi_{il}\rangle\bar{\alpha}_{jl}\left[\Psi(\bar{\alpha}_{jl}+\bar{\beta}_{jl}) - \Psi(\bar{\alpha}_{jl}) + \bar{\beta}_{jl}\Psi'(\bar{\alpha}_{jl}+\bar{\beta}_{jl})(\langle\ln\beta_{jl}\rangle - \ln\bar{\beta}_{jl})\right]$$

$$a_j^* = a_j + \sum_{i=1}^{N}\langle Z_{ij}\rangle, \quad v_{jl}^* = v_{jl} - \sum_{i=1}^{N}\langle Z_{ij}\rangle\langle\phi_{il}\rangle\ln X_{il}$$

where $\Psi(\cdot)$ is the digamma function: $\Psi(x) = \frac{d}{dx}\ln\Gamma(x)$. Note that, $\widetilde{\mathcal{R}}$ and $\widetilde{\mathcal{F}}$ are the lower bounds of $\mathcal{R} = \langle\ln\frac{\Gamma(\alpha+\beta)}{\Gamma(\alpha)\Gamma(\beta)}\rangle$ and $\mathcal{F} = \langle\ln\frac{\Gamma(\lambda+\tau)}{\Gamma(\lambda)\Gamma(\tau)}\rangle$, respectively. Since these expectations are intractable, we use the second-order Taylor series expansion to find their lower bounds as proposed in [22]. The solutions to the hyperparameters of $Q(\boldsymbol{\beta})$, $Q(\boldsymbol{\lambda})$ and $Q(\boldsymbol{\tau})$ can be computed similarly as for $\boldsymbol{u}^*$ and $\boldsymbol{v}^*$. The expected values in the above formulas are given by

$$\langle Z_{ij}\rangle = r_{ij}, \quad \langle W_{ilk}\rangle = m_{ilk}, \quad \langle\phi_{il}\rangle = f_{il}, \quad \langle 1-\phi_{il}\rangle = 1 - f_{il} \tag{18}$$

$$\langle \ln \pi_j \rangle = \Psi(a_j^*) - \Psi(\sum_{j=1}^{M} a_j^*), \quad \langle \ln \alpha \rangle = \Psi(u^*) - \ln v^* \tag{19}$$

$$\langle \ln \beta \rangle = \Psi(p^*) - \ln q^*, \quad \langle \ln \lambda \rangle = \Psi(g^*) - \ln h^*, \quad \langle \ln \tau \rangle = \Psi(s^*) - \ln t^*, \tag{20}$$

$$\langle \ln \eta_{kl} \rangle = \Psi(c_k^*) - \Psi(\sum_{k=1}^{K} c_k^*), \quad \langle \ln \epsilon_{l_1} \rangle = \Psi(b_1^*) - \Psi(b_1^* + b_2^*), \quad \langle \ln \epsilon_{l_2} \rangle = \Psi(b_2^*) - \Psi(b_1^* + b_2^*) \tag{21}$$

$$\bar{\alpha}_{jl} = \langle \alpha_{jl} \rangle = \frac{u_{jl}^*}{v_{jl}^*}, \quad \bar{\beta}_{jl} = \langle \beta_{jl} \rangle = \frac{p_{jl}^*}{q_{jl}^*}, \quad \bar{\lambda}_{kl} = \langle \lambda_{kl} \rangle = \frac{g_{kl}^*}{h_{kl}^*}, \quad \bar{\tau}_{kl} = \langle \tau_{kl} \rangle = \frac{s_{kl}^*}{t_{kl}^*} \tag{22}$$

To verify the correctness of model selection in this variational framework, we adopt a variational version of the Deviance Information Criterion (DIC) [23] as the selection criterion. The DIC has the advantage that it is straightforward to compute and does not require to specify the number of unknown parameters as other model selection criteria (e.g. BIC). The DIC is defined as

$$\mathrm{DIC} = 2\mathcal{D} - 2\ln p(\mathcal{X}|\breve{\Theta}) \tag{23}$$

where $\breve{\Theta}$ is the posterior mean and we have

$$\ln p(\mathcal{X}|\breve{\Theta}) = \sum_{i=1}^{N} \ln \left[\sum_{j=1}^{M} \breve{\pi}_j \frac{\Gamma(\sum_{l=1}^{D} \breve{\alpha}_{jl})}{\prod_{l=1}^{D} \Gamma(\breve{\alpha}_{jl})} \prod_{l=1}^{D} X_{il}^{\breve{\alpha}_{jl}-1} \right] \tag{24}$$

$$
\begin{aligned}
\mathcal{D} \approx & -2 \int Q(\Theta) \ln \left[\frac{Q(\Theta)}{p(\Theta)} \right] d\Theta + 2\ln \left[\frac{Q(\breve{\Theta})}{p(\breve{\Theta})} \right] \\
= & 2\Big[\sum_{j=1}^{M} \sum_{l=1}^{D} (u_{jl}^* - u_{jl})(\ln \breve{\alpha}_{jl} - \langle \ln \alpha_{jl} \rangle) + (p_{jl}^* - p_{jl})(\ln \breve{\beta}_{jl} - \langle \ln \beta_{jl} \rangle) \Big] \\
& + 2\Big[\sum_{k=1}^{K} \sum_{l=1}^{D} (g_{kl}^* - g_{kl})(\ln \breve{\lambda}_{kl} - \langle \ln \lambda_{kl} \rangle) + (s_{kl}^* - s_{kl})(\ln \breve{\tau}_{kl} - \langle \ln \tau_{kl} \rangle) \Big] \\
& + 2\Big[\sum_{j=1}^{M} (a_j^* - a_j)(\ln \breve{\pi}_j - \langle \ln \pi_j \rangle) + \sum_{i=1}^{N} \sum_{j=1}^{M} r_{ij}(\langle \ln \pi_j \rangle - \ln \breve{\pi}_j) \Big] \\
& + 2\Big[\sum_{k=1}^{K} \sum_{l=1}^{D} (c_k^* - c_k)(\ln \breve{\eta}_{kl} - \langle \ln \eta_{kl} \rangle) + \sum_{i=1}^{N} \sum_{k=1}^{K} \sum_{l=1}^{D} m_{ikl}(\langle \ln \eta_{kl} \rangle - \ln \breve{\eta}_{kl}) \Big] \\
& + 2\Big[\sum_{l=1}^{D} (b_1^* - b_1)(\ln \breve{\epsilon}_{l_1} - \langle \ln \epsilon_{l_1} \rangle) + (b_2^* - b_2)(\ln \breve{\epsilon}_{l_2} - \langle \ln \epsilon_{l_2} \rangle) \\
& + \sum_{i=1}^{N} \sum_{l=1}^{D} f_{il}(\langle \ln \epsilon_{l_1} \rangle - \ln \breve{\epsilon}_{l_1}) + (1 - f_{il})(\langle \ln \epsilon_{l_2} \rangle - \ln \breve{\epsilon}_{l_2}) \Big]
\end{aligned}
\tag{25}
$$

where $\check{\pi}_j = a_j^*/\sum_{j=1}^{M} a_j^*$, $\check{c}_k = c_k^*/\sum_{k=1}^{K} c_k^*$, $\check{\epsilon}_{l_1} = b_1^*/(b_1^*+b_2^*)$, $\check{\epsilon}_{l_2} = b_2^*/(b_1^*+b_2^*)$ $\check{\alpha}_{jl} = u_{jl}^*/v_{jl}^*$, $\check{\beta}_{jl} = p_{jl}^*/q_{jl}^*$, $\check{\lambda}_{kl} = g_{kl}^*/h_{kl}^*$, $\check{\tau}_{kl} = s_{kl}^*/t_{kl}^*$. Since the variational solutions are coupled together through the expected values of the other factor, these solutions can thus be obtained iteratively by using an EM-like algorithm with a guaranteed convergence. The complete learning algorithm is summarized in Algorithm 1.

Algorithm 1. Variational GD mixture with feature selection

1: Choose the initial number of components M and K.
2: Initialize the values for hyper-parameters a, c, b, u, v, p, q, g, h, s and t.
3: Initialize the values of r_{ij} and m_{ikl} by K-Means algorithm.
4: **repeat**
5:   *The variational E-step*: use the current values of model parameters to evaluate the moments in (18) to (22).
6:   *The variational M-step*: update the variational solutions through (14) to (17).
7: **until** Convergence criteria is reached.
8: Calculate the current DIC value.
9: Detect the optimal number of components M and K by eliminating the components with small mixing coefficients close to 0.

---

4 Experimental Results

In this section, two challenging applications namely face detection and face expression recognition are highlighted using the proposed variational GD mixture model (*varGDMM*) and the bag-of-visual words representation. In all of our experiments, we initialize the number of components M and K to 15 and 10, respectively. In order to provide broad non-informative prior distributions, the initial values of the hyperparameters u, p, g and s for the conjugate priors are set to 1, and v, q, h, t are set to 0.01. Furthermore, hyperparameters a, b, and c are initialized to 0. For comparison, we have also applied three other approaches for these two applications: the EM-based approach to learn GD mixture models with feature selection (*GDMM*) proposed in [7], the variational Gaussian mixture model with feature selection (*varGMM*) proposed in [8] and the Gaussian mixture model with feature selection (*GMM*) as learned in [16].

4.1 Face Detection

Experimental Methodology. The methodology that we have adopted for face detection can be summarized as follows. First, SIFT descriptors are extracted from each image using the Difference-of-Gaussians (DoG) interest point detector [20]. Next, a visual vocabulary $\mathcal{V}$ is constructed by quantizing these SIFT vectors into visual words w using K-means algorithm and each image is then represented as the frequency histogram over the visual words. Based on our experiments, the optimal performance can be obtained when $\mathcal{V} = 800$. Then, we apply the probabilistic latent semantic analysis (pLSA) model [11] to the bag of visual words representation which allows the description of each image as a D-dimensional vector of proportions where D is the number of aspects (or learnt topics) [4]. Finally, we employ the proposed $varGDMM$ as a classifier to detect human faces by assigning the testing image to the group (face or non-face) which has the highest posterior probability according to Bayes' decision rule.

Fig. 1. Sample images of the Caltech face data set and the Caltech background data set: the first row contains face samples, the second row represents background samples

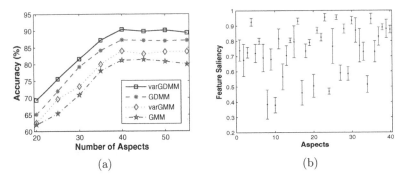

Fig. 2. (a) Detection accuracy vs. the number of aspects; (b) Feature saliency for each aspect

Table 1. The face detection rate (%) using different methods

Method	Detection accuracy %
$varGDMM$	90.39
$GDMM$	87.28
$varGMM$	83.97
GMM	81.15

The data set that we have used for face detection is the Caltech face data set[1]. It contains 452 front human face images which are recorded under natural conditions, i.e. varying illumination, expressions and complex background. Each image has a size of 896×592 pixels. For non-face images, the Caltech background data set was adopted and it contains 452 background images with size of 378×251 pixels [2]. Sample images from the Caltech face and the Caltech background data sets are displayed in Fig. 1.

Each of the two data sets (face and non-face) is randomly divided into two halves: one for training and the other for testing. We evaluated the detection performance of the proposed algorithm by running it 20 times. The experimental results of face detection using different algorithms are presented in Table 1. This table clearly shows that our algorithm outperforms the three other approaches by providing the highest detection rate (90.39%). Notice that, the two GD mixture algorithms perform better than the algorithms using Gaussian mixtures. This is actually as expected since images are represented by vectors of proportions which are more appropriately modeled by the GD rather than the Gaussian. Moreover, the choice of the number of aspects also influences the accuracy of detection as shown in Fig. 2 (a). Based on this figure, the highest accuracy can be obtained when the number of aspects is set to 40. The corresponding feature saliencies of the 40-dimensional aspects obtained by *varGDMM* are illustrated in Fig. 2 (b). As shown in this figure, the features have different relevance degrees and then contribute differently to the classification task.

4.2   Face Expression Recognition

In this experiment, the application of facial expression recognition is highlighted. A similar methodology was adopted in this application as for the face detection. The major difference is that in this application, instead of using SIFT features, we adopted a high gradient component analysis method as proposed in [18] to extract facial expression features (32-dimensional) in spatio-temporal domain from each image. The resultant local texture features obtained by the high gradient component analysis have shown appealing results on face expression recognition in the past [18,19]. The data set that we used for this experiment is the well-known Japanese Female Facial Expression (JAFFE) data set [21]. It includes 213 images of 7 facial expressions (6 basic facial expressions: anger, disgust,

Anger   Disgust   Fear   Happiness   Sadness   Surprise   Neutral

Fig. 3. Sample images from the JAFFE data set

[1] http://www.robots.ox.ac.uk/~vgg/data.html
[2] http://www.vision.caltech.edu/html-files/archive.html

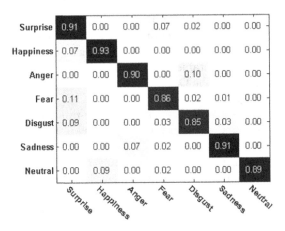

Fig. 4. Confusion matrix obtained by *varGDMM*

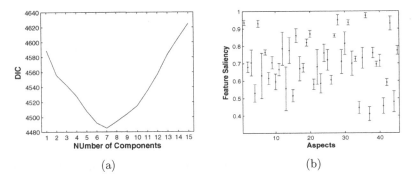

(a)   (b)

Fig. 5. (a) The number of image categories detected by *varGDMM* with DIC; (b) Feature saliency for each aspect

fear, happiness, sadness and surprise + 1 neutral) posed by 10 Japanese female models aged 20 to 40. Sample images from this data set with different facial expressions are shown in Fig. 3.

In this experiment, we randomly divided the data set into two partitions: one for constructing the visual vocabulary, another for testing. We evaluated the

Table 2. The average classification accuracy and the number of components $(\hat{M})$ computed by different algorithms

	varGDMM	*GDMM*	*varGMM*	*GMM*
$\hat{M}$	6.95	6.85	6.90	6.80
Accuracy (%)	89.28	87.43	85.25	82.94

performance of the proposed algorithm by running it 20 times. The confusion matrix for the JAFFE data set calculated by the *varGDMM* is shown in Fig. 4.

The average classification accuracy of the facial expression images and the average number of components obtained by each algorithm are shown in Table 2. As we can see, our algorithm provides higher classification accuracy than the other algorithms while detecting the number of categories more accurately. Fig. 5 (a) displays the DIC values as a function of the number of facial expression categories and shows clearly that the optimal number of categories is actually 7. Additionally, the number of components for the mixture model representing irrelevant features are estimated as 2. In this experiment, the optimal number of aspects is 45. The feature saliencies of the 45-dimensional aspects calculated by *varGDMM* are illustrated in Fig. 5 (b). Obviously, different features are assigned with different degrees of importance. For instance, there are four features (features number 34, 37, 41 and 44) that have saliencies lower than 0.5, and then provide less contribution in clustering. This is because these aspects are associated to all categories and have less discrimination power. By contrast, six features (features number 1, 5, 28, 31, 36 and 43) have high relevance degrees with feature saliencies greater than 0.9. In addition, according to Fig. 5 (b), the standard deviations of the features with high relevance degrees are generally smaller than those for the irrelevant features, which means that our algorithm for selecting relevant features is consistent.

5   Conclusion

In our work, we have proposed a variational framework for finite GD mixture models. By adopting the proposed algorithm with the pLSA model and the bag-of-words representation, we have developed a powerful approach for dealing with the problems of face detection and facial expression recognition. The effectiveness and the efficiency of the proposed approach are demonstrated through the experimental results.

References

1. Amit, Y., Trouvé, A.: POP: Patchwork of Parts Models for Object Recognition. International Journal of Computer Vision 75, 267–282 (2007)
2. Attias, H.: A Variational Bayes Framework for Graphical Models. In: Proc. of Advances in Neural Information Processing Systems (NIPS), pp. 209–215 (1999)
3. Bishop, C.M.: Pattern recognition and machine learning. Springer (2006)
4. Bosch, A., Zisserman, A., Muñoz, X.: Scene Classification Via pLSA. In: Leonardis, A., Bischof, H., Pinz, A. (eds.) ECCV 2006, Part IV. LNCS, vol. 3954, pp. 517–530. Springer, Heidelberg (2006)
5. Bouguila, N., Ziou, D.: A Hybrid SEM Algorithm for High-Dimensional Unsupervised Learning Using a Finite Generalized Dirichlet Mixture. IEEE Transactions on Image Processing 15(9), 2657–2668 (2006)
6. Bouguila, N., Ziou, D.: High-Dimensional Unsupervised Selection and Estimation of a Finite Generalized Dirichlet Mixture Model Based on Minimum Message Length. IEEE Transactions on Pattern Analysis and Machine Intelligence 29, 1716–1731 (2007)

7. Boutemedjet, S., Bouguila, N., Ziou, D.: A Hybrid Feature Extraction Selection Approach for High-Dimensional Non-Gaussian Data Clustering. IEEE Transactions on Pattern Analysis and Machine Intelligence 31(8), 1429–1443 (2009)
8. Constantinopoulos, C., Titsias, M.K., Likas, A.: Bayesian Feature and Model Selection for Gaussian Mixture Models. IEEE Transactions on Pattern Analysis and Machine Intelligence 28(6), 1013–1018 (2006)
9. Csurka, G., Dance, C.R., Fan, L., Willamowski, J., Bray, C.: Visual Categorization with Bags of Keypoints. In: Workshop on Statistical Learning in Computer Vision (ECCV), pp. 1–22 (2004)
10. Fasel, B., Luettin, J.: Automatic Facial Expression Analysis: A Survey. Pattern Recognition 36(1), 259–275 (1999)
11. Hofmann, T.: Unsupervised Learning by Probabilistic Latent Semantic Analysis. Machine Learning 42(1/2), 177–196 (2001)
12. Hwang, W.S., Weng, J.: Hierarchical discriminant regression. IEEE Transactions on Pattern Analysis and Machine Intelligence 22(11), 1277–1293 (2000)
13. Jaakkola, T.S., Jordan, M.I.: Computing Upper and Lower Bounds on Likelihoods in Intractable Networks. In: Proc. of the Conference in Uncertainty in Artificial Intelligence (UAI), pp. 340–348 (1996)
14. Jordan, M.I., Ghahramani, Z., Jaakkola, T., Saul, L.K.: An Introduction to Variational Methods for Graphical Models. Machine Learning 37, 183–233 (1999)
15. Kotsia, I., Pitas, I., Zafeiriou, S., Zafeiriou, S.: Novel Multiclass Classifiers Based on the Minimization of the Within-Class Variance. IEEE Transactions on Neural Networks 20(1), 14–34 (2009)
16. Law, M.H.C., Figueiredo, M.A.T., Jain, A.K.: Simultaneous Feature Selection and Clustering Using Mixture Models. IEEE Transactions on Pattern Analysis and Machine Intelligence 26(9), 1154–1166 (2004)
17. Li, S.Z., Zhu, L., Zhang, Z., Blake, A., Zhang, H., Shum, H.: Statistical Learning of Multi-view Face Detection. In: Heyden, A., Sparr, G., Nielsen, M., Johansen, P. (eds.) ECCV 2002, Part IV. LNCS, vol. 2353, pp. 67–81. Springer, Heidelberg (2002)
18. Lien, J.J.J.: Automatic Recognition of Facial Expressions Using Hidden Markov Models and Estimation of Expression Intensity. Ph.D. thesis, Robotics Institute, Carnegie Mellon University (April 1998)
19. Lien, J.J., Kanade, T., Cohn, J., Li, C.C.: Subtly Different Facial Expression Recognition and Expression Intensity Estimation. In: Proceedings of 1998 IEEE Computer Society Conference on Computer Vision and Pattern Recognition, pp. 853–859 (June 1998)
20. Lowe, D.G.: Distinctive Image Features from Scale-Invariant Keypoints. International Journal of Computer Vision 60(2), 91–110 (2004)
21. Lyons, M., Akamatsu, S., Kamachi, M., Gyoba, J.: Coding Facial Expressions with Gabor Wavelets. In: Proceedings of Third IEEE International Conference on Automatic Face and Gesture Recognition 1998, pp. 200–205 (April 1998)
22. Ma, Z., Leijon, A.: Bayesian Estimation of Beta Mixture Models with Variational Inference. IEEE Transactions on Pattern Analysis and Machine Intelligence 33(11), 2160–2173 (2011)
23. McGrory, C.A., Titterington, D.M.: Variational Approximations in Bayesian Model Selection for Finite Mixture Distributions. Computational Statistics and Data Analysis 51, 5352–5367 (2006)
24. Saul, L.K., Jordan, M.I.: Exploiting Tractable Substructures in Intractable Networks. In: Advances in Neural Information Processing Systems (NIPS), pp. 486–492 (1995)

Advanced Architecture of the Integrated IT Platform with High Security Level

Gerard Frankowski and Marcin Jerzak

Institute of Bioorganic Chemistry of the Polish Academy of Sciences,
Poznań Supercomputing and Networking Center,
ul. Z. Noskowskiego 12/14, 61-704 Poznań,
http://www.pcss.pl
{gerard.frankowski,marcin.jerzak}@man.poznan.pl

Abstract. "Advanced architecture of the Integrated IT Platform with high security level" is a development project within the area of the homeland security. The results of the project are intended to be used by a Polish Law Enforcement Authority. The goal of the project is to develop the architecture of a network and application IT environment for secure, reliable, efficient and scalable processing and storing data by current and future applications and services. Additionally, a clear migration path from the current infrastructure must be provided. Together with the increasing demand on the network bandwidth, expected growth of the multimedia mobile applications and with particular respect to strict security requirements it raises numerous IT security challenges. In this paper we would like to describe how security was addressed within the project and how external research, our own experience, and results of ther similar research projects (including those from the homeland security area) have been combined to provide the maximum possible protection level. We hope that the described approach may be used at least as an initial pattern in other similar initiatives.

Keywords: IT security, defense-in-depth, network, infrastructure, cloud computing, project, LEA, homeland security.

1 Introduction

1.1 The Platform and the Project

The Integrated IT Platform with high security level is the subject of a 2-year development project, funded by the Polish Ministry of Science and Higher Education within the confines of the 11th Contest of Development Projects for Homeland Security. The project is realized by the scientific-industrial consortium led by Poznan Supercomputing and Networking Center that is also the project research partner. The consortium is complemented by three industrial partners: Alma, Talex and Verax.

The main goal of the project is to develop the architecture of a network and application IT environment for secure, reliable, efficient and scalable processing

A. Dziech and A. Czyżewski (Eds.): MCSS 2012, CCIS 287, pp. 107–117, 2012.

and storing data by current and future applications and services utilized by one of the Polish Law Enforcement Authorities (called "LEA" or "the final user"). The final output will be the pilot installation based upon the Polish NREN — the PIONIER network, normally operated by PSNC.

As the proposed platform architecture is composed of multiple layers, a number of work packages have been differentiated, each dedicated to a particular layer: Advanced Data Management System, Data Repositories, Application Data Processing Platform and Network Infrastructure. Additionally, two vertical work packages have been added: Architecture Management and Security. The structure of the work packages is shown in Fig. 1.

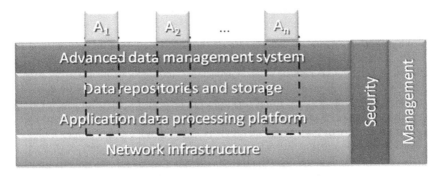

$A_1, A_2, ..., A_n$ - current and future applications

Fig. 1. Architecture of the Integrated Platform

1.2 Research Challenges

The most significant research challenge that has been faced concerns combining the most recent technologies, like e.g. virtualization, cloud computing, Master Data Management etc. with the current state of the IT infrastructure of the final user. The destination platform must not require the final user to rebuild all used applications and services. Where necessary, a clear migration path must be built. Additionally, there are numerous constraints and requirements, regarding e.g. heterogeneity and complexity of the current services, the large number of users (clients), the increasing demand for network bandwidth (especially for mobile clients and users), and particularly security requirements, also with respect to the Polish national law.

Another great challenge was to identify the final requirements of the Integrated Platform. That task was difficult as the Platform is expected to conform with the IT solutions that will be on the market within 3-5 years, while the final users may not have sufficient knowledge on them yet. There are also cases where it may be predicted with a high probability that a particular application will be extremely useful for the LEA within 5 years, but currently the LEA does not consider that solution at all. Therefore, to some extent, the requirements have been identified basing on known (deployed or being deployed) solutions used by the LEAs from different countries, as well as on some external research results. The challenges especially concern mobile multimedia applications and systems, which was especially significant when designing the network architecture. For instance, due to the results of the vendor research ([1]) it has been decided that special emphasis must be put on efficient broadband mobile networks for the purposes of computerized mapping technology/GPS, mobile broadband networks, or vehicle and personal location systems. An example of an application taken into account to calculate potential future requirements of the advanced platform, is a "Brazilian Robocop" [2]. The results of projects like Proteus [3] or INDECT [4] have also been investigated for that purpose.

2   Security of the Platform

2.1   Security Challenges

From the security point of view, the challenges associated with applications that are expected to be deployed within the next 3-5 years, especially concern security of mobile clients, confidentiality, integrity and accountability of the data sent through wireless networks (especially GSM), and adjusting security measures to the available resources in order to not decrease availability of the data and services under the acceptable level. Security always involves some cost and e.g. encrypting the mobile communication channel may cause exhausting the mobile client processor time and therefore extend the transmission time.

Therefore it was essential to include security from the beginning of the project design process. For each element of the platform, security recommendations have been discussed separately and the solutions have been prepared individually. The coexistence of various platform elements and their combined security requirements have also been globally studied.

Additionally, an assumption has been made that the security of the integrated platform must not be based only on innovative systems and solutions proposed within the same project. Those solutions might. And, finally, the described project is formally not a research one. That is why it has been decided to combine several well proven methodologies and solutions (some of which, however, had been the subject of research in other, earlier R&D projects). That approach assures high security level, while innovatory solutions might appear failed, which would expose the infrastructure to danger.

What is yet to be mentioned, the proposed security solutions could not directly involve the current infrastructure of the LEA, but the pilot installation that is the result of the project. The above fact results in another challenge — the designed security solutions should be consistent with those currently applied by the final user, enable cooperation, or offer an acceptable migration plan.

It must be noted that not all details of the proposed security solutions may be publicly disclosed.

2.2 Defense-in-Depth Principle

According to the defense-in-depth principle, it is beneficiary for complex solutions to provide security on different levels. Apart from strictly technical facets, there are also crucial organizational and procedural steps that ought to be taken at various stages of the infrastructure design. Other security measures include administration and management areas as well. The following paragraphs explain in more details what security measures have been proposed.

The defense-in-depth approach may be justified economically. It is obvious that breaking multiple protection layers is more difficult for an attacker — in terms of security economy the attack cost increases, while the attacker will only try to penetrate the system if the expected gain is higher than the cost [5]. A side note should be made here — in the case of the infrastructure expected to be used by a LEA, the gain estimate should be drastically higher than usual as the potentially stolen data may be extremely precious. This justifies accepting defence-in-depth strategy even more as this is the optimal way to increase the attack cost by multiplying technical obstacles that the attacker would have to overcome.

2.3 Embedding Security into the Project Life

Introduction. It is now generally recognized that IT security must not be an extra feature, added to the project in its final stage, but it should rather be embedded from the beginning of the project life. Research shows (e.g. [6]) that security errors are more expensive to be repaired in a late stage of the project life, especially after the results have already been deployed. Even the biggest systems and software vendors tend to implement software development life cycles oriented in some degree towards security — e.g. Microsoft SDL [7].

It is worth noting that embedding security within the whole project lifecycle tends to be a common practice in currently run R&D projects. Examples may be a large European GN3 project [8] or the Polish NGI project — PL-Grid [9]. Although the mentioned projects are more oriented towards producing software, the general rule that integrating security with all project, development and deployment activities helps to avoid different types of security threats still applies. Such IT security handling assumptions have already been presented as suggested practices [10].

Therefore it has been decided within the project that security consultations should be undertaken from its beginning. Within the described project, there is no dedicated software to be implemented. That is why the rules applying to software development life cycles do not quite apply, but there are still the project stages where making a conceptual mistake with security implications would have a significant impact on the further project stages, with the potential need of an additional lifecycle iteration. The project stages are presented in Fig. 2. They will also be shortly described in the following subparagraphs with appropriate reference to performed security-oriented activities.

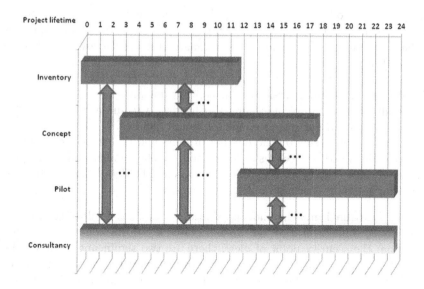

Fig. 2. The project stages with external and internal consultancies

As it may be seen from the picture above, the stages are overlapping, which allows to make iterations. The iterations may be caused e.g. by introducing new, promising IT technologies or by altering the requirements by the final users.

The Inventory of the Current Infrastructure. The project work has started from performing inventory activities concerning the current IT infrastructure of the LEA. It must be noted that the project does not assume making any direct changes to the current LEA infrastructure, but the requirement of providing a clear migration path towards the proposed Integrated Platform could not be fulfilled otherwise. The implications for security are therefore clear — it has been necessary to identify the used security systems to decide whether they will conform with the proposed solutions, how they should be supported, complemented or used for cooperation. An example may be Intrusion Detection Systems: identifying different types of IDS solutions will be useful to determine whether the proposed MetaIDS system may exchange IDMEF messages with those systems.

Integrated Platform Concept. This is the most important project stage and it produces one of the main results of the projects. The concept uses the results gained from the inventory stage in order to propose solutions that will assure both using the most recent solutions and technologies and the possibility easy migration. From the security point of view, this is also the place for providing continuous security consultancy for particular work packages.

The process of elaborating the Integrated Platform concept is itself iterative, based also on bilateral talks with the representatives of the final user. Additionally, some particular factors on the final user's side might have some impact on the concept (e.g. modified plans of the IT infrastructure development). Security specialists within the project have performed a number of consultancies, either with the representatives of other work packages (the results are described in chapter 2.4) or with the final users. There have been some additional requests from the final users to explain certain security facets more thoroughly, which will be realized in the final version of the concept document.

The result of the concept stage are also hardware requirements for the Integrated IT Platform. For instance, the minimum bandwidth for the backbone has been calculated, basing on the number of users and the envisaged profile of using applications and services, as at least 10Gb/s. The number and geographic dispersion of data centers have alo been proposed. The detailed requirements, however, are not to be disclosed on the current stage.

The Pilot Installation. The pilot installation is the final result of the project and the direct implementation of the Integrated Platform concept described earlier. From the security point of view, there are two significant factors to be considered. Firstly, it will be necessary to verify whether the recommendations that have been provided during the concept stage are implemented correctly. Secondly, the final pilot infrastructure will undergo independent security tests. Independence does not mean here that an external security specialist will be hired, but that a person(s) from the project consortium, who did not take any part in the project work before, will perform the assessment. It will be therefore unbiased and more objective. The tests will consist of the penetration testing stage (due to the specifics of the final user — the tests will be conducted from the internal network, but with different starting roles), followed by clear-box assessments (e.g. reviews of the system and application configurations).

2.4   Security Support for Particular Work Packages

Introduction. The following chapter will present how the security features and recommendations have been embedded into particular horizontal layers of the Platform. Applying security to multiple layers of the network infrastructure conforms with defense-in-depth principle.

Network Architecture. Within the network architecture work packages, besides providing general suggestions on how to design, build and administer the network (both on the technical and procedural level) securely, some particular security areas have been identified. As it has been described earlier, they focus mainly on security of mobile devices, data stored on them and being sent through the physically untrusted channel (e.g. wirelessly). For instance, the communication model has been proposed with respect to the resilience of the network. As for the wired networks, particular facets of IPv6 protocols have also been shortly described, especially those concerning the possibility of a potential DDoS attacks [11].

Application Data Processing Platform. Two important security facets have been addressed for this work package: security of virtualization and of the cloud computing model. While the migration towards virtualization has already been accepted by the LEA, independently of the conducted project, the cloud computing model seems to be some sort of a technical and conceptual breakthrough, although potential applications utilizing technical benefits are unexpectedly large. Nevertheless, security of the data stored in the cloud (even a private cloud, as within the LEA only particular users may have access to some data that might be potentially transferred to the cloud to make sophisticated computations) is the main concern.

The virtualization has been generally found as the solution that not only brings business benefits, but also — when properly applied and configured — increases the security level itself. It provides an additional separation layer between the attacker and the physical server resources. Moreover, virtualization simplifies resource management, while management mistakes may also be the cause of security vulnerabilities. Several general security recommendations have been issued, like e.g. assigning the group of virtual machines processing a similar sort of data to the same physical host or appropriate separation between data sent by virtual machines and those sent on the physical host level.

The situation is different in the case of cloud computing that also brings significant business benefits but with a direct cost on security. In order to present the potential threats, research has been done on clouds security (e.g. [12]) and the identified threats have been applied to the concept of private or hybrid (e.g. shared by different LEAs) cloud. On the one hand, threats like insecure interfaces, service abuse, interoperability problems or internal attacks, have less meaning within the private cloud or do not expose an additional attack vector. On the other hand, restrictions with access to particular data still may cause problems.

Because using a cloud, even the private one, is a new approach for the LEA, securing that sort of cloud has become one of the main areas of work for the security task. A detailed security report is being developed, concerning the appropriate security criteria for ensuring the private cloud, which are issues associated with the network layer, the data center, operational security and internal regulations [13].

That is why the second facet of clouds security described thoroughly in the concept report was data anonymity. One of the proposed potential applications of the cloud computing model was performing computationally intensive operations oriented towards discovering knowledge in the LEA's databases containing millions of records (e.g. discovering connections, looking for anomalies). However, it may be that only a single user is permitted by law to have access to particular data, while, when being processed in the private cloud, they might have to be read by another user.

The goal of the data anonymity is to make the unauthorized access to data unusable — an attacker cannot draw any useful conclusion from the read data. A review of data anonymity techniques has been made — including general approaches, like generalization, swapping, modifying values or encryption, or more formal ones with known computational cost, like k-anonymity. This simple and formal technique allows to have a database formatted in a way that will not allow the attacker to unambiguously identify any confidential or sensitive data. It works by generalizing the database records which are identified as having a sensitive meaning until every row is identical to at least k-1 other rows [14].

For each method, a short explanation has been provided on how the method may impact the expected computations and whether it may be applied at all. For instance, generalization will still allow to make computations but with a decreased accuracy level (e.g. the results will be about "tall people", not about "people 189 cm tall"), but it cannot always be applied (e.g. the IMEI number or personal ID cannot be generalized as they are expected to uniquely identify a thing or a person).

The scope of the project does not allow to implement particular methods, which is merely why the general examples have been provided. However, the prepared materials allow e.g. the LEA to issue requirements for applications that they are going to deploy in the future.

We have derived a research idea to provide a formalized description of data that should be anonymized. A set of attributes and their values might be designed to describe particular columns in the database as well as the whole table. An attribute for a column containing personal IDs might e.g. state that they must be translated into another, unique IDs, while the one for a column that stores the people's age could allow generalization to a 5 years' interval. An attribute for the whole table could e.g. determine the level of k-anonymity (namely, the value of k). However, research work of this type is again outside the scope of the Integrated IT Platform project. Therefore we will describe that approach in general in the concept document. It may be used e.g. to build requirements for vendors of future LEA's applications.

Storage Facilities. Within that work package, besides general recommendations (like encrypting the backup), several detailed suggestions have been provided. As an example, the problem of storing data on SSD drives, that will probably become a standard in the nearest future, has been raised. Due to the properties of that technology (relatively high ratio of damaged sectors and features like wear-leveling), it is in general impossible to securely erase a particular file or the whole disk contents from the software level (the research shows that, independently on the used erasing pattern, it may be possible to retrieve at least 4% of the original file contents [15]). Usually there are problems with the disk controller software that might be used for erasing data, and finally the demagnetizer will also not help. According to the mentioned research, the only possible workaround is the whole drive encryption before using it. Additionally, suitable procedures of physical destruction of unused disks should be prepared and applied.

Advanced Data Management System. The Integrated Platform proposes the Master Data Management model. So far the general security recommendations have been prepared, concerning the role-based authorization models for accessing data and general maintaining the databases securely. However, additional consultancy is taking place, especially considering the data shared by the LEA with external entities.

2.5   Additional Security Solutions

MetaIDS — Intrusion Detection System. MetaIDS has been developed within the confines of the Polish Platform of Homeland Security project "Knowledge and Security Management in High Security Level Services" [16]. It will only be deployed within the Integrated Platform as an auxiliary security tool, especially to protect the Unix/Linux-based servers. MetaIDS is composed of the central management system and a number of sensors (agents) working with low privileges on protected hosts. The system is able to detect attack patterns defined as sequences of events that separately do not have to be suspicious. Another significant feature of the system is the possibility of exchanging the threat information using the IDMEF standard [17]. Therefore it may better support other IDS systems that are already used by the LEA.

As it is virtually impossible to have an IDS that is able to detect every threat, (n + 1) different IDSs assure better coverage of all attack patterns than n IDSs, especially if those solutions work differently. Besides exchanging of IDMEF messages, an example of cooperation between MetaIDS and other IDS systems might be building an "IDS cascade" on the network meeting point, where several IDSs consecutively control data travelling through the point.

We have clear plans to perform further research on the MetaIDS system in order to extend its capabilities. The plans concentrate on building models of suspicious behaviour (understood not only as actions performed within the system, but also as e.g. events of appearing particular entries in the log files etc.), using the concepts of Markov processes, Petri nets or genetic algorithms. However, those extensions may only be a subject of another R&D project and may be then incorporated into the Integrated IT Platform in its further stage as add-ons for the current MetaIDS system.

SARA — System for Automatic Reporting and Administration. SARA is a system for static security control, developed within the confines of the PL-Grid national project. It allows to combine information on software and systems security vulnerabilities stored (with the help of the standards like CPE, CVE and CVSS) in the National Vulnerability Database with the manually provided description of the IT infrastructure. Similarly to the MetaIDS system, it will only be deployed within the Integrated Platform as an auxiliary security tool. It is worth noting that the LEA uses its dedicated software (not present anywhere else on the market), so the feature of SARA that enables defining and using custom CPE entries describing non-public applications and additional, internal vulnerability repositories, should help to maintain the appropriate security level of the whole infrastructure [18].

Security Policies and Procedures. Technical security should be supported with security policies and procedures. It is obvious that the LEA has its Security Management Information System and that the authors of the project cannot impact it. The prepared security procedures (like other security measures) concern only the pilot installation though. Additionally it has been expressed by the LEA that an experience to see how security procedures are prepared by external, independent entities, would be appreciated.

Additionally, it is assumed to provide a set of security configuration checklists concerning the used technologies. The checklists will be delivered at the very end of the project. Thus they will be as up-to-date as possible.

2.6   The ICICLE System

Another proposed solution within the project is the ICICLE (Intelligent Cybercrime Information ColLEctor and Integrator) system. Like the MetaIDS system, that solution has been developed within the confines of the "Knowledge and Security Management in High Security Level Services" project [16]. The ICICLE's goal is not to increase the IT security level, but it may be used to improve the overall homeland security. The system supports communication between LEAs and ISP/ASPs in exchanging retained data. It introduces a consistent XML format for data exchange, based on ETSI Lawful Interception Technical Committee work [19]. Another significant feature of the system is the ability of generating cascades of queries (i.e. an answer to the query A, if conforms with the defined formats, allows to automatically generate and send further queries, derived from the answer to A). ICICLE is an engine that may work with arbitrary GUI applications, provided that appropriate communication interfaces are prepared. Therefore different LEAs may use their dedicated solutions for cases management together with ICICLE.

Currently the ICICLE system cannot be implemented due to delays in implementing appropriate law acts, which makes it impossible to test the system in real conditions. For the Integrated Platform project, ICICLE will be deployed as an exemplary service embedded within the platform. We have developed automated tetsing suites for ICICLE, and we are integrating the system with the PKI used by the LEA. Finally, we prepare ICICLE for handling a different data exchange format. Legal authorities are going to use a similar format to the one proposed, but in a binary form, therefore appropriate interfaces would have to be designed and implemented for the project deployment.

3   Summary

The project of the Integrated Platform is still changing, as the concept and particular assumptions (including security issues) are under constant consultations with the final user. However, the overall proposed approach has been generally accepted by the user. The consortium is of the opinion that the combination of applying security measures in horizontal work packages and the additional vertical protection solutions allow to both properly secure the pilot installation and support the deployment of the Integrated Platform by the final user in the optimal way. We hope that the described approach may be used at least as an initial pattern in other similar initiatives.

References

1. Motorola Solutions Media Center. Public safety organisations plan next generation services, `http://mediacenter.motorolasolutions.com/content/detail.aspx?ReleaseID=14136`
2. Yapp, R.: Brazilian police to use 'robocop-style' glasses at world cup, `http://www.telegraph.co.uk/news/worldnews/southamerica/brazil/8446088/Brazilian-police-to-use-Robocop-style-glasses-at-World-Cup.html`
3. Proteus. integrated mobile system for counterterrosrism and rescue operations, `http://www.projektproteus.pl/?lg=_en`
4. Indect: intelligent information system supporting observation, searching and detection for security of citizens in urban environment, `http://www.indect-project.eu`
5. Odlyzko, A.: Economics, Psychology, and Sociology of Security. In: Wright, R.N. (ed.) FC 2003. LNCS, vol. 2742, pp. 182–189. Springer, Heidelberg (2003)
6. Morana, M.M.: Building security into the software life cycle, `http://www.blackhat.com/presentations/bh-usa-06/bh-us-06-Morana-R3.0.pdf`
7. Microsoft security development lifecycle, `http://www.microsoft.com/security/sdl`
8. Geant project, `http://www.geant.net`
9. Pl-grid project, `http://www.plgrid.pl/en`
10. Balcerek, B., Frankowski, G., Kwiecień, A., Smutnicki, A., Teodorczyk, M.: Security Best Practices: Applying Defense-in-Depth Strategy to Protect the NGI_PL. In: Bubak, M., Szepieniec, T., Wiatr, K. (eds.) PL-Grid 2011. LNCS, vol. 7136, pp. 128–141. Springer, Heidelberg (2012)
11. Jerzak, M., Czarniecki, L., Frankowski, G.: Ddos attacks in the future internet. In: Towards and Beyond Europe 2020: The Significance of Future Internet for Regional Development. Future Internet Event Report (in preparation)
12. ENISA. Cloud computing: Benefits, risks and recommendations for information security, `http://www.enisa.europa.eu/act/rm/files/deliverables/cloudcomputing-risk-assessment/at_download/fullReport`
13. Winkler, V.(J.R.): Securing the cloud - cloud computer security, techniques and tactics, pp. 195–209. Elsevier Inc. (2011)
14. Sweeney, L.: k-anonymity: a model for protecting privacy. International Journal on Uncertainty, Fuzziness and Knowledge-based Systems 10(5), 557–570 (2002)
15. Spada, F.E., Swanson, S., Wei, M., Grupp, L.M.: Reliably erasing data from flash-based solid state drives, `http://static.usenix.org/events/fast11/tech/full_papers/Wei.pdf`
16. Knowledge and security management in high security level services, `http://ppbw.pcss.pl/en`
17. Wojtysiak, M., Jerzak, M.: Distributed intrusion detection systems: Metalds case study. Computational Methods in Science in Technology, Special Issue 2010, 135–145 (2010)
18. Frankowski, G., Rzepka, M.: SARA – System for Inventory and Static Security Control in a Grid Infrastructure. In: Bubak, M., Szepieniec, T., Wiatr, K. (eds.) PL-Grid 2011. LNCS, vol. 7136, pp. 102–113. Springer, Heidelberg (2012)
19. ETSI Lawful Interception Technical Committee. Etsi ts 102 657 v1.9.1 - retained data handling: handover interface for the request and delivery of retained data

Video Detection Algorithm Using an Optical Flow Calculation Method

Andrzej Głowacz[1], Zbigniew Mikrut[2], and Piotr Pawlik[2]

[1] AGH University of Science and Technology,
Department of Telecommunications, Kraków, Poland
aglowacz@agh.edu.pl
[2] AGH University of Science and Technology,
Department of Automatics, Kraków, Poland
{zibi,piotrus}@agh.edu.pl

Abstract. The article presents the concept and implementation of an algorithm for detecting and counting vehicles based on optical flow analysis. The effectiveness and calculation time of three optical flow algorithms (Lucas-Kanade, Horn-Schunck and Brox) were compared. Taking into account the effectiveness and calculation time the Horn-Schunck algorithm was selected and applied to separating moving objects. The authors found that the algorithm is effective at detecting objects when they are subject to binarisation using a fixed threshold. Thanks to the specialized software the results obtained by the algorithm were compared with the manual ones: the total vehicle detection and counting rate achieved by the algorithm was 95,4%. The algorithm is capable to analyse about 8 frames per second (Intel Core i7 920, 2.66 GHz processor, Win7x64).

Keywords: vehicle detection, vehicle counting, optical flow, video detector, traffic analysis.

1   Introduction

One of the aims of the INSIGMA [19] project is the analysis of situations taking place at traffic intersections. The analysis should be conducted automatically on the basis of a sequence of camera images. The subject of the analysis will be multi-lane roads, or in certain cases individual lanes. The analysis will aim to define the following:

— Length of traffic queue (in meters or number of waiting vehicles),
— Number of vehicles leaving the intersection,
— Vehicle speed.

The process of analysing sequences of digital images is usually divided into several stages. They are:

1. Preliminary detection of objects: example methods include using differential images or motion fields, generating background and subtracting the current image, and searching for defined features,

A. Dziech and A. Czyżewski (Eds.): MCSS 2012, CCIS 287, pp. 118–129, 2012.

2. Segmentation, generally achieved through thresholding, which divides the observed scene into objects and background,
3. Analysis of the detected objects: labelling, framing, integration or division of objects, defining their key features and other parameters.

Chapter 2 presents three algorithms for calculating optical flow which form the basis of detection of moving vehicles. The comparison was conducted in terms of the accuracy of object detection and calculation time. Chapter 3 presents the segmentation algorithm attempting to solve the problems of integration and division of objects corresponding to vehicles.

Calculations were carried out on video sequences recorded from an upper storey of a student accommodation building at a traffic intersection in Kraków. The results of the video detection algorithm are presented in Chapter 4, using the example of counting the total number of vehicles.

2   Optical Flow Calculation Algorithms

The segmentation process was based on the optical flow method, allowing the detection of moving objects. The method was previously studied at the Biocybernetics Laboratory [15, 10, 1, 17], although no practical application was found. Tests were concluded at the object segmentation stage due to the relatively long time taken to calculate optical flow using computers available at the time (analysis took approx. 30 times longer than real time in spite of having been implemented using the C programming language and in spite of optimisation attempts). It seems that better conditions are currently in place for restarting the experiments: significantly faster, multi-core computers are now available, as well as additional methods of equalising calculations using graphics processing units (GPUs) or field programmable gate arrays (FPGA).

An alternative to the optical flow method is an algorithm for detecting static and dynamic objects by generating and removing background; this method was used in the video detector in Kraków [14, 2, 13]. By applying a suitable reduction to visual information (up to a few lines of pixels, defined along each lane), the video detector software operated in real-time, even on an industrial PC104+ (Celeron 400 MHz, 128MB SDRAM). The video detector operated in an inductive loop simulation mode, since it was assumed at the time that it will replace inductive loops and will be able to transmit information to the traffic light controller in the same way.

The authors foresee several advantages in using methods based on calculating optical flow, such as disassociating the results of vehicle detection from atmospheric conditions, and the ability to detect vehicles whose colour is similar to the background. Additionally information about the optical flow direction will be useful in the future research, especially when the vehicles could be occluded by the ones moving in opposite direction. However using optical flow method forces the development of algorithms that track the detected vehicles and remember their position, in particular when they are stationary.

The methods for computing optical flow can be divided into three basic groups:

1. Gradient methods, based on the analysis of the derivatives of image intensity values,
2. Frequency methods, based on filters operating in the frequency domain,
3. Correlation methods searching the image space.

Regardless of the differences between these methods, the majority are based on three phases of the calculations [4, 5, 3]. They are:

— Preliminary processing using appropriate filters in order to obtain the desired signal structure and improve the signal-to-noise ratio.
— Calculating basic measurements such as partial derivatives after a certain time period or local correlation areas,
— Integrating the measurements in order to calculate a 2-dimensional optical flow.

Based on papers [5, 11, 7], whose authors conducted comparative tests of methods used to compute optical flow, and on results obtained in [15], it was found that the suitability of two first order gradient methods should be tested in the first instance: the Lucas-Kanade method working on a local scale [12], and the Horn-Schunck method operating on a global scale [8, 9]. It is interesting to compare the two methods, since they are based on similar entry data (derivative order); however, the calculation method producing the final result is different in both cases.

The means of implementing both methods have been described in detail in [15]. Results obtained by using both methods are affected by certain parameters, which are as follows:

— The means of computing partial derivatives: the numerical differentiation method used, affecting the properties of the gradient and calculation precision,
— The threshold criterion parameter τ for the Lucas-Kanade method, affecting the size and position of the area for which the 2-dimensional optical speed is being computed, which in turn affects the precision of the calculations,
— The α parameter for the Horn-Schunck method, regulating the effects of limiting the fluidity of optical speed, and as such affecting the continuity of the optical flow,
— The number of iterations for the Horn-Schunck method affecting the propagation of information on optical flow and providing for the detection of greater speeds allowing for greater displacement.

The third tested method (Brox's algorithm) is regarded as the most precise; however, it is also the most complex to calculate. It was implemented in MATLAB on the basis of [6, 16] by Visesh Chari [18]. The method achieves a global reduction of the functional, very similar to the one proposed by Horn and Schunck [8], although a double linearisation of the components of the functional is carried out instead of single. In addition, calculations are performed on the given number of levels of the image pyramid: optical flow computed on lower resolution images is used to initialise calculations on higher resolution images using the warping technique.

2.1   Algorithm Testing

Preliminary tests involved the selection of a short film, documenting traffic flow at an intersection in Kraków. The recording was taken from high above (a balcony in a student accommodation block) in early spring (this is relevant, since trees which would usually partially obscure the view were bare), in light drizzle.

The optical flow was calculated using three algorithms:

- The Lucas-Kanade algorithm (for the $\tau = 32$ parameter),
- The Horn-Schunck algorithm in two variants: for parameters $\alpha = 31$ and $iter = 8$, the central 4-point difference was used as the differentiation method; for parameters $\alpha = 33$ and $iter = 9$, a first order difference was used to calculate the derivative,
- The Brox algorithm [6, 16], which utilised warping and an image pyramid.

For each calculation step, the vectors of the optical flow were written to file as two components: horizontal and vertical. Further calculations were conducted using MATLAB scripts. Information on the vectors' direction was not used at this stage; only vector lengths were calculated and binarised on several levels.

Figure 1 depicts the results obtained for one of the film frames. It shows the binarised optical flow vector lengths; thresholds were selected for each of the three methods to allow for the best detection of moving objects. The results presented are representative of the tested methods and the entire film.

2.2   Results of Algorithm Comparison

The Brox method produces objects that are well filled in. It is extremely sensitive to even the slightest camera movements (cf. Fig. 1 for threshold 0.2). The method tends to merge objects located close together (for example, two vehicles near the lower edge of the frame were merged and could not be separated even at a threshold of 0.4). In order to detect slowly moving objects, it is necessary to use a lower threshold (0.2); however, this introduces interference caused by camera movement.

The Lucas-Kanade method also generated a lot of interference at a threshold of 0.2, although it was less significant than for the Brox method. Detection is most effective at a threshold of 0.4, where more objects are detected than for the Brox method; however, they do not always form a single whole (they are divided).

When using the Horn-Schunck method, interference is produced at a threshold of 0.2; however, it is point interference, which is less common than the other results discussed here. Objects are well filled out at this threshold.

The analysis becomes more complicated when the road surface is wet; vehicle images are reflected in it, increasing their size. On one hand vehicles appear larger than they really are, and on the other this leads to the integration of separate objects.

Tiny vibrations of the camera cause significant interference in the optical flow. The problem is visible in Fig. 1 for the lowest thresholds, and even more obvious in other fragments of the film.

Table 1 presents basic information on the tests conducted. In conjunction with the results depicted in Fig. 1, it indicates the Horn-Schunck method as the most effective at implementing the tasks outlined in the Introduction.

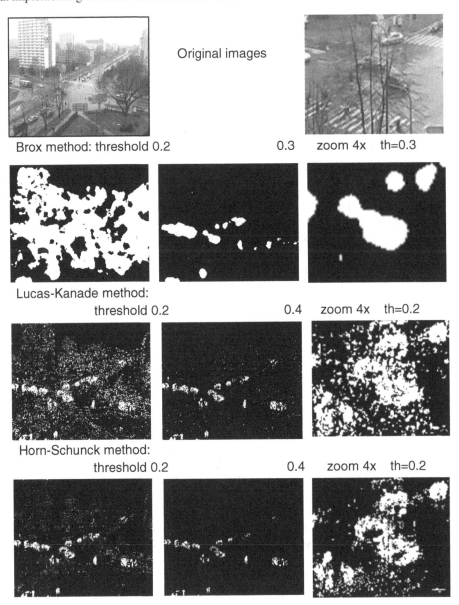

Fig. 1. Effects of the optical flow binarisation calculated using three methods

Table 1. Optical flow computation: results of preliminary experiments* (in the column „Range" the minimal and maximal values of the optical flow vector lengths are presented)

Method	Image size [pix]	Mean calculation time for a single frame** [s]	Range (for frames with no vibration)	Implementation language
Horn-Schunck $\alpha = 31$, *iter* = 8	640×480	0.09	[0; 2.67]	C
Horn-Schunck $\alpha = 33$, *iter* = 9	640×480	0.10	[0; 3.36]	C
Lucas-Kanade $\tau = 32$	640×480	0.09	[0; 6.64]	C
Brox	640×480	36.00	---	MATLAB
	320×240	7.70	[0; 0.88]	MATLAB

* analysis of every third film frame
** for an Intel Core i7 920, 2.66GHz processor (W7x64)

3   Vehicle Detection Algorithm

The input of the described algorithm is a binary image obtained on the basis of optical flow calculated using the Horn-Schunck method (cf. Chapter 2). The purpose of the algorithm is to isolate moving objects (such as vehicles) and to track them on subsequent film frames. The algorithm takes into account (to a varying degree) problems of a single object splitting into several parts, several objects merging into one, and objects stopping temporarily.

It was found that after the optical flow is binarised, there is no guarantee of imaging a vehicle into a single visual object. As a result of vehicles having uniform areas (which do not generate changes in the optical flow) or becoming occluded by fixed background objects, the vehicle can split into two or more objects. In turn, vehicles occluding one another while moving along adjacent lanes may result in images of two or more vehicles merging into a single visual object. As a result, following segmentation, algorithms combining objects that were split incorrectly or dividing those that were merged incorrectly were implemented.

The algorithm operates on a list of objects (representations of vehicles) containing data such as surface area, bounding box, and several recent locations of the centre of gravity. Vehicle identification involves the car being correctly assigned to an object in the representation list in the given film frame. This process also uses the preceding film frame.

It is initially assumed (at the first frame) that a single object represents precisely a single vehicle. Basic parameters representing the vehicle – surface area and the bounding box– are calculated. Additionally, small objects are attached to significantly larger ones with which they share appropriately large parts of the bounding boxes. Unique identifiers are then assigned to representations describing the vehicles;

subsequent parameters describing the vehicles (e.g. centre of gravity) are marked after assigning the identifiers.

The second and subsequent frames are analysed on the basis of information obtained in the previous step of the algorithm. The start of the analysis runs identically to the start of the first step: an auxiliary list of representations is created with the assumption that a single item on the list corresponds to a single vehicle. The auxiliary list is then adjusted to match the list from the previous step (henceforth known as the main list). This involves assigning records from the auxiliary list to records from the main list such that they both describe the same vehicle (on two subsequent frames).

The process must take into account the following instances:

a)   A single record in the auxiliary list corresponds to a single record in the main list,
b)   Two or more records in the auxiliary list correspond to a single record,
c)   A single record in the auxiliary list corresponds to two or more records in the main list,
d)   Several records in the auxiliary list correspond to several records in the main list,

where the term "corresponds" is understood as the records containing a sufficiently large intersection on two consecutive frames. The first three cases are shown in Figure 2 as an activity diagram illustrating the discussed algorithm.

The first case is also the simplest: a record from the auxiliary list is assigned a unique identifier corresponding to the record from the main list; the car is represented by a single visual object from the scene.

In the second case, a single object is divided into two or more objects. All records from the auxiliary list are assigned to an identifier corresponding to the record from the main list; this describes a situation where segmentation results in a single vehicle being represented by several objects. In order to minimise the risk of incorrectly merging several object into a single object, centres of gravity of these objects are analysed; in the event of exceeding the distance limit, objects that are too "distant" receive new identifiers (these objects represent new vehicles).

The third case is the opposite of the second, and concerns the merging of representations of several vehicles into a single visual object. In order to avoid losing control over the motion of each object (vehicle), additional segmentation of the "common" visual object is conducted by morphological dilation of the shared part (formed by subtracting objects/vehicles from the previous frame). Dilation is directional: each object fragment is expanded in the direction of its motion, defined by the vector of the changing position of the centre of gravity. If the vehicle is stationary (the vector is zero), the dilation is performed in all directions.

The fourth case (several visual objects to several vehicles) occurs very rarely. It is not discussed here, since the outlay required to study it exceeds the frequency of its occurrence many times over.

After balancing the connections between the records from both lists, records from the auxiliary list without counterparts in the main list are retrieved. They are regarded as representations of new vehicles, and assigned new identifiers.

In turn, records from the main list without counterparts in the auxiliary list represent vehicles which have left the control area or which have stopped. They are marked accordingly and removed after a certain period.

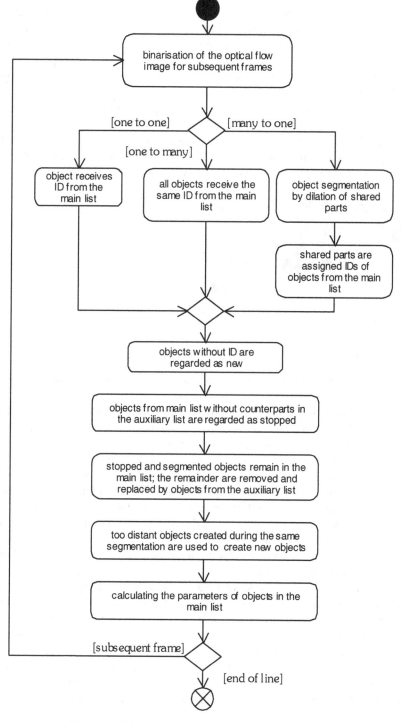

Fig. 2. Activity diagram for the proposed algorithm

The stage of harmonising the lists aims to assign records from the auxiliary list with suitable and unique identifiers. The next stage is the re-indexing of objects from the current frame with the assigned identifiers. The main list is updated last. Records used in the dilation process (added to other objects), and records corresponding to vehicles that have stopped or moved outside the area of interest are kept. The remaining records are replaced by records from the auxiliary list. This stage is followed by calculating the parameters of all objects in the scene once again, taking into account their new indexes (identifiers) and the process is repeated from the beginning.

The algorithm helps create a list of all currently tracked vehicles. It can be represented graphically by indexing the binary image with the vehicles' unique identifiers. Viewing films where algorithms have been visualised this way shows that when vehicles are not obscured, there are no problems with tracking them (cf. Fig. 3).

Fig. 3. An example of correct tracking of objects that do not obscure one another

The division of two lanes of traffic into one lane going straight on and one being used for turning is also working correctly. This corresponds to initially recognising two overlapping vehicles as a single object, which is subsequently correctly split into two, once the vehicles move a suitable distance apart (cf. Fig. 4).

In Fig. 3, the measuring line for counting vehicles turning right is marked in black. Specialized software used for manually counting vehicles driving by, with a memory function for the moment of counting, has also been created. A comparison of the algorithm performance (solid line) with actual data (dashed line) is shown in Figure 5.

The algorithm described above detects vehicles at points very close to manually-recorded times. In Figure 5 three algorithm errors are marked by arrows. Thanks to the specialized software it was possible to find the causes of errors. The algorithm is currently unable to make a distinction between vehicles travelling side by side at similar speeds in adjacent lanes, and they are treated as a single object (arrow no 1). Furthermore, there are cases of incorrect merging of objects into a single representation of a vehicle (cf. Fig. 4, Fig.5 - arrows no 2 and 3): two long vehicles (a truck and a bus) of uniform color were divided by the algorithm into pieces. This part of the algorithm requires further work. Nevertheless the vehicle count results are very good. For four measuring lines the obtained counting rates (for ~5 min film) were 91%, 93%, 98% and 100% (with total false positive/negative rates FPR=FNR= 2,3%).

Fig. 4. An example of tracking an object (grey, in the bottom left corner) formed when two vehicles moving in different directions enter the intersection

4   Summary and Conclusions

The article presents the concept and implementation of an algorithm for detecting and counting vehicles based on optical flow analysis. The effectiveness and calculation time of three algorithms were compared. The Horn-Schunck algorithm was selected and applied to separating moving objects. The authors found that the algorithm is effective at isolating objects when they are subject to binarisation using a fixed threshold. The authors constructed a basic algorithm for vehicle detection; its operation and results are presented in Chapter 3. The total counting rate achieved by the algorithm was 95,4%.

Summing up the operation of the proposed algorithm, the following statements can be made:

— It operates correctly when vehicles do not obscure one another,
— Vehicles moving along lanes one of which is used for turning only are initially merged as a result of overlapping; however, they are then separated correctly,
— Occasionally, a single vehicle is divided into several object (this concerns specific vehicles of uniform colour),
— Vehicles moving side by side at similar speeds in adjacent lanes are not distinguished,
— The algorithm is capable to analyse about 8 frames per second (Intel Core i7 920, 2.66GHz processor, Win7x64).

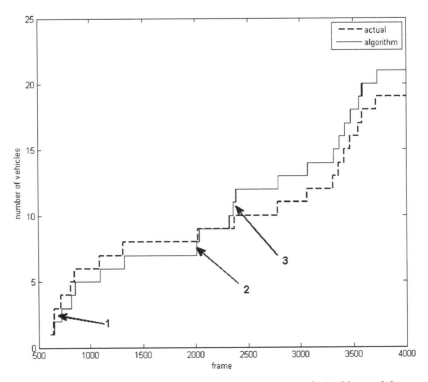

Fig. 5. A comparison of results of vehicle detection (and counting) with actual data

Acknowledgements. The work was co-financed by the European Union from the European Regional Development Fund, as part of the Innovative Economy Operational Programme INSIGMA no. POIG.01.01.02-00-062/09. The authors extend thanks to their student team of Szymon Bigaj and Jacek Kościow for drafting the first version of the software for manually counting vehicles.

References

1. Adamski, A., Bubliński, Z., Mikrut, Z., Pawlik, P.: The image-based automatic monitoring for safety traffic lanes intersections. In: Piecha, J. (ed.) Transactions on Transport Systems Telematics, Wyd. Politechniki Śląskiej, Gliwice, pp. 92–102 (2004)
2. Adamski, A., Mikrut, Z.: The Cracovian prototype of videodetectors feedback in transportation systems. In: Piecha, J. (ed.) Trans. on Transport Systems Telematics, Wyd. Politechniki Śląskiej, Gliwice, pp. 140–151 (2004)
3. Beauchemin, S.S., Barron, J.L.: The Computation of Optical Flow. ACM Computing Surveys 27(3), 433–467 (1995)
4. Barron, J.L., Beauchemin, S.S., Fleet, D.J.: On Optical Flow. In: 6th Int. Conf. on Artificial Intelligence and Information-Control Systems of Robots, Bratislava, Slovakia, September 12-16, pp. 3–14 (1994)

5. Barron, J.L., Fleet, D.J., Beauchemin, S.S.: Performance of optical flow techniques. Int. Journal of Computer Vision 12(1), 43–77 (1994)
6. Brox, T., Bruhn, A., Papenberg, N., Weickert, J.: High Accuracy Optical Flow Estimation Based on a Theory for Warping. In: Pajdla, T., Matas, J(G.) (eds.) ECCV 2004. LNCS, vol. 3024, pp. 25–36. Springer, Heidelberg (2004)
7. Galvin, B., McCane, B., Novins, K., Mason, D., Mills, S.: Recovering Motion Fields: An Evaluation of Eight Optical Flow Algorithms. In: Proc. of the British Machine Vision Conference, BMVC (1998)
8. Horn, B.K.P., Schunck, B.G.: Determining optical flow. Artificial Intelligence 17, 185–204 (1981)
9. Horn, B.K.P., Schunck, B.G.: Determining optical flow: a retrospective. Artificial Intelligence 59, 81–87 (1993)
10. Kotula, K., Mikrut, Z.: Detection and segmentation of vehicles based on a hierarchical "optical flow" algorithm. Trans. on Transport Systems Telematics, 34–46 (2006)
11. Liu, H., Hong, T., Herman, M., Camus, T., Chellappa, R.: Accuracy vs. Efficiency Trade-offs in Optical Flow Algorithms. Computer Vision and Image Understanding (CVIU) 72(3), 271–286 (1998)
12. Lucas, B.D., Kanade, T.: An iterative image registration technique with an application to stereo vision. In: Proc. 7th Intl. Joint Conf. on Artificial Intelligence (IJACAI), Vancouver, August 24-28, pp. 674–679 (1981)
13. Mikrut, Z.: Road Traffic Measurement Using Videodetection. Image Processing and Communications 3(3-4), 19–30 (1997)
14. Mikrut, Z.: The Cracovian Videodetector - from Ideas to Embedding. In: Proc. Int. Conf. Transportation and Logistics Integrated Systems ITS-ILS 2007, Kraków, pp. 29–37 (2007)
15. Mikrut, Z., Pałczyński, K.: Image sequences segmentation based on optical flow method. Automatyka AGH 7(3), 371–384 (2003) (in Polish)
16. Sand, P., Teller, S.: Particle video. In: IEEE Computer Vision and Pattern Recognition, CVPR (2006)
17. Tadeusiewicz, R.: How Intelligent Should Be System for Image Analysis? In: Kwasnicka, H., Jain, L.C. (eds.) Innovations in Intelligent Image Analysis. SCI, vol. 339, pp. V – X. Springer, Heidelberg (2011)
18. Chari, V.: High Accuracy Optical Flow Using a Theory for Warping, http://perception.inrialpes.fr/~chari/myweb/Software/ (accessed March 20, 2012)
19. INSIGMA Project. AGH UST, Kraków, http://insigma.kt.agh.edu.pl (accessed February 4, 2012)

INSTREET - Application for Urban Photograph Localization

Michał Grega, Seweryn Łach, and Bogusław Cyganek*

AGH University of Science and Technology
Department of Telecommunications
al. Mickiewicza 30, 30-059 Krakow
Poland
grega@kt.agh.edu.pl

Abstract. The paper proposes a solution to the problem of geolocation of a photograph based on comparison of its content to a geolocated database of street view images. The proposed algorithm allows the pinpointing of the location where a photograph was taken. In order to solve this problem we have proposed an algorithm based on MPEG-7 features. We have conducted large-scale tests (4% of area of the city of Krakow, Poland) that prove that the algorithm is characterized by high accuracy at the cost of high computing power required.

Keywords: geolocation, geotagging, localisation of photographs, criminal investigations, MPEG-7.

1 Introduction

Some cameras available today are equipped with a GPS module. This module allows the geographic coordinates of the place where the photograph was taken to be added to its metadata (EXIF). However, if such information is either unavailable or has been removed from the photograph metadata, the photograph's location may be either difficult or impossible to determine.

This problem is one often encountered by police officers during their investigations. If the result of the investigation starts depending on the geolocation of the photograph, the problem becomes crucial. In this case, the only way to geolocate the photograph is to hire analysts whose job is to search manually for the most probable site. This work is incredibly tedious and has a low probability of success.

In this paper, we propose a fully automated method for geolocating photographs taken in urban scenery. Our algorithm is based on low-level image features and uses MPEG-7 descriptors for the task. As a source database of reference images, data from a street view application is utilized. We show that our algorithm is able to pinpoint the location of the photograph based on a single

---

* This work was supported by the European Commission under the Grant INDECT No. FP7 218086.

A. Dziech and A. Czyżewski (Eds.): MCSS 2012, CCIS 287, pp. 130–138, 2012.

characteristic detail at the cost of the high computing power required for the task.

The problem considered in this paper has been approached in a small number of papers. Hays and Efros [2] propose a system, similar to the one proposed here, that attempts to obtain geolocation based on a single image. In their paper, the authors utilize geolocated images obtained from an image sharing service. Seven different features are compared in terms of accuracy of localization (where an accuracy of 200km is considered satisfactory). Unlike the authors of [2], we focus on street view data and try to pinpoint the location of the photograph exactly. In their next paper [3], the authors focus on using a larger database and optimizing the geolocation algorithm. Work presented by Zhang and Kosecka [4] is closest in concept to that presented here. The authors attempt to solve the problem by utilizing the SIFT [6] algorithm. A great effort is made in order to compensate for perspective transforms caused by the different angles the photographs may be taken at. In our work we attempted to use the SURF algorithm [5] (enhanced and more efficient when compared to SIFT), but it was proven to be less accurate than our MPEG-7 approach. The importance of the topic is further confirmed by a call published by the US National Intelligence Office: Intelligence Advanced Research Projects Activity (IARPA) for a solution to the presented problem (Solicitation IARPA-BAA-11-05, May 2011).

The rest of the paper is structured as follows. Section 2 presents the architecture of the proposed solution. Section 3 presents the results of the experiments. The paper is concluded in Section 4.

2   Solution Architecture

In order to be able to geolocate a photograph we have followed the path that is usually followed by human specialists. We have focused on architectural details that are characteristic to a specific place. We have proposed an algorithm that, while being complex and demanding high computing power, is, at the same time, fully automated. The algorithm is presented in Figure 1.

In brief, our algorithm accepts a fragment of the localized photograph as an input. This rectangular fragment is chosen by the user and should contain a characteristic architectural detail of the photograph. For instance, it could be an unusually shaped window in a building, or a piece of a wall of a characteristic colour, texture and/or shape. Low level descriptors of this fragment are compared against low-level descriptors of photographs coming from a street view service.

A street view service is a popular Internet application that combines interactive maps with panoramic photographs made at street level. Examples of such services are StreetView from Google, NORC (which covers cities in mid-eastern Europe not covered by Google StreetView) and Microsoft Streetside. Photographs from these services are of high resolution and quality on one hand, and on the second – geolocated, and therefore create a perfect reference database for the INSTREET application.

Our approach assumes the utilization of a cascade of one texture descriptor (Edge Histogram) and four colour descriptors. Afterwards, a weighted rank

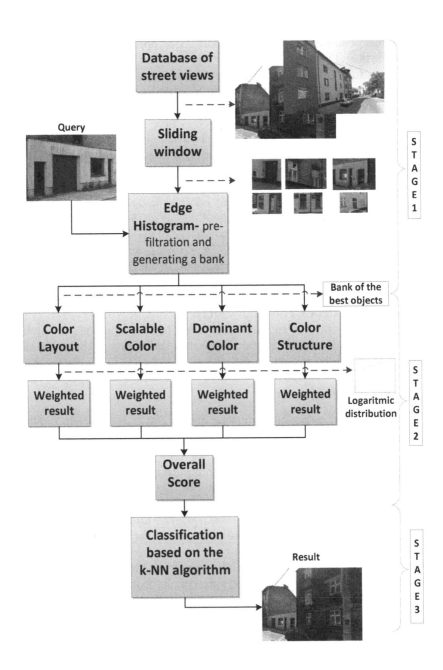

Fig. 1. Architecture of the INSTREET

algorithm is used. Finally the k-nearest neighbor algorithm (k-NN) returns the candidate panoramas. The best candidate street view panoramas with their location are chosen by the algorithm and presented to the user to do the final overview and geolocation of the photograph. The following sections describe the subsequent stages of the INSTREET algorithm.

2.1 Stage 1 — Inputs and Pre-filtering

As mentioned before, the proposed algorithm requires the user to choose a characteristic object (a.k.a. region of interest) from the photograph (Figure 2) to be geolocated in the process. It is very important that the user marks a homogeneous set of the most characteristics objects in the photograph, for example a gateway-door-window set. This operation has a crucial role in the classification process. The tests conducted showed that if the selected object was too detailed (for example a letterbox) or too general (a whole photo) the overall quality of the result decreases drastically.

Fig. 2. Selection the region of interest

The user is also required to mark on the map the region to be searched. Our algorithm is optimized to work in the urban environment. The larger the area the user chooses, the more street view panoramas have to be searched and the longer the search time.

The panoramas from the selected region are divided into subimages using the sliding window algorithm. The mode of operation of this algorithm is characterized by two parameters: an area of overlap between two adjacent blocks, and a size of sliding window in consecutive iterations. In the system presented, the best results were achieved for 90% of overlap area and 10% resized down the size of sliding windows per iteration. According to these parameters, one panoramic view is divided into about 30,000 subimages, while one street may consist of 40-80 views. This causes the INSTREET to be a very time-consuming system.

The pre-filtering is realised by using the edge histogram algorithm from the MPEG-7 standard [1]. The edge histogram is an algorithm that enables an image to be reduced to a vector that represents the numbers and directionality of edges in an image. Moreover, it enables a distance to be computed between two vectors, which leads to an estimation of visual similarity between the two images. The edge histogram distinguishes five types of edges in local image regions, four directional (vertical, horizontal, 45 degree, 135 degree) and one non-directional.

This information is provided for every subimage. Subimages in the process of Edge Histogram calculations are defined by dividing the image into 16 non-overlapping parts. As a result, the edge histogram returns a vector of values, also referred to as a descriptor value. A distance between two images is computed in compliance with the MPEG-7 standard [1]. This distance represents the visual similarity between the query and the analyzed image.

At this stage in the algorithm, the characteristic object chosen by the user is compared using the edge histogram descriptor with subimages created by the sliding window algorithm. A list of best candidate subimages is created and continuously updated during the process of pre-filtering. This list is referred to as a bank of results. Such an approach allows for optimized memory management and for more time efficient operation of the application.

2.2   Stage 2 – Colour-Based Filters

At this stage, the INSTREET compares the candidates in the bank with the user-selected region using colour-based MPEG7 descriptors. These descriptors are the Color Layout, Scalable Color, Dominant Color and Color Structure. For each of these descriptors, all candidates in the bank are compared against the user-selected region. For each of the descriptors the results are ordered from the best to the worst match.

Using four different descriptors caused a problem with the final classification. It was resolved by using a weighted classification - the ranks algorithm, which is based on a logarithmic distribution. The algorithm proceeds as follows. Each result from each descriptor is assigned a rank ranging from 1 to the size of the bank. The rank represents the similarity between the user-selected region and the result, where 1 is the most similar. Every image has four ranks assigned, one for each descriptor. The INSTREET penalises results for being low ranked by the descriptors using the logarithmic distribution. This distribution was chosen from among others after a set of experiments. The overall score for one object is expressed by the following equation:

$$R(i) = \sum_{j=1}^{4} \log[X(i,j)] \tag{1}$$

where:
$R(i)$ – overall rank for i-object
$X(i,j)$ – result for ith object in jth-descriptor (CL,SC,DC,CS)

As a result, the final output of this stage of the algorithm is a set of candidate images with an associated overall score.

It is worth mentioning, that the presented algorithm will not be effective in case if the exemplary photograph is made during night time due to differences in colours between day and night photos. In order to compensate for different lighting conditions a colour normalization algorithm is now under research.

2.3   Stage 3 — Final Classification

The final stage of the algorithm is to utilize the k — nearest neighbor $(k - NN)$ algorithm. This is a method for classifying objects based on the closest distance in a feature space. After the logarithmic classification, a set of nearest pictures is available. The role of the $k - NN$ algorithm is to assign objects to their own model (a street view). As a consequence it is possible to determine which class of objects is the most numerous in the $k - NN$ classification.

The algorithm outputs the best candidate panoramas. We calculate the performance of the algorithm as the number of results the user has to browse in order to find the desired location. The lower the number the better.

3   Experiment and Results

This section describes the practical experiments performed with the INSTREET algorithm that were conducted in order to assess both accuracy and time performance.

3.1   Data

The database of panoramas used in the experiment includes 4,000 views from Krakow (third largest city in Poland, 1 million inhabitants) streets and covers about 4% of the whole area of the city. It can be estimated that the whole dataset for Krakow would consist of approx. 100,000 views. The queries used in the experiments were real, random photos from Krakow streets.

3.2   Search Results

Exemplary results are presented in Figure 3. INSTREET is able to recognize the correct sites from a set of 4,000 views by reducing the number of panoramas that have to be inspected by two orders of magnitude. In most cases the appropriate localization was placed in the first ten images. In the worst case, from 4024 views the INSTREET algorithm was able to select 37 candidates which is 0,92% of all the street views in order to recognize the correct localization (Tab. 1). In addition, the panoramas clustered near to the searched one have a similar context, taking into account the visual characteristics of the object being sought. This fact confirms that INSTREET is working properly and at an appropriate quality.

Fig. 3. An exemplary result which shows several first responses

Table 1. Result for the worst tested case

Number of street views	Number of selected images	[%]
4024	37	0.92

3.3   Time Performance

All tests were carried out on a PC with the following specifications:

- Processor 2x2.2 [GHz] AMD Turion 32 bit
- Memory RAM: 2 [GB]
- OS: Linux Ubuntu 10.04

As the system presented has to process a huge amount of data, the weak point of the INSTREET application is the time performance. Searching through the area described above takes about 60 hours of computing time. It should be noted, however, that the search process is fully automated and can be performed without user intervention. In order to improve the time efficiency, the idea of using downsized panoramas was investigated. It was necessary to check whether smaller sized panoramas would have an impact on the search result. Two options were tested:

- Panoramas downsized 4 times per dimension
- Panoramas downsized 2 times per dimension

In the first case, downsizing accelerated the algorithm execution time by nearly 3 times (Tabl .2). On the other hand, this operation caused meaningful changes in the final classification. Comparing results for downsized and original panoramas, it was easy to conclude that the downsized panoramas cause a serious decrease in the discriminating strength of the algorithm. The number of panoramas in a result set grew up by nearly 100%.

In the second case, the execution time was decreased by about 30%. Moreover, the influence on the results was insignificant.

Table 2. An effect of view size on the processing time

View size	Time [hours]	Acceleration
Original	66	x1
1/2 Original	44	x1.5
1/4 Original	24	x2.75

3.4 SURF Algorithm in the INSTREET

During the work on the INSTREET algorithm, we have attempted to utilize the Speeded Up Robust Feature (SURF) algorithm [5] instead of MPEG-7 descriptors. This method uses matches between key points in the query and the training image. Significantly, only the most characteristic features of the image are taken into account. Low contrast candidate points are discarded as the most meaningful points are situated on the edges or at high contrast points.

The question of whether the SURF algorithm is suitable for use in the IN-STREET application was tested. At first it was necessary to check the quality of retrieval, in order to compare the results with MPEG-7. Our tests proved that the SURF algorithm is faster that the MPEG-7 based algorithm. In a few cases, the qualitative results were comparable to the MPEG-7. However, the SURF has problems with perspective transformations, especially when the geolocated photograph is taken from a substantially different angle than the panorama in the database.

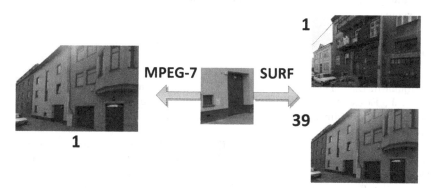

Fig. 4. Two tested approaches — an example of results

The case of a different angle is presented in Figure 4. The selected characteristic site comes from a picture which was taken from the left-hand side while the associated street view has a perspective from the right-hand side. In consequence, the SURF algorithm was able to geolocate the image only as the 39th candidate, when MPEG-7 made a correct hit with the first candidate. In all the cases tested, the MPEG-7 algorithm was able to provide proper geolocation within the ten first results returned to the user. SURF algorithm may be considered as a fast and rough method in cases where speed of search is more important than accuracy.

4   Conclusions and Further Work

In this paper we have proposed an algorithm for geolocating photographs in an urban environment. Our algorithm requires the user to select a region of interest in the photograph, which may be a characteristic architectural detail, for example. Based on this rectangular image, our algorithm runs through a database of geolocated panoramas made at the street view level. Using MPEG-7 descriptors our algorithm proposes the geolocated panoramas to the user that are most likely to be made in the location of the query photograph.

Our algorithm, while being very time consuming, is able to find the photograph's location with high accuracy. In most cases, the search location was proposed to the user within the first ten results. In the worst case, the search location was within 1% of the returned results.

In further work, we plan to create a GUI for the algorithm in order to create a prototype application. We will also work on speeding up the most time-consuming operation, which is the calculation of descriptor values by delegating this task to a GPU instead of the CPU. It is also considered to employ a super-computing cluster for the task.

Having created a GUI and optimized the execution time, we would like to assess the exact accuracy of the algorithm as a function of the geographical area that is analysed. Currently such tests are impossible due to the algorithm's very long execution.

References

1. Manjunath, B.S., Salembier, P., Sikora, T.: Introduction to MPEG-7. Wiley (2002)
2. Hays, J., Efros, A.A.: IM2GPS: estimating geographic information from a single image. In: Proceedings of the IEEE Conf. on Computer Vision and Pattern Recognition, CVPR (2008)
3. Kalogerakis, E., Vesselova, O., Hays, J., Efros, A.A., Hertzmann, A.: Image Sequence Geolocation with Human Travel Priors. In: Proceedings of the IEEE Internaltional Conference on Computer Vision Recognition, ICCV (2009)
4. Zhang, W., Kosecka, J.: Image Based Localization in Urban Environments. In: International Symposium on 3D Data Processing, Visualization and Transmission, 3DPVT 2006, North Carolina, Chapel Hill (2006)
5. Bay, H., Ess, A., Tuytelaars, T., Van Gool, L.: SURF: Speeded Up Robust Features. Computer Vision and Image Understanding (CVIU) 110(3), 346–359 (2008)
6. Lowe, D.G.: Object recognition from local scale-invariant features. In: Proceedings of the International Conference on Computer Vision 1999, vol. 2, pp. 1150–1157 (1999)

Real Data Performance Evaluation of CAISS Watermarking Scheme

Piotr Guzik, Andrzej Matiolański, and Andrzej Dziech

Department of Telecommunication,
AGH University of Science and Technology,
Mickiewicza 30, 30-059, Cracow, Poland
{guzik,matiolanski,dziech}@kt.agh.edu.pl

Abstract. In this paper we present real data Bit Error Rate (BER) performance tests of recently proposed corelation-and-bit-aware improved spread spectrum (CAISS) watermarking scheme [8]. Our tests were performed in DCT domain. The results show significant improvement as compared with traditional spread spectrum technique applied to the same and identically prepared data. Tests performed under medium JPEG compression and fixed Peak Signal-to-Noise Ratio (PSNR) level indicate that appropriate choice of CAISS parameters results in over a two orders of magnitude smaller BER as compared with spread spectrum technique without side information about correlation. We also compared CAISS with improved spread spectrum scheme and found that CAISS can perform approximately two times better than ISS (in terms of BER) after medium JPEG compression.

Keywords: Watermarking, Spread Spectrum, DCT domain.

1 Introduction

Digital watermarking is a technique that allows embedding multimedia objects with additional information which is stored among object's bits. Watermark can be used to hide data in a signal in such a way that it is not perceived by human but may be retrieved with use of dedicated algorithm. This technology can be used in copyright protection, hidden communication and many other applications.

Watermarking system consists of two parts. The first one is encoder which embeds host signal with some additional information. The second, called decoder, is able to detect watermark and retrieve the information. There are many techniques of watermark data embedding. Two main classes of watermarking schemes are quantization-based techniques [1] and spread sprectrum (SS) techniques. SS scheme was introduced by Cox [3]. That technique allows spreading the information over the entire host signal. The information may be simply added to the host signal (additive SS [2]) or it may multiply the signal (multiplicative SS [7]). Usually it is assumed that the original host signal is not known at the decoder side (blind decoding) and Pearson correlation is used to detect watermarks. Such

A. Dziech and A. Czyżewski (Eds.): MCSS 2012, CCIS 287, pp. 139–147, 2012.
© Springer-Verlag Berlin Heidelberg 2012

a decoding scheme is vulnerable to any correlation between the host signal and the embedded information. Malvar and Florencio [4] proposed improved spread spectrum (ISS), the method for reducing negative effect of above-mentioned correlation on decoding performance. In ISS embedding scheme strength depends on the correlation between the signals. Valizadeh and Wang [8] proposed another method - correlation-and-bit-aware spread spectrum that additionally modifies watermark embedding strength with respect to the embedded bit sign and the correlation sign.

2 CAISS Embedding Schema Description

In the traditional spread spectrum, bit message b is embedded into a host signal with an amplitude $A > 0$. The information bit that is embedded takes a value from binary set $\{-1, +1\}$. Every bit is spread over the host signal using security key $s = [s_1, s_2, ..., s_N]^T$, where s_i is taken randomly from binary set $\{-1, +1\}$. Watermaked signal is then described by the following equation:

$$r_i = x_i + s_i A b, \quad i = 1, 2, ..., N. \tag{1}$$

At the decoder side the information bit is estimated as a sign of Pearson correlation coefficient between received signal and the security key:

$$b = sign\{z\} \tag{2}$$

where:

$$z = \frac{\sum_{i=1}^{N}(r_i - \bar{r}) \sum_{i=1}^{N}(s_i - \bar{s})}{\sqrt{\sum_{i=1}^{N}(r_i - \bar{r})^2 \sum_{i=1}^{N}(s_i - \bar{s})^2}} \tag{3}$$

An interference between the host signal and the embedded bit introduces noise effect at the decoder side and is a source of errors durign decoding. One of possible method to reduce interference effect of the host signal was proposed by Malvar and Florencio [4]. In the ISS method they introduced a free parameter λ_h that is used to modulate the embedding strength in the presence of interference between signals. Malvar and Flornecio shown that optimal λ_h for gaussian signals is $1/N$, where N is the size of a vector representing digital signal. For convenience, we decided to use $\lambda = \lambda_h N$. The interference is calculated here as a dot product of the signals. Watermarked signal is obtained as follows:

$$r = x + sAb - \lambda ss^T x/N \tag{4}$$

The idea of reducing distortions from interference between host and embedded signals in SS was extended by Valizadeh and Wang [8]. At first they proposed correlation-aware spread spectrum (CASS) scheme where the bit message is inserted into host signal with two different amplitude values. The value of the modulation amplitude depends on the sign of correlation between the key signal

s and the host signal x and bit b to embed. If the correlation sign is equal to b, smaller amplitude is choosen:

$$r = \begin{cases} x + sA_1, & \text{if } s^T x \geq 0, b = +1 \\ x - sA_2, & \text{if } s^T x \geq 0, b = -1 \\ x - sA_1, & \text{if } s^T x < 0, b = -1 \\ x + sA_2, & \text{if } s^T x < 0, b = +1 \end{cases} \tag{5}$$

In the CAISS method Valizadeh and Wang combined CASS with ISS. Again, theoretically optimal λ_h value is $1/N$ for gaussian signals and again we use $\lambda = \lambda_h N$ for convenience:

$$r = \begin{cases} x + sA_1, & \text{if } s^T x \geq 0, b = +1 \\ x - sA_2 - \lambda s(s^T x)/N, & \text{if } s^T x \geq 0, b = -1 \\ x - sA_1, & \text{if } s^T x < 0, b = -1 \\ x + sA_2 - \lambda s(s^T x)/N, & \text{if } s^T x < 0, b = +1 \end{cases} \tag{6}$$

3   Testing Methodology

All of presented results were obtained from tests performed on BOWS2 original data-base [5]. Since this database is a large one (10000 images) most of our tests were performed only on n_{used} images. The procedure of choosing images was as follows: first we sorted all images with respect to their pixel value variances and then picked every $m - th$ image were $m = 10000/n_{used}$. We excluded from our analysis 3 images that were vastly overexposed, namely: 1126.pgm, 1258.pgm and 1478.pgm. In CAISS scheme watermark embedding strength depends on correlation between the watermark and the content it is embedded into. This implies that for the same embedding parameters, the distorion depends on the content. This property produces visible artifacts while embedding watermarks in a block manner in spatial domain. Though PSNR of whole image may be lower than the desired level, some of the blocks are distorted much more than others and thus resulting image is not acceptable. It is clearly seen in Fig 1. To cope with that problem, we embedded watermarks in DCT domain. At the begining we deleted high- and low-frequency coefficients defined by first $K/2$ and last $K/2$ values sorted in zig-zag manner. The rest were randomly permuted - that assures us that each and every block will follow the same distribution. In the next step those coefficients were divided into blocks of length L.

Since the watermark is added to the signal it may result in pixel values greater than 255 or smaller than 0. We operate on 8-bit images thus all of pixel values are clipped to that 8-bit range of 0 to 255. Additionally in order to reduce this problem we restrict the intensity range of input images to the range of 5 to 250. Thus if the amplitude of watermark pattern does not exceed 5, clipping is avoided.

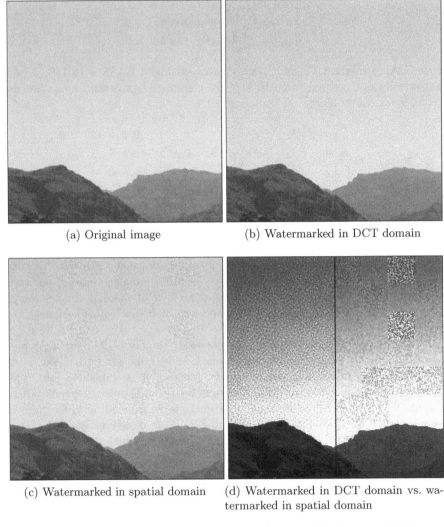

(a) Original image   (b) Watermarked in DCT domain

(c) Watermarked in spatial domain   (d) Watermarked in DCT domain vs. watermarked in spatial domain

Fig. 1. CAISS watermarking in DCT and spatial domain. All of watermarked images have PSNR of 36.99 dB. Left part of Fig 1d is a left half of DCT domain watermarked image whilst right part is another half of an image watermarked in spatial domain. The contrast of an image is enhanced in order to show the difference between artifacts in DCT and spatial domain.

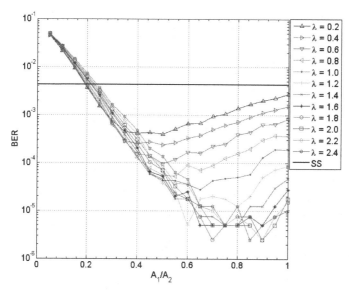

Fig. 2. Bit Error Rate vs. changing A_1 to A_2 ratio for different λ values in CAISS scheme obtained for 500 images, PSNR fixed to 37.0 ± 0.3 dB, JPEG quality level 75% and 800 bits embedded into each image. Black horizontal line represents traditional spread spectrum performance.

In Fig 1b all of 800 embedded bits were correctly decoded after 75% JPEG compression while there were 12 erroneously decoded bits in Fig 1c though only 256 bits were embedded.

We noticed that the dependence between the distortion and the content is a cause of wide non-gaussian PSNR distributions for given parameters settings. Thus we modified embedding strength by changing $A1$ and $A2$, preserving A_1/A_2 ratio to keep PSNR in desired range.

4   Results

In [8] authors simulated theoretical performance of CAISS method using artificial Gaussian signals. Real images usually significantly deviates from Gaussian distribution. We use DCT coefficients as a host signal for above-mentioned reasons, thus we are dealing with values that are approximated by Laplacian probability distribution functions [6]. We started our evaluation from calculating BER performance, varying λ and A_1/A_2 ratio. The results are shown in Fig 2. Every value was obtained after embedding 500 images with 800 bits of watermark - one bit in one block of size $L = 256$. All of images were of size 512×512 thus for each image we left $K = 57344$ (21.9%) DCT coefficients unmodified. After embedding, every image was saved as JPEG file with quality level 75%. For A_1/A_2 ratio smaller than ~ 0.2, CAISS behaves worse than traditional spread spectrum regardless of λ value. That is clearly visible in Fig 3 where we plotted the same

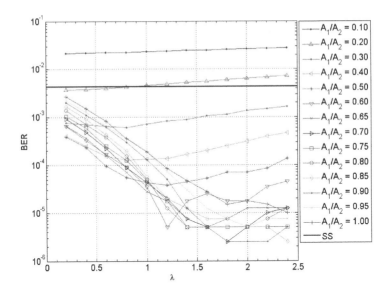

Fig. 3. Bit Error Rate vs. changing λ for different A_1 to A_2 ratio values in CAISS scheme obtained for 500 images, PSNR fixed to 37.0 ± 0.3 dB, JPEG quality level 75% and 800 bits embedded into each image

results but from different perspective. Bit Error Rate decreases rapidly with increasing A_1/A_2 ratio. For small λ values the decrease stops around $A_1/A_2 \approx 0.4$. With increase of λ the minimum shifts towards larger A_1/A_2 values. Also minimum value of BER drops pretty quickly. Finally for $\lambda \approx 2$ and $A_1/A_2 \approx 0.9$, BER is ≈ 1000 times smaller than for traditional spread spectrum. One should however be aware that this result is only approximate since there were only a few bits erroneously decoded with best parameters settings. Further increase of λ or A_1 to A_2 ratio does not improve BER performance or even slightly worsen the results.

Next we compared CAISS with ISS. To be fair in our comparison, we decided to choose best λ parameter for ISS scheme. As previously, we chose 500 images and embedded them with 800 bits of watermark for every checked λ value. The results are presented in Fig 4. Increasing λ makes BER to drop quickly and deep minimum is definitely present around $\lambda \approx 1.5$. BER values for $\lambda = 1.4$ and $\lambda = 1.6$ are noisy since they both were obtained from only 6 erroneously decoded bits. Finally we decided to use three different values of λ: 1.4, 1.5 and 1.6 in further tests. To get more realiable data for CAISS and ISS comparison we performed another test on all 9997 images (as indicated earlier, three were excluded) from BOWS2 original data-base. Such tests are computationally demanding thus we only chose a few suboptimal parameters settings. The results of this test are presented in Tab 1. Numbers involved in BER calculation (n) are this time much higher, and calculated BERs re more realiable. It is clear that CAISS outperforms ISS by a factor of ~ 2. At the same time BER for CAISS is ~ 400

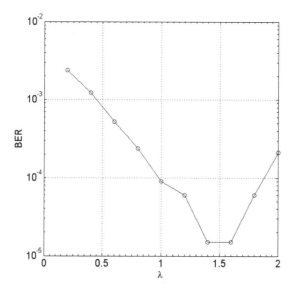

Fig. 4. Bit Error Rate vs. λ in ISS scheme, obtained for 500 images, PSNR fixed to 37.0 ± 0.3 dB, JPEG quality level 75% and 800 bits embedded into each image

Table 1. Comparison between traditional spread spectrum (SS), and correlation aware spread spectrum techniques. Results obtained for 9997 images with PSNR fixed to 37.0 ± 0.3 dB, JPEG quality level 75% and 800 bits embedded.

Scheme	λ	A_1/A_2	n	BER
SS	-	-	33322	0.0042
ISS	1.4	-	218	0.000028
ISS	1.5	-	174	0.000022
ISS	1.6	-	152	0.000019
CAISS	2.0	0.85	82	0.000010
CAISS	2.0	0.90	103	0.000013

times smaller than for SS. Choosen ISS and CAISS settings were only suboptimal and it is probably possible to get slightly lower BER for different parameters settings for both considered methods. At last, we examined BER behaviour of our suboptimally set CAISS and ISS schemes under changing JPEG quality level. The results are shown in Fig 5. Here we used 500 images for SS and 2000 images for ISS and CAISS. The embedding procedure was the same as previously. We may notice, that both CAISS and ISS with proposed settings outperforms traditional spread spectrum by about two orders of magnitude. Still, BER for CAISS is consequently ~ 2 times smaller than for ISS. Only for quality level 85% ISS seems to perform better but that is probably a matter of very small count of erroneously decoded bits. For JPEG quality level larger than 85% we did not note any bit errors in CAISS or ISS scheme, hence $BER \leq 10^{-6}$.

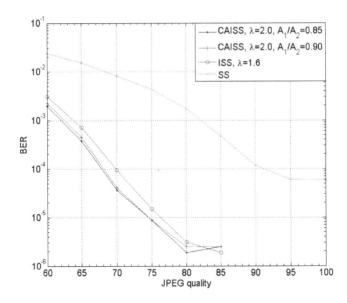

Fig. 5. BER vs. JPEG quality level for considered watermarking schemes

5   Conclusions

In this paper BER performance of corelation-and-bit-aware spread spectrum (CAISS) watermarking was compared with traditional spread spectrum approach. It was also compared with a performance of CAISS precursor - improved spread spectrum (ISS). Tests were carried in DCT domain, thus we were dealing with non-gaussian signals. Results indicate that considered technique is robust for medium JPEG compression and it outperformes traditional spread spectrum technique by more than two orders of magnitude in terms of BER. However that holds only for assorted scheme parameters. At the same time CAISS outperformes ISS by a factor of about two. For JPEG quality level as low as 65% we were obtainig BER considerably better than 0.1% while embedding 800 bits into an image of size 512×512. All of our tests were prepared with PSNR fixed at convenient level of 37.0 ± 0.3 dB, thus we confirmed that this technique should be considered as a best choice for watermarking aplications that benefits from spread spectrum algorithms properties.

Acknowledgment. Research funded within the project No. POIG.02.03.03-00-008/08, entitled "MAYDAY EURO 2012- the supercomputer platform of context-depended analysis of multimedia data streams for identifying specified objects or safety threads". The project is subsidized by the European regional development fund and by the Polish State budget.

References

1. Chen, B., Wornell, G.W.: Quantization index modulation methods for digital watermarking and information embedding of multimedia. J. VLSI Signal Process. Syst. 27(1/2), 7–33 (2001)
2. Cheng, Q., Huang, T.S.: An additive approach to transform-domain information hiding and optimum detection structure. IEEE Transactions on Multimedia 3(3), 273–284 (2001)
3. Cox, I.J., Kilian, J., Leighton, F.T., Shamoon, T.: Secure spread spectrum watermarking for multimedia. IEEE Transactions on Image Processing 6(12), 1673–1687 (1997)
4. Malvar, H.S., Florencio, D.A.: An improved spread spectrum technique for robust watermarking. In: 2002 IEEE International Conference on Acoustics, Speech, and Signal Processing (ICASSP), vol. 4, pp. IV–3301–IV–3304 (May 2002)
5. opcjonalny opcjonalny. Bows2-original data-base. opcjonalny, opcjonalny opcjonalny. opcjonalny
6. Reininger, R., Gibson, J.: Distributions of the two-dimensional dct coefficients for images. IEEE Transactions on Communications 31(6), 835–839 (1983)
7. Valizadeh, A., Wang, Z.J.: A framework of multiplicative spread spectrum embedding for data hiding: Performance, decoder and signature design. In: IEEE Global Telecommunications Conference, GLOBECOM 2009, November 30-December 4, pp. 1–6 (2009)
8. Valizadeh, A., Wang, Z.J.: Correlation-and-bit-aware spread spectrum embedding for data hiding. IEEE Transactions on Information Forensics and Security 6(2), 267–282 (2011)

Towards Robust Visual Knife Detection in Images: Active Appearance Models Initialised with Shape-Specific Interest Points

Marcin Kmieć[1], Andrzej Głowacz[2], and Andrzej Dziech[2]

[1] Department of Automatics
AGH University of Science and Technology
al. Mickiewicza 30, 30-059 Krakow, Poland
[2] Department of Telecommunications
AGH University of Science and Technology
al. Mickiewicza 30, 30-059 Krakow, Poland
{kmiec,aglowacz}@agh.edu.pl,
dziech@kt.agh.edu.pl

Abstract. In this paper a novel application of Active Appearance Models to detecting knives in images is presented. Contrary to popular applications of this computer vision algorithm such as face segmentation or medical image analysis, we use it not only to locate an instance of an object that is known to exist in the analysed image. Using an interest point typical to knives we try to answer the question, whether a knife is or is non-existent in the image in question. We propose an entire detection scheme and examine its performance on a sample test set. The work presented in this paper aims at creating a robust visual knife detector that will find application in computerised monitoring of the public using CCTV.

Keywords: Active Appearance Models, Knife Detection, Computerised Video Surveillance, Harris Interest Point Detector.

1 Introduction

One application of object detection in images is computer-aided video surveillance. A CCTV operator usually monitors multiple video feeds at the same time, which is a complex and challenging task in terms of allocating attention effectively. One study suggests that detection rates for operators monitoring 4, 9, and 16 screens oscillate around 83%, 84% and 64% respectively and will drop significantly after an hour [1]. Therefore, the need to automate the process is obvious.

This study focuses on automatic detection of knives in images. The idea of software-based knife detection has a practical application in surveillance of the public using CCTV. Should a knife be detected, an alarm is raised and the human operator can immediately focus their attention on that very scene and either confirm or reject

A. Dziech and A. Czyżewski (Eds.): MCSS 2012, CCIS 287, pp. 148–158, 2012.

that detection. Although people will almost always outperform software algorithms in object detection in images, in the long run the computer could be of significant assistance to the human CCTV operator when it comes to dealing with up to tens of different simultaneous video feeds for many hours a day.

As such, knives are a very wide class of objects of immense diversity. Moreover, they easily reflect light, which reduces their visibility in video-sequences. Therefore automatic knife detection in images poses a challenging task. In this paper, a novel application of the well-established active appearance models is presented. So far, they have been extensively used for medical image interpretation [2 3 4] and for matching and tracking of faces [5 6]. The novelty of this work is twofold. First, to the best knowledge of the authors, no research in the area of knife detection is conducted except for their own, and second, active appearance models have so far not been used to detect objects belonging to a general class. They have been used in what is referred to as detection as in [7] but that is not what is meant by detection in computer vision in the strict sense of the word. By detecting, for example a face in an image, we mean answering the question of whether there is or there is not a face in the given image. This process can be characterised by two parameters, the positive and false-positive detection rates. As in the case of [7], before what is referred to as detection is performed, an assumption exists that the object is somewhere in the image, and the task is to precisely locate it. For instance given an image of face, finding the nose is not a task of detection since it is assumed that all faces have noses. It is rather the task of localisation. It can be characterised by the level of localisation accuracy, but not by positive and negative detection rates. In this paper a method for detection of objects, in this case knives, using Active Appearance Models is introduced. It aims at answering the question of whether a knife is or is not existent in the image in question. The effectiveness of this approach has been evaluated on high-quality knife images, the results of which have paved the way for performing evaluation on video-surveillance sequences.

2   Related Work

Although as stated earlier not many scientists dedicate their research to the task of knife detection, computerised monitoring of CCTV images in general has become a popular research topic at present. Typical applications include external intruder detection, people and vehicle counting, or crowd flux analysis. All of the above utilize information that is conveyed through consecutive frames of a video sequence. In the case of knife detection we have focused on detecting knives in single frames based on their shape and appearance. One possible way of making use of information coming from a video-sequence would be to initially locate a detection candidate in a frame, then using optical-flow to locate that very object a number of frames later and perform knife detection on that object again. That would reduce the number of false positives and might be the next step towards a robust knife detection scheme.

The work presented in this paper follows our research on using Histograms of Oriented Gradients in knife detection [8]. Combining two independent detection techniques will surely improve the overall performance of a knife detection scheme.

3   Hardware Framework Description

In computerised monitoring of CCTV the human operator monitors a number of screens linked to strategically placed cameras looking for illegal activities or potentially dangerous situations. His work is aided by a computerised system, in which images coming from all cameras undergo automatic analysis. Should a threat be detected by the system, the human operator gets alerted and can focus his attention on the selected screen. At this point it is important to stress that such a system should be characterised by a very low false detection rate, otherwise the poor operator's job would be unbearable. On the other hand, if positive and negative rates were respectively high and low enough, the number of cameras the operator monitors at once could be increased.

The subject of our detection scheme is rather difficult to view through an ordinary CCTV camera. Knives are on average smaller in size than the subjects of popular detection algorithms such as faces [9], pedestrians [10] or cars [11]. They easily reflect light and what is even more troublesome, a knife remains virtually invisible if one looks directly at its edge as shown in Figure 1b. Although in this paper we focus on the algorithmic aspect of this novel detection scheme, we feel it is important to explain how image acquisition hardware should be set up in order to ensure a high detection rate.

First, the area under surveillance should be covered with more than one camera in order to avoid situations where knives are visible to a very limited extent, similarly to Figure 1b. The more cameras the better, but a reasonable number of three cameras positioned at 120 degree intervals should allow for sufficient image acquisition quality. Second, the cameras deployed need to have a megapixel resolution. This is a necessary condition if we think of the relatively small size of knives on average.

4   Active Appearance Models

Active Appearance Models (AAM) were introduced [12] in 1998 as a generalisation of the popular Active Contour Model (Snake) and Active Shape Model algorithms. They are a learning-based method, which was originally applied to interpret face images. Due to their generic nature and high effectiveness in object locating numerous applications in medical image interpretation followed. The typical medical application is finding an object, usually a body organ in a medical image of a specific body part, such as locating the bone in a magnetic resonance image (MRI) of the knee [13], the left and right ventricles in a cardiac MRI [14] or the heart in a chest radiogram [15].In general, AAM can be described as a statistical model of the shape and pixel intensities (texture) across the object. The term appearance means the combination of shape and texture, while the word 'active' describes the use of an algorithm that fits the shape and texture model in new images. In the training phase, objects of interest are manually labelled in images from the training set with so called landmark points to define their shape. A set of three annotated knives has been depicted in Figure 2.

(a) Knife well visible   (b) Knife almost invisible

Fig. 1. Sample video-surveillance images containing knives

In AAM objects are thus defined by the pixel intensities within their shape polygon and the shape itself. Principle Component Analysis (PCA) is then employed to model variability between objects in images from the training set. Before that happens, all shape polygons need to be aligned to the normalised common mean. This is achieved by the means of Generalised Procrustes Analysis [16], a method widely used in statistical shape analysis. The algorithm outline can be summarised in four steps:

1. Arbitrarily choose the reference shape (typically the first object's shape).
2. Superimpose all other shapes to the reference shape.
3. Now calculate the mean shape of the superimposed shapes.
4. If the distance of the mean shape to the reference one is larger than a threshold, set mean shape as the reference shape and go back to point 2. Otherwise return the mean shape.

The distance in point 4 is simply the square root of the sum of squared distances between corresponding points in two considered shapes and is called Procrustes distance. Superimposing two shapes includes translating, rotating and uniformly scaling objects and is also a four step process:

1. Calculate the centres of gravity (COG) both of the mean and the shape being superimposed to the mean.
2. Rescale the shape being superimposed so that it has size equal to the mean.
3. Align the position of the two COGs.
4. Rotate the shape being superimposed so that the Procrustes distance is minimal.

By the means of PCA performed on the shape data we obtain the following formula which can approximate any shape example:

$$x = \bar{x} + P_s b_s$$

where $\bar{x}$ is the mean shape, P_s are the eigenvectors of the shape dispersion and b_s is a shape parameters set. To build the statistical model of appearance we need to warp all images in the training set so that their landmark points match the mean shape. Texture warped to match the mean shape is referred to as appearance in the original paper [12]. Let us denote grey level information from the shape normalised image

Fig. 2. Images of knives with landmark points annotated

within the mean shape as g_{im}. To reduce lighting variation one needs to normalise pixel intensities with an offset $\bar{\psi}$ and apply a scaling ξ:

$$g = (g_{im} - \psi)/\xi$$

The values of the two parameters are obtained in course of a recursive process, details of which can be found in [12]. Once pixel intensities of samples in the training set have been normalised a statistical model of pixel intensities can be defined:

$$g = \bar{g} + P_g b_g$$

where $\bar{g}$ is the normalised mean pixel intensity vector, P_g are the eigenvectors of the pixel-intensity dispersion and b_g is a parameter set. For each image in the training set a concatenated vector b is generated:

$$b = \begin{bmatrix} W_s b_s \\ b_g \end{bmatrix} = \begin{bmatrix} W_s P_s^T (x - \bar{x}) \\ P_g^T (g - \bar{g}) \end{bmatrix}$$

Now another PCA is applied on the concatenated vector, the result of which is:

$$b = P_c c, \ P_c = \begin{bmatrix} P_{c,s} \\ P_{c,g} \end{bmatrix}$$

c is a parameter set that controls both the shape and the appearance and P_c are the eigenvectors. New object instances can be generated by altering the c parameter, shape and pixel intensities for a given c can be calculated as follows:

$$x = \bar{x} + P_s W_s P_{c,s} c, \quad g = \bar{g} + P_g P_{c,g} c,$$

W_s is a matrix of weights between pixel distances and their intensities.

5   Corner Detection

It is a characteristic of all knives that the tip of the knife is distinct, lies at the intersection of two edges and can be regarded as a corner that is a point that has two different, dominant edge directions in its neighbourhood. Multiple corner detectors are used in computer vision; for the purpose of tip of the knife detection we have used the Harris corner detection algorithm [17]. This popular interest point detector is invariant to geometric transformation and to a large extent to illumination changes as well as resistant to image noise. It is based on the simple principle that edges and corners change noticeably more than other elements of the image with a window shifted a little in all directions. The difference in pixel intensities caused by a window shift (Δ_x, Δ_y) is measured by the means of the local auto-correlation function [17]:

$$c(x, y) = \sum_W [I(x_i, y_i) - I(x_i + \Delta_{x,y_i} + \Delta y)]^2$$

(1)

where $I(x_i, y_i)$ denotes gray-scale pixel intensity at the given location in the Gaussian window W centred on (x, y). By approximating the shifted image with a Taylor series, it can be proved [17] that expression (1) is equal to

$$c(x, y) = \begin{bmatrix} \Delta_x & \Delta_y \end{bmatrix} C(x, y) \begin{bmatrix} \Delta_x \\ \Delta_y \end{bmatrix}$$

where is C a matrix that summarises the predominant directions of the gradient in the specified neighbourhood of the point (x, y). The eigenvalues of the matrix C form a description the pixel in question, if they are both high they indicate a corner.

In Figure 3 three knife images with corners detected are shown. The number of detected corners depends on the threshold. What we are interested in is to detect the tip of the knife. Therefore the threshold is set reasonably low to ensure that the tip of the knife will be detected as a corner, even at the price of more corners being detected.

6   Detection Scheme and Results

As stated in the introduction, Active Appearance Models have been used in locating objects in images. The most notable example of such application is locating face elements i.e. determining the exact position of the eyes, mouth and nose given an image of a face. In this case the assumption is that there always is a face in the analysed image. Should the AAM be performed on a none-face image, it would converge to the nearest local minimum. This is still theoretically correct, but makes no sense from a practical point of view. Moreover, Active Appearance Models are sensitive to the initial location of landmark points in the image. Even if there is a face somewhere in a large image, for the algorithm to correctly segment it into elements, the initial location of its landmark points needs to be roughly around the face region. A common technique for face segmentation with AAM is the use of Viola and Jones's face detector [9] to initialise the AAM as in [18].

Fig. 3. Interest points in knife-images detected by the Harris Corner Detector

We have trained an AAM for knife localisation using 6 images of knives. Adding additional images to the training set has brought little if any improvement to the AAM performance. We believe that this is due to the rather simple shape we are dealing with, in contrast to for example the shape of the human face.

The first and foremost assumption for our detection scheme is that if there is a knife in the 64×128 pixel detection window, it is vertically centred and its tip faces up. This can be achieved by sliding the detection window across the image at all possible scales, in a manner similar to how the HOG detector works [10]. In addition, the image undergoing detection needs to be rotated 360 degrees at a fixed angle step.

The results of running the Harris corner detector of knife images fulfilling the above conditions have been presented in Figure 3. We can see that the tip of the knife is highly likely to be designated as a corner. A general outline of our knife detection scheme is presented below:

---

1. Select corners that fulfil the following location rule; they lie within 20% of detection window's width from the window's vertical symmetry axis and within 25% of detection window's height from the top of the window. Put the selected points in the knife-tip candidate set. Choose the first point and remove it from the set.
2. Set the AAM's landmark corresponding to the knife-tip at the chosen point.

3. Run the knife-trained AAM, calculate the minimum area bounding box of the landmark-points that have now converged to a new location. The box's longer symmetry axis shall be considered the knife candidate's pseudo symmetry axis.

4. Calculate the percentage of landmark points that lie on the edges found in the image undergoing detection by the Canny edge detector.

5. Set the AAM's landmark point corresponding to the knife tip to the following three locations: three pixels to the left, above and to the right of the initial knife-tip and perform steps 3 and 4 on them.

6. The detection's result is positive if the following conditions have been met:

I. In all four cases (the original corner and the three points in its neighbourhood) landmark points of the model converge to the same location.

II. In all four cases the knife-candidate's pseudo symmetry-axes have a similar skew-angle, that corresponds to the assumed knife's vertical orientation.

III. Most of the landmark points lie on edges detected by the Canny edge detector.

If the result is positive stop the detection procedure. If at least one of the above conditions is not met the detection's result is negative. In this case continue to the next point.

7. If the knife-tip candidate set is not empty, choose the next point, remove it from the set and go back to point 2.

The idea behind this detection scheme is that if the appearance (and the shape) the AAM has been trained to locate exists in the image, the model will converge to it if reasonably initialized even from slightly varying locations (see point 5). In other words, the AAM will converge to the same location, which is evaluated with the min-area bounding box and skew-angle. If the appearance AAM is searching for does not exist, the model will converge to some random locations. Moreover, to make the positive detection condition even stronger, we assume that most of the landmark points lie on an edge, to eliminate the case where there are no edges visible (therefore no knife exists). The exact number of landmarks that actually lie on an edge can be chosen only by a heuristic rule. We have chosen this threshold to be 70% of the points on each of the two edges of the knife-blade. Additionally, since the knife is assumed to be aligned vertically in the detection window, the knife-candidate's skew-angle has to correspond to this assumption.

The positive test set consisted of 40 images in normalised 64x128 pixel format that depicted knives vertically centred with the tip facing up. This position of the knife is assumed in our detection-scheme and corresponds to the detection window being slid across a large image at all scales and orientations. Out of 40 images depicting knives, 37 have been correctly labelled as containing knives, whereas in 3 cases the result of detection was negative. Sample images of correctly classified images are presented in Figure 4a. They depict knives localised by the AAM, initialised at the tip of the knife, as described above. The negative test set consisted of 40 images randomly cropped from images that did not contain knives, samples from this set are

a) Positive detection results.   b) Negative detection results.

Fig. 4. Sample detection results

Table 1. Characteristics of the proposed AAM based detection scheme

Number of Images in Positive Test Set	Number of Correctly Classified Positive Images	Classification Accuracy	Number of Images in Negative Test Set	Number of Wrongly Classified Negative Images	False Positive Rate
40	37	92,5%	40	0	0%

shown in Figure 4b. None of the images has been falsely labelled as containing a knife. The detection accuracy and the false positive rates have been summarised in Table 1.

The approach to detecting knives in images described in this paper is at an early stage. It has been tested on a positive and a negative sets of images, both of which do not come from video surveillance sequences. The positive test set is made up of images of knives rotated and cropped from photographs found online, whereas the negative test set originates from randomly cropped images that did not contain knives. If our method of detecting knives was to fail on a set of images with knives clearly visible, further research in this direction would be probably ill-founded. Since the obtained results are promising, the next step will be tests performed on genuine video surveillance sequences, so that knives are less clearly visible than in our initial test set, with frames not containing knives also coming from CCTV footage.

We aim at creating a knife-detector that is capable of working in real-time. An apparent drawback of all rotation non-invariant detection methods is the need to perform the detection procedure on images rotated at a fixed angle step [19]. It could prove useful to, instead of rotating an image, use AAMs trained to fit knives positioned at different angles. This way the time needed to rotate images is saved. Limiting the region of interest in images to certain areas that are more likely to contain knives, for example by using a people detector or optical flow to indicate parts of the image in motion, could on one hand speed up the very detection process, but those techniques could themselves be computationally expensive and an optimal balance is required here.

7   Summary and Conclusions

The proposed knife detection scheme is novel in terms of applying Active Appearance Models to the problem of object detection. So far AAM have been used to locate instances of an object in images which were known to contain it, such as medical images containing a body organ or images of face, where the task was to locate particular face elements. In our work we have used the fact that the knife-blade has a very specific interest point, which can be easily detected as a corner, i.e. the tip of the knife. This point can be used to initialise the AAM. We then use the fact, that the AAM initialised from similar locations would always converge to the same object, provided that this object is existent in the detection window. Moreover, a likely knife-candidate should be centred vertically in the detection window. Combining the above observations, we get a set of rules that allow the decision on whether or not the object in question is a knife. The efficiency of this approach is at a reasonable level. In the future, combining AAM with a HOG based knife detector is planned in order to obtain a robust knife detection scheme.

Acknowledgements. This work has been co-financed by the European Regional Development Fund under the Innovative Economy Operational Programme, INSIGMA project no. POIG.01.01.02-00-062/09.

References

1. Tickner, A.H., Poulton, E.C.: Monitoring up to 16 synthetic television pictures showing a great deal of movement. Ergonomics 16(4), 381–401 (1973)
2. Beichel, R., Bischof, H., Leberl, F., Sonka, M.: Robust active appearance models and their application to medical image analysis. IEEE Transactions on Medical Imaging 24(9), 1151–1169 (2005)
3. Jiang, Y., Zhang, Z., Cen, F., Tsui, H.T., Lau, T.K.: An enhanced appearance model for ultrasound image segmentation. In: Proceedings of the 17th International Conference on Pattern Recognition, ICPR 2004, August 23-26, vol. 3, pp. 802–805 (2004)
4. Leung, K.Y.E., van Stralen, M., van Burken, G., van der Steen, A.F.W., de Jong, N., Bosch, J.G.: Automatic 3D left ventricular border detection using active appearance models. In: 2010 IEEE Ultrasonics Symposium (IUS), October 11-14, pp. 197–200 (2010)
5. Edwards, G.J., Taylor, C.J., Cootes, T.F.: Interpreting face images using active appearance models. In: Proceedings of Third IEEE International Conference on Automatic Face and Gesture Recognition, 1998, April 14-16, pp. 300–305 (1998)
6. Kim, D., Sung, J.: A Real-Time Face Tracking using the Stereo Active Appearance Model. In: 2006 IEEE International Conference on Image Processing, October 8-11, pp. 2833–2836 (2006)
7. Leung, K.Y.E., van Stralen, M., van Burken, G., van der Steen, A.F.W., de Jong, N., Bosch, J.G.: Automatic 3D left ventricular border detection using active appearance models. In: 2010 IEEE Ultrasonics Symposium (IUS), October 11-14, pp. 197–200 (2010)
8. Kmieć, M., Glowacz, A.: An Approach to Robust Visual Knife Detection (manuscript in submission)

 9. Viola, P., Jones, M.: Rapid Object Detection Using a Boosted Cascade of Simple Features. In: Proc. IEEE Computer Society Conference on Computer Vision and Pattern Recognition (CVPR), Kauai, USA (2001)
10. Dalal, N., Triggs, B.: Histograms of oriented gradients for human detection. In: IEEE Computer Society Conference on Computer Vision and Pattern Recognition, CVPR 2005, June 25, vol. 1, pp. 886–893 (2005)
11. Buch, N., Orwell, J., Velastin, S.A.: 3D Extended Histogram of Oriented Gradients (3DHOG) for Classification of Road Users in Urban Scenes, Kingston University, UK (2009)
12. Cootes, T.F., Edwards, G.J., Taylor, C.J.: Active Appearance Models. In: Burkhardt, H., Neumann, B. (eds.) ECCV 1998. LNCS, vol. 1407, pp. 484–498. Springer, Heidelberg (1998)
13. Cootes, T.F., Taylor, C.J.: Statistical Models of Appearance for Computer Vision. Tech. report, University of Manchester (2001)
14. Mitchell, S., Lelieveldt, B., Geest, R., Schaap, J., Reiber, J., Sonka, M.: Segmentation of cardiacMR images: An active appearance model approach. In: Medical Imaging 2000: Image Processing, San Diego CA, SPIE, vol. 1, pp. 224–234. SPIE (2000)
15. van Ginneken, B., Stegmann, M.B., Loog, M.: Segmentation of anatomical structures in chestradiographs using supervised methods: A comparative study on a public database. Medical Image Analysis (2004)
16. Stegmann, M.B., Gomez, D.D.: A brief introduction to statistical shape analysis. Technical report, Informatics and Mathematical Modelling, Technical University of Denmark, DTU (March 2002)
17. Derpanis, K.G.: The Harris Corner Detector (October 2004)
18. Cristinacce, D., Cootes, T.F.: A comparison of shape constrained facial feature detectors. In: Proceedings of Sixth IEEE International Conference on Automatic Face and Gesture Recognition, 2004, May 17-19, pp. 375–380 (2004)
19. Tadeusiewicz, R., Korohoda, P.: Computer Analysis and Image Processing, FPT, Cracow (1997)

Object Detection and Measurement Using Stereo Images

Christian Kollmitzer

Electronic Engineering
University of Applied Sciences Technikum Wien
Höchstädtplatz 4, A-1200 Vienna Austria
kollmitz@technikum-wien.at

Abstract. This paper presents an improved method for detecting objects in stereo images and of calculating the distance, size and speed of these objects in real time. This can be achieved by applying a standard background subtraction method on the left and right image, subsequently a method known as subtraction stereo calculates the disparity of detected objects. This calculation is supported by several additional parameters like the center of object, the color distribution and the object size. The disparity is used to verify the plausibility of detected objects and to calculate the distance and position of this object. Out of position and distance the size of the object can be extracted, additionally the speed of objects can be calculated when tracked over several frames. A dense disparity map produced during the learning phase serves as additional possibility to improve the detection accuracy and reliability.

Keywords: Computer Vision, Stereo Vision, Foreground Segmentation, Disparity Map, Subtraction Stereo.

1 Introduction

In surveillance systems and autonomous mobile robots cameras are common sensors for detecting objects. The standard one-camera system is able to detect objects by differentiating these objects from the background by motion, brightness or color. This leads to misinterpretations and unclear situations, especially when the object is similar to the background, occluded or when the background changes rapidly, like at changing illumination.

In a setup, where a static camera observes a scene, objects have to be distinguished from the background. The evaluated system learns over time to find out, when background or foreground is present. The improvement of the robustness of this detection requires additional information to normal color cameras. This additional information can be provided by a second camera, which leads to depth information and allows calculating the position of objects in the scene.

All computational processes are evaluated by means of a computer system using an i5 processor running with a frequency of 2.4GHz. The target is a detection- frame-rate of 20 fps at a resolution of 640x480 pixels. All algorithms use the libraries of OpenCV [7] and run in Windows7.

A. Dziech and A. Czyżewski (Eds.): MCSS 2012, CCIS 287, pp. 159–167, 2012.
© Springer-Verlag Berlin Heidelberg 2012

2   Camera Rig

The used camera set is constructed in a way that two identical cameras are mounted on a stable bar with a base distance of 413 mm (Fig. 1).

Base Distance: The resolution of the distance (Z) measurement is based on the horizontal distance (T) between the left and right camera.

$$Z = \frac{f.T}{(x^l - x^r) - offset} \qquad (1)$$

offset = distance between the left and right horizontal camera center in pixels; f = focus in pixels; T = base distance in mm; Z = distance in mm. The pixel unit is present in the numerator and denominator; therefore the resulting unit is mm.

$$\Delta Z = \frac{Z^2}{fT} \Delta d \qquad (2)$$

The resolution of the distance measurement can be improved by increasing the base distance (T) or by increasing the focus (f; tele lens). Δd = minimal disparity = 1 pixel. In this setting the base distance was chosen in a way that distances up to 20m can be measured. The focus is 530 pixels, which gives a field of view which is suitable for surveillance of indoor and outdoor areas without camera movement.

Camera Resolution: The resolution is a tradeoff between exact object detection and computational cost. As a standard resolution 640x480 was chosen with the ability to use also 1280x1024 for higher accuracy.

Fig. 1. Camera rig

3   Image Calibration and Rectification

The cameras have to be calibrated and the acquired images have to be rectified to achieve congruent images, which allow disparity calculation. The calibration for cameras which cover a higher distance range is better performed in two steps. First each camera has to be calibrated separately by means of chessboards, which are presented several times and lead to intrinsic parameters, which cover misalignment of the camera chip, focal distance and distortion of the lenses. In a second step the rig is calibrated again by presenting a chessboard in several positions. By this the extrinsic parameters like camera distance and rotation to each other is examined.

With these calibration parameters both images are rectified, which leads to horizontally aligned images, which ease the disparity calculation (Fig.2). In this case the search for identical points in both images is reduced to a horizontal search. After calibration the focus setting of the cameras should not change, therefore all automatic focus adjustments of cameras have to be turned off.

Fig. 2. Rectified images

4 Disparity Calculation

Disparity calculation is a matching problem, the position of identical points in the left and the right image has to be detected. Due to rectification the search can be limited to horizontal lines. Several algorithms have been already presented and evaluated. The computational effort (matching cost) for different methods and algorithms have been evaluated in [1].

Better algorithms find more correct corresponding points but typically the matching cost is higher. In this evaluation the disparity algorithm "semi global block matching" is used [5].

The implementation of this algorithm has a high calculation cost and is depending on the resolution of the images.

Disparity calculation with "semi global block matching" mode per frame:

- Resolution 320x240:   39ms
- Resolution 640x480:   210ms
- Resolution 1280x1024:   920ms

For a real time application calculation times should be smaller than 50ms to achieve a frame rate of 20 fps. This method has been used in the further evaluation during the learning phase of 100 frames and gives a dense disparity map for the background used as reference background with a resolution of 320x240 pixels. An even denser disparity map can be achieved by averaging all images during the learning phase and forming a stable background (Fig.3).

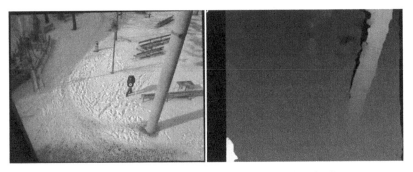

Fig. 3. Left image and corresponding background disparity image

5   Background Registration

During this evaluation different algorithms have been used, starting with a method called "modified codebook" [4] [5]. The quality of detecting objects and separating these objects from background is good but the computational effort is very high.

Codebook background registration time per frame:

- Resolution 320x240:   49ms
- Resolution 640x480:   230ms
- Resolution 1280x1024:   1700ms

The combination of disparity algorithm "semi global block mode" and background registration with "modified codebook" results in a calculation time of 440ms per frame at the intended resolution of 640x480 pixels. With these algorithms the average frame rate is about 2fps, which is not acceptable for surveillance purposes.

This problem has been solved by evaluating a different type of background-registration and disparity calculation. For the background registration an adaptive median background subtraction was used [6]. During a learning phase of 40 images the median of each image pixels history is calculated and used as background. During the detection phase this background is subtracted from the actual image; a threshold function is used to distinguish between foreground and background.

6   Subtraction Stereo

The reduction of the matching cost for the disparity calculation can be achieved by a method known as "subtraction stereo" [2]. In this case the disparity calculation is not applied to the whole left and right image but to the stereo images after the background registration. Thus only areas which have changed and are different from background are used for disparity. This can be done in several ways. One is proposed in a paper of K. Umeda [2] calculating the horizontal distance of detected foreground objects. Evaluating this method results in noisy position calculation, caused by varying object sizes, due to background registration.

This method is modified in this evaluation in a way, that only the left image undergoes a background registration. After smoothing foreground pixels with a procedure known as "connecting components" [7] foreground pixels are clustered and objects identified. The bounding box of the identified object is used to cut out a template of the rectified left image. The next step is to search image data of the right image for this template. The search can be limited to a rectangle, which lies left of the object-position in the left image on the same horizontal line. The search process uses normed correlation.

$$result(x,y) = \frac{\sum_{a,b}(template(a,b).right(x+a,y+b))}{\sum_{a,b} template(a,b)^2.\sum_{a,b} right(x+a,y+b)^2} \qquad (3)$$

The best correspondence can be found by searching the maximum within the result area. This position marks the center of the object in the right image and the distance between the object centers in the right and left image represent the disparity.

Out of the disparity and the position of the object in the image the 3D position of the object can be calculated. If the center of the object is tracked over several images an x/y diagram of the projected path can be drawn (Fig. 7).

The object center position is calculated (d = disparity; T = distance of the cameras; f = focus of the cameras, X,Y,Z = position of the object center in 3D space)

$$z = \frac{f.T}{d} \qquad (4)$$

$$x = \frac{z.d}{f} \qquad (5)$$

$$y = \frac{z.dy}{f} \qquad (6)$$

The calculated position is inserted in the display (Fig.4). Additionally certain points can be selected by the user in the right and left image and thus determine the coordinates of these points. This allows to measure distances and positions of reference points (Fig.5) within the field of view.

Fig. 4. Left and right image with object detection and position

Fig. 5. Distance measurement

7   Object Tracking

The center position of registered object (Fig.6) is tracked and allows drawing the way of tracked objects in an x/y diagram (Fig.7).

Fig. 6. Frame of object tracking video with position of center

This diagram (Fig.7) shows that the person first walked in z direction away from the camera (blue) and then walked towards the camera in a wiggly line (red). There are still positions which are not correctly recognized, this has to be improved by filtering. The resolution changes with distance; at 20m distance the resolution is 75cm.

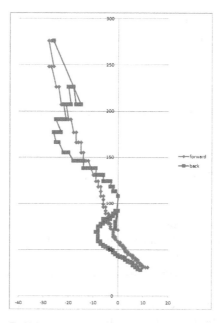

Fig. 7. Object center tracking projection (x/y in dm)

8   Object Measurement

Out of the center coordinates the position of the center can be calculated. The speed calculation uses the center positions, determines the distance between neighboring frames centers and measures time between frames (Fig.8). Height and width of objects can be calculated as estimation or with more computational effort in detail.

In the estimation it is assumed, that the object is farther away and all points of the object have nearly the same distance to the camera. If the calculation has to be more detailed, all individual measuring points have to undergo disparity detection, similar to the center disparity detection.

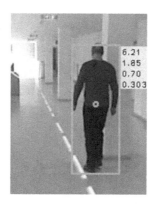

Fig. 8. Detected object with distance, height, width and speed

The error of the distance measurement has been verified by distance measurement with a laser measuring device and lies below 10%.

9   Algorithm

In the complete algorithm left and right images are acquired and during a learning phase of 100 images a standard dense disparity map of the background is produced with a running average process and the "semi global block matching" method.

Moving objects are detected in both images with the method "adaptive median background subtraction", detected pixels are evaluated belonging to the object with the method "connected components". This produces two images holding detected objects.

A second disparity map is produced out of the two images with the detected objects. This disparity map is calculated by evaluating the horizontal difference of the center of objects in the left and right image.

A third disparity map is calculated by using the method described in 6 (subtraction stereo).

The first disparity map represents the background and can be used to measure the distance between the objects and the background.

The second disparity map decides, if the object is visible in both images and if a detailed detection is reasonable.

The third disparity map is used to calculate the objects properties like center, size, and speed. (Fig.9)

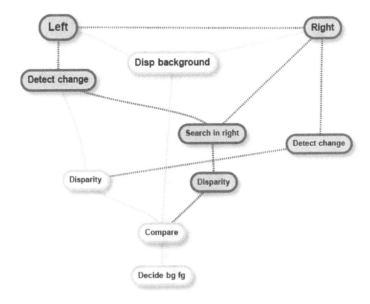

Fig. 9. Object detection algorithm

Computational cost of this algorithm for up to three objects is 50ms per frame at a resolution of 640x320 pixels, which is sufficient for real time use. The calculation time depends on the object size and on the number of objects.

10   Discussion

These methods allow a stable localisation of objects. Based on this work, improvements like better shadow removal [3], multi-object-tracking, identifying multiple objects out of a crowd are under consideration. To achieve better distance resolution the cameras should be mounted with greater distance which complicates calibration. As calibration is crucial for calculation accuracy, methods for automatic calibration and recalibration should be developed.

Acknowledgments. The work reported in this article has been done within the framework of the European FP7-SEC Project INDECT (http://www.indect-project.eu)

References

1. Hirschmüller, H., Scharstein, D.: Evaluation of Stereo Matching Costs on Images with Radiometric Differences. IEEE Transactions on Pattern Analysis and Machine Intelligence (2008)
2. Umeda, K., et al.: Subtraction Stereo - A Stereo Camera System that focuses on Moving Regions. In: Proceedings of SPIE-IS&T Electronic Imaging, 7239 Three-Dimensional Imaging Metrology (2009)
3. Terabayashi, K., et al.: Improvement of Human Tracking in Stereoscopic Environment Using Subtraction Stereo with Shadow Detection. International Journal of Automation Technology 5(6), 924–931 (2011)
4. Kollmitzer, C., Weichselbaum, J., Hager, C.: Background modeling by combining codebook method and disparity maps. In: Proceedings of MCSS (2010)
5. Kim, K., Chalidabhongse, T.H., Harwood, D., Davis, L.S.: Real-time foreground-background segmentation using codebook model. Real-Time Imaging (2005)
6. Lo, B., Velastin, S.: Automatic congestion detection system for underground platforms. In: Proceedings of 2001 International Symposium on Intelligent Multimedia, Video, and Speech Processing, Hong Kong, pp. 158–161 (May 2001)
7. Bradsky, G., Kaehler, A.: Learning OpenCV, Computer Vision with the OpenCV Library (Book Style). O'Reilly Media, Sebastopol (2008)

Multiple Sound Sources Localization in Real Time Using Acoustic Vector Sensor

Józef Kotus

Multimedia Systems Dept., Gdansk University of Technology,
Narutowicza 11-12, 80-233 Gdansk, Poland
joseph@sound.eti.pg.gda.pl

Abstract. Method and preliminary results of multiple sound sources localization in real time using the acoustic vector sensor were presented in this study. Direction of arrival (DOA) for considered source was determined based on sound intensity method supported by Fourier analysis. Obtained spectrum components for considered signal allowed to determine the DOA value for the particular frequency independently. The accuracy of the developed and practically implemented algorithm was evaluated on the basis of laboratory tests. Both synthetic acoustic signals (pure tones and noises) and real sounds were used during the measurements. Real signals had the same or different spectral energy distribution both on time and frequency domain. The setup of the experiment and obtained results were described in details in the text. Taking obtained results into consideration is important to emphasize that the localization of the multiple sound sources using single acoustic vector sensor is possible. The localization accuracy was the best for signals which spectral energy distribution was different.

Keywords: sound detection, sound source localization, audio surveillance.

1 Introduction

In previous work the sound source localization methods based on sound intensity computed in time domain were presented [1-8]. Those techniques worked with broadband signals received from multichannel acoustic vector sensor. Single, dominant sound source was localized properly by means of those methods [1-6], [8]. When more than one sound source produced the acoustic energy simultaneously, determination their positions was very difficult. Quite different approach to multiple sound sources localization in real time using the acoustic vector sensor was presented in this study [9, 10]. Term of the multiple sound sources in this research means that two sources produced the acoustic energy simultaneously from different directions. Main differences depends on computation the sound intensity components in the frequency domain. Fast Fourier Transform was applied for this purpose. Obtained spectrum coefficients for considered signal allowed to determine the DOA values for the particular frequency independently. Developed algorithm was designed to analyze the acoustic signals in real time. It was presented in details in section 2. Due to the

A. Dziech and A. Czyżewski (Eds.): MCSS 2012, CCIS 287, pp. 168–179, 2012.

computation of sound intensity for particular frequency independently, the method should work properly for signals which are different in the frequency domain. It is important to emphasize that localization process of sound source takes into consideration the dynamic of acoustic energy emission. It means that even if the particular sound sources have the same average spectral energy distributions they can still be localized properly. It will took place if they will produce the energy in different parts of time (sound sources should be incoherent). Applied method, used signals and organization of measurement tests were described in details in section 4. Obtained results were presented in section 5.

2   Multiple Sound Sources Localization Algorithm

Proposed, implemented and practically evaluated algorithm to multiple sound sources localization is based on calculation sound intensity level in frequency domain. It is quite different approach in opposite to methods which rely on sound intensity calculation in the time domain. The block diagram of the algorithm was presented in Fig. 1. Calculation process was divided into six functionally different steps. Particular phases were described in details below.

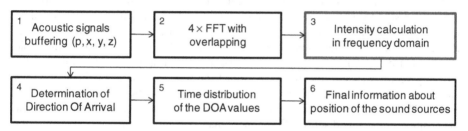

Fig. 1. The block diagram of the proposed algorithm. Red rectangle indicate the most important part of the developed method.

The multichannel acoustic vector sensor produces the following signals: sound pressure p and three orthogonal particle velocity components u_x, u_y, u_z. In the first step each signals were buffered and prepared to FFT calculation. The Hanning window was applied in this case [11]. Next the 4096 point FFT calculation for each signals were performed, sampling frequency was equal to 48 kHz (frequency resolution: 11.7 Hz). Such parameters provide sufficient spectral resolution. It is important to emphasize that the proper overlapping technique is crucial for the next steps. The overlap degree from 0% to 75% was tested (it was explained and illustrated in the results section). It means that in the next FFT frame from 4096 to 1024 samples were new. On the other hand for 4096 signal samples from 1 to 4 FFT frames were calculated. The FFT calculation were performed for each acoustic components separately. In block 3 the sound intensity in frequency domain was computed. Intensity component for x direction was defined as (eq.1):

$$I_x(i) = X_p(i) \cdot \overline{X_{ux}(i)} \tag{1}$$

where:

$I_x(i)$ – sound intensity component for x direction for i-th frequencies,

$X_p(i)$ – coefficients of complex spectrum for i-th frequencies for acoustic pressure signal,

$\overline{X_{ux}(i)}$, conjugated spectrum coefficients for particle velocity into x direction.

Intensity components for y and z direction were computed in the same way. The final intensity vector was given by equation 2 [12]:

$$\vec{I} = I_x\vec{e}_x + I_y\vec{e}_y + I_z\vec{e}_z \qquad (2)$$

It is the most important part of whole algorithm. The direction of arrival for particular frequency was obtained as a result of this block (with given spectral resolution, 11.7 Hz in this case). After that the direction of arrival for each frequencies were determined. The values of azimuth and elevation angles were calculated based on transformation from Cartesian to spherical coordinate system. Obtained DOA values were used to compute its time distribution. The time distribution collect the intensity values for given direction with 1 degree quantization. To reduce the noise level in the computed time distribution characteristic not all intensity vectors were used. During the computation process the noise floor for particular frequencies were calculated. Only the intensity vectors which value exceeded the noise floor increased by the defined threshold were used to accumulate the time distribution for given direction. In presented algorithm the threshold was equal to 10 dB (for this value the noise influence can be neglected, moreover in future practical implementation the threshold value could be specified by the user). In the accumulating process the value of the intensity vector was additionally used. In fact the time distribution indicate the total value of intensity vectors which occurred for given direction in considered time period. The maximum value observed in the time distribution characteristics indicate the position of the sound source. When more than one sound source was present in the acoustic field in considered time period, the additional maximum value on time distribution characteristic can occur. At the end of the calculation process (block 6), the final information about the position of particular sound sources position was indicated. The final time distribution of DOA values was smooth by means of weighted moving average. Angle values for particular peaks were obtained using local difference calculation. The difference between the actual and next DOA value was computed. When the difference changes the sign from positive to negative, the local maximum should occur. If the difference is greater than assumed threshold it means that it indicate the possible position of sound source.

3   Evaluation of the Proposed Algorithm

For simplify and reduce the complexity of the description of localization accuracy only azimuth angle was taken into consideration. Elevation angle was neglected. The sound source localization accuracy (α_{err}) was defined in that case as a difference between the computed direction of arrival angle (α_{AVS}) and Ground Truth angle value (α_{GT}) (it indicate the real position of the sound source) for considered acoustic event. This parameter was given by the equation (3).

$$\alpha_{err} = \alpha_{AVS} - \alpha_{GT} \tag{3}$$

The evaluation process relied on determination the α_{err} value for every sound source. To proper determination of this parameter both α_{AVS} and α_{GT} should be known. The proposed sound sources localization algorithm returns the result as a value of the angular direction of arrival for particular sound sources (α_{AVS}). The multichannel loudspeaker system placed in the anechoic chamber was used to simulate the sound sources, therefore the α_{GT} values were easy to determine. The prepared measurement system, used test signals and realized scenarios were described in this section.

3.1   Measurement System

Setup of the measurement equipment employed in the experiment is presented in Fig. 2. Placement of speakers and angles (α) between them and the USP probe were presented in Figs. 2 and 3.

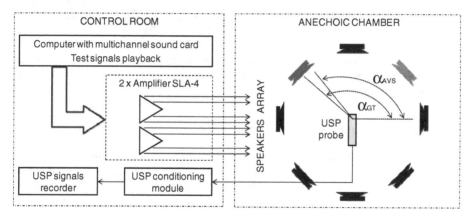

Fig. 2. Setup of measurement system, red and green color of the loudspeakers presents one of the scenario, when two different sounds were played

In an anechoic chamber 8 REVEAL 601p speakers, an USP probe were installed. The USP probe was fixed about 1.5 meters above the floor. In the control room a PC computer with Marc 8 Multichannel audio interface was used to generate test signals. Signals from the USP probe were recorded using MAYA44 USB audio interface. Two SLA-4 type 4-channel amplifiers were employed to power the speakers. The angular width of the speakers ($\Delta\alpha$) was also measured and illustrated. Real angle values between the speakers and USP Probe were used as a reference data during the evaluation process (see α_{GT} in eq. 3). The localization results of particular reference signals emitted in sequence by particular loudspeakers were presented in Fig. 3. Right picture presents the interior of anechoic chamber, red circle indicate the position of used acoustic vector sensor (USP probe). Used loudspeaker system can also be noticed. The anechoic chamber simulate the conditions of free acoustic field. Reflections coming from different surfaces can be neglected. It is important to emphasize two essential things. First the distance between the USP probe and loudspeaker is rather small. The dimensions of the sound source was taken into consideration in localization accuracy estimation. Second, the used loudspeakers only simulates the real sound source. In practice, the sound source that produce considered type of acoustic signal can have different dimensions. That facts can be important to estimate the correctness of the sound sources localization in real conditions.

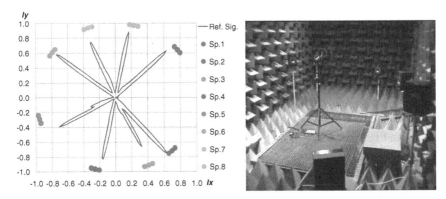

Fig. 3. The localization of particular reference signals emitted in sequence by particular loudspeakers. Right picture presents the interior of anechoic chamber, red circle indicate the position of used acoustic vector sensor (USP probe)

3.2   Test Signals and Measurement Scenarios

Both synthetic acoustic signals (such as pure tones and narrow band pink noises) and real sounds were used during the measurements. Pure tones has the following frequencies: 250Hz, 500Hz, 800Hz, 1000Hz, 1250Hz, 2000Hz, 4000Hz. Noises signals were prepared on the basis of pink noise broadband signal filtrated by means one third octave band filters. Center frequencies of particular filters were the same like for tonal signals. Signal amplitudes have been normalized therefore they had the same acoustic energy. Selected real signals used during the tests were different in the frequency and time domain. Exemplary recordings several type of sound sources such like: speech, scream, car horn, shot and broken glass were prepared and applied during the tests. The main assumption of the prepared measurement scenarios was simultaneous presentation signals which have the same or different energy distribution both in time and frequency domain. Main aim of such scenarios was evaluation of localization accuracy as a function of characteristic of the sound source. For this purpose during one session the particular sound source was used two times. First the single sound source from one direction was played, next second sound source played from other direction and finally both sound sources were played simultaneously from their directions. For such arrangement of the sound source reproduction, the localization accuracy of single and multiple sound sources can be determine. Another parameter which was taken into consideration was the cohesion of sound sources. It was verified by reproduction different or the identical sounds from given directions. For identical sound sources we do not notice any differences between the sound sources both on time and frequency domain. It means that proposed algorithm in such conditions cannot work properly. The example frequency characteristic of selected test signals and measure scenario were presented in Figure 4. For every measurement session two calibration sequence were used, before and after the test signals presentation. Pure tone about 1 kHz frequency and one second duration time was played by particular loudspeakers sequentially. The hardware setup and measurement session were proper when the localization results of the particular loudspeakers for both calibration sequences were the same.

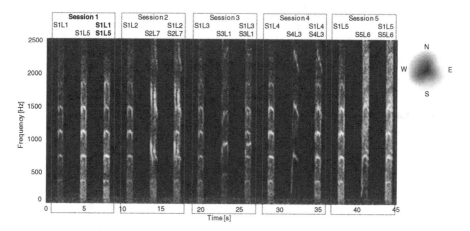

Fig. 4. Example frequency characteristic of selected test signals (scream) and measure scenario (dotted red rectangles). During session 1 identical sound signals were presented from different directions (coherent sound sources). In other sessions different scream signals were used.

4   Results

4.1   Calculation of Time Distribution of DOA Values

During evaluation of the developed algorithm a several methods used to calculation time distribution of DOA were taken into consideration. First method (H1) assumed that the DOA values for particular direction was increased by one if the frequency component in that direction was present. This approach worked well for signals whose time duration was longer than 0.5 [s]. For impulse sound sources (shot, broken glass) it was difficult to determine the main direction of arrival properly. In second method (H2), level of intensity vector for particular frequency of spectrum was taken into account. It means that the time distribution for given angle value indicate the total value of intensity vectors which occurred for given direction in considered time period. Thanks to this modification, the selectivity of obtained results increased rapidly. But for impulse sounds sources many additional local peaks were still observed. In some cases uncertainty of final decision about position of particular sound sources was high, because two or more local maximum were observed.

Application of weighted moving average solved that problem (H3). Averaged period was equal to 7. Weighted coefficients were calculated based on Hanning window. In Figs. 5 and 6 normalized time distribution of DOA values (Hnv[%]) for impulse sound sources obtained by means of particular method were presented. In Fig. 7 the polar representation of the DOA values were show.

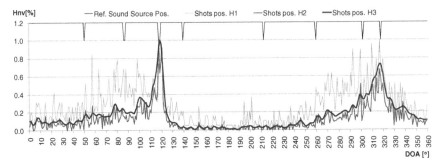

Fig. 5. Normalized time distribution of DOA values for impulse sound sources obtained by means of particular method – two shot sounds emitted from different directions

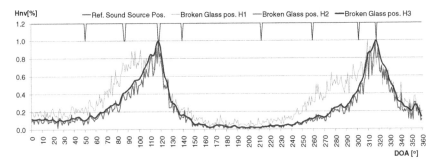

Fig. 6. Normalized time distribution of DOA values for impulse sound sources obtained by means of particular method – two broken glass sounds emitted from different directions

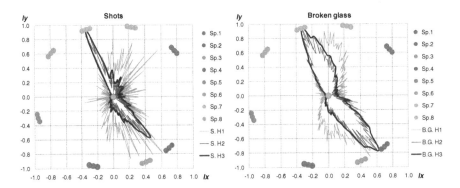

Fig. 7. Polar representation of DOA values for considered method. Left plot – shot, right plot – broken glass. Increase of localization accuracy for particular sound sources is clearly visible

The green line - H1 indicates the method 1, red line - H2 and blue line - H3, method 2 and 3 respectively. The black peaks indicate the reference position of the sound sources. If the black peak is common with the peak of the DOA it means that the position of sound source was indicated properly. Sound sources were simulated by speakers 2 and 6. In the left plot the localization results for shot sounds were depicted. Localization results for two broken glass played simultaneously were shown in right plot. It is clearly to notice that indication of localization for impulse sound sources for method H1 and H2 is ambiguous. Weighted moving average produce smooth DOA characteristic. Two main peaks can be observed. Automatic precise localization of the sound source position were possible in consider case.

4.2   Length of Overlap

Another issue that was considered in evaluation process was the length of overlap used in the FFT calculations. Three different overlap values were tested: 0%, 50% and 75%. It was explained in Fig. 8. Particular FFT frames were marked with different color. It means that in the next FFT frame from 4096 to 1024 samples were new. On the other hand for 4096 signal samples from 1 to 4 FFT frames were calculated. The FFT calculation were performed for each acoustic components separately. In Figs. 9 and 10 time distribution of DOA values for different overlapping lengths were presented. The results for impulse sounds like shots or broken glass were shown.

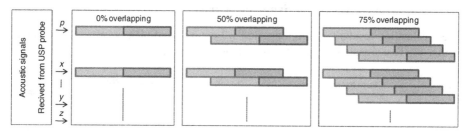

Fig. 8. Tested overlapping configuration. Green rectangles – current frame, red – next frame.

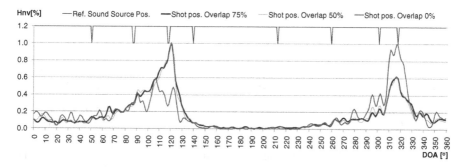

Fig. 9. Time distribution of DOA values for different overlapping level for two shots presented at the same time

For longer acoustic events the difference was very low and it was not presented here. On the basis of obtained DOA characteristics optimal length of overlap was specified. It was equal to 50%. Lower value can cause the loss of data, especially for impulse sounds. Such situation can be observed in Fig. 9.

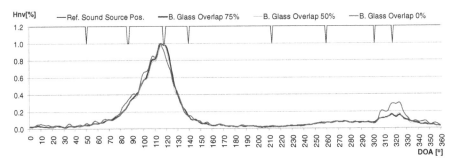

Fig. 10. Time distribution of DOA values for different overlapping level for two broken glass presented at the same time

On the other hand greater overlap value not delivered more data during the time distribution calculation. High similarity of green and blue curves proves this statement. 50% overlap was finally applied in the designed localization method.

4.3   Localization Results for Synthetic Signals

The localization results for pure tones and 1/3 octave band pink noise were presented in Fig. 11. Test signals were emitted from loudspeaker 1 and 6.

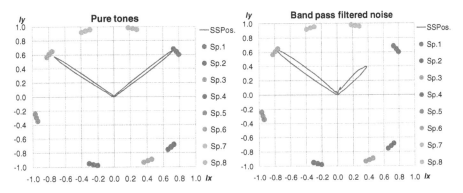

Fig. 11. Localization results for synthetic signals. Left plot - pure tones 1000Hz and 1250Hz, right plot - 1/3 octave band filtered pink noise, the centre frequency 1000 Hz and 1250 Hz.

For signals different in the frequency domain the localization accuracy was the best. Particular sound sources were localized perfectly. But when the identical, time synchronized, signals were presented from different directions, the proper localization of sound sources was insufficient and in some cases completely impossible.

4.4   Localization Results for Real Signals

Below results of localization accuracy obtained by means of real signals were presented. In Fig. 12 car horn and scream were played from speaker 3 and 4. In this case the localization was precise for different signals. High inaccuracy was noticed for coherent sound sources.

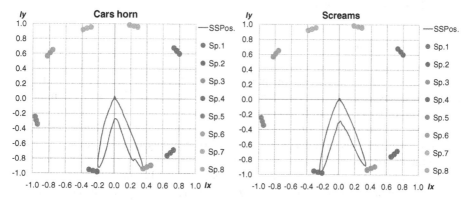

Fig. 12. Localization results for real acoustic signals presented simultaneously by speakers 3 and 4. Left plot - car horn, right plot - scream.

In Fig. 13 the localization results of impulse sound were shown. The localization was proper for signals different in frequency domain.

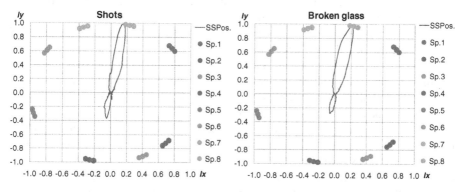

Fig. 13. Localization results for real acoustic signals presented simultaneously by speakers 4 and 8. Left plot - shot, right plot - broken glass.

Additionally, the localization accuracy goes down rapidly if sound sources were close to each other and amplitude one of them was dominant. It is important to emphasize, that for impulse sound sources the localization was based on few FFT frames (it depended on length of the considered acoustic event). Angular resolution of multiple sound sources localization also depended on the type of analyzed signal. Length of particular signals has crucial role for final accuracy. The best results were obtained for signals lasting more that 1 second. The type of given sound source is main difficulty for proper blind localization process in real time. If the analysis will be done offline during the situation reconstruction process, user can precisely select interest part of signal and do the localization process more accurate. Proposed algorithm can be used as interesting and useful tool during the offline forensic audio analysis. In such case all of described parameters could be selected and changed manually adequate to considered signal.

5   Conclusions

Method and preliminary results of multiple sound sources localization in real time using the acoustic vector sensor were presented in this study. Term of the multiple sound sources in this research means that two sources produced the acoustic energy simultaneously from different directions. The several properties of the developed algorithm were discussed in details on the basis of specially prepared tests conducted in laboratory conditions. First was selection the best method for computation the time distribution of DOA values. Second was discussion about length of overlap. Hanning weighted moving average and 50% length of overlap were optimal and gave the greater localization accuracy. The multiple sound sources localization can be done by means single acoustic vector sensor and sound intensity computation in frequency domain.

Localization accuracy and angular resolution depended on length of the analyzed signals and local differences both in time and frequency domain. The best results were obtained for signals longer than 1 second and different in time and frequency domain. For shorter signals the decrease of accuracy and angular resolution were observed. Moreover for coherent signals (both in time and frequency domain) the proposed algorithm did not work properly. The type of given sound source is main difficulty for proper blind localization process in real time. In such case the information about the position of detected sound sources is presented immediately. Additional types of information about the sound source like beginning, end and length of activity can be also obtained and to presented.

Method can be applied to analysis both fixed or moving sound sources. Their trajectory can be tracked independently. The described method can be useful in a surveillance systems to monitor and visualize the acoustic field of specified region. The direction of arrival can be used to control the Pan-Tilt-Zoom (PTZ) camera to automatically pointing it towards the direction of the detected sound source.

It is important to emphasize that the proposed method can be used as interesting and useful tool also during the offline forensic audio analysis. The described algorithm can be also used as visualization technique called spectrogram direction of arrival. In such case all of described parameters could be selected and changed manually adequate to considered signal.

In future work the method will be examined in real disturbance conditions. Additional improvements of functionality as spatial filtration into the defined direction and integration with other DSP method such as adaptive detection and automatic classification of sound events will also be implemented (for this reason the Hanning window was applied in FFT calculation). The comparison of the presented solution with traditional methods based on microphone arrays will be done.

Acknowledgements. Research is subsidized by the European Commission within FP7 project "INDECT" (Grant Agreement No. 218086).

References

1. Kotus, J.: Application of passive acoustic radar to automatic localization, tracking and classification of sound sources. Information Technologies 18, 111–116 (2010)
2. Czyżewski, A., Kotus, J.: Automatic localization and continuous tracking of mobile sound source using passive acoustic radar. Military University of Technology (2010)
3. Kotus, J., Kunka, B., Czyżewski, A., Szczuko, P., Dalka, P., Rybacki, R.: Gaze-tracking and acoustic vector sensors technologies for PTZ camera steering and acoustic event detection. In: 1st International Workshop: Interactive Multimodal Pattern Recognition in Embedded Systems (IMPRESS 2010), Bilbao, Spain, September 1, pp. 276–280 (2010)
4. Kotus, J., Łopatka, K., Kopaczewski, K., Czyżewski, A.: Automatic Audio-Visual Threat Detection. In: MCSS 2010: IEEE International Conference on Multimedia Communications, Services and Security, Krakow, Poland, May 6-7, pp. 140–144 (2010)
5. Łopatka, K., Kotus, J., Czyżewski, A.: Monitoring of audience of public events employing acoustic vector sensors. In: ISSET 2011, May 19-21 (2011)
6. Kotus, J., Łopatka, K., Cżyzewski, A.: Detection and Localization of Selected Acoustic Events in 3D Acoustic Field for Smart Surveillance Applications. In: Dziech, A., Czyżewski, A. (eds.) MCSS 2011. CCIS, vol. 149, pp. 55–63. Springer, Heidelberg (2011)
7. Hawkes, M., Nehorai, A.: Acoustic vector-sensor beamforming and Capon direction estimation. IEEE Trans. Signal Processing 46, 2291–2304 (1998)
8. Hawkes, M., Nehorai, A.: Wideband source localization using a distributed acoustic vector sensor array. IEEE Trans. Signal Processing 51, 1479–1491 (2003)
9. de Bree, H.-E., Wind, J., Sadasivan, S.: Broad banded acoustic vector sensors for outdoor monitoring propeller driven aircraft. In: DAGA 2010, Conference Proceedings, Berlin, Germany, March 15-18 (2010)
10. de Bree, H.-E., Wind, J., de Theije, P.: Detection, localization and tracking of aircraft using acoustic vector sensors. In: Inter Noise 2011 Proceedings, Osaka, Japan, September 4-7 (2011)
11. Smith, S.W.: The Scientist and Engineer's Guide to Digital Signal Processing. California Technical Publishing (1997)
12. Basten, T., de Bree, H.-E., Tijs, E.: Localization and tracking of aircraft with ground based 3D sound probes. In: ERF33, Kazan, Russia (2007)

New Efficient Method of Digital Video Stabilization for In-Car Camera

Aleksander Lamża and Zygmunt Wróbel

Department of Biomedical Computer Systems, Institute of Computer Science, Faculty
of Computer and Materials Science, University of Silesia in Katowice,
ul. Bedzinska 39, 41–200 Sosnowiec
{aleksander.lamza,zygmunt.wrobel}@us.edu.pl

Abstract. The problem of image stabilization in video sequences is important in many applications, including cameras mounted on vehicles. This type of vision systems can be used in vehicles that are the equipment of police and other road services as well as in driving assist systems. The stability of video aquired while driving directly affect the quality resulting in performance of further processing. In this paper, the new approach to digital video stabilization was presented. The proposed method is based on block-matching algorithm, namely Gray-Coded Bit Matching Plain. Modification of the algorithm improves performance by reducing the size of the analyzed block. This was achieved by the introduction of additional information of acceleration in the vertical axis aquired from accelerometer coupled with camera.

Keywords: image stabilization, accelerometer, gray-coded bit plain matching, in-car camera.

1 Introduction

Nowadays, monitoring of the environment and tracking objects through cameras finds increasing numbers of applications. There are two main areas of applications: cameras permanently located on buildings, masts etc. and portable cameras or ones located on vehicles. The first solution is used mostly to monitor the streets, shops, public utility places or crossroads. The second solution is used mostly for automobiles as part of driver assistance systems, especially in Automated Guided Vehicles (AGVs) and mobile robots. Usually the video sequence, besides being recorded and possibly stored, is processed and analyzed in order to allow identification and tracking of selected objects, such as people or their faces, cars, registration plates etc., detecting roadsigns and dangerous situations that might occur in the streets (in case of systems mounted in cars) etc.

A camera located in a car is constantly subjected to vibrations resulting from the car driving on bumpy streets and by the engine of the car itself, which cannot be always easily eliminated by proper mounting. This strongly influences the quality of the data stream and decreases the effectiveness of the algorithms that process analyze the data, often making it impossible to recognize the selected

A. Dziech and A. Czyżewski (Eds.): MCSS 2012, CCIS 287, pp. 183–190, 2012.

objects correctly. Image stabilization techniques may be divided into two groups: optical and digital methods. Optical image stabilization is a method in which a prism or one of the lenses is moved in response to the processed signal coming from the gyroscope. Such solutions are usually used in photography and filming equipment, mostly because od the considerable size of the lens equipped with a stabilization system of that sort. For small on-vehicle cameras the second method – digital image stabilization (DIS) – is applied.

2   Digital Image Stabilization Algorithms

The main difference between digital and optical image stabilization methods is the place where the compensation of the movement occurs. In optical methods, the optical system of the camera is modified, while in digital methods, the changes are introduced at the level of image after it is recorded by the camera. The simplest solution is to move a window inside the image recorded by the camera (Fig. 1). However, this requires the width of the margin to be established based on the maximum amplitude of vibrations to be eliminated.

a)   b)

Fig. 1. The movement of a window inside an image according to the computed vibration vector: (a) window at time t; (b) window at time $t + 1$ moved by a vector m

Sometimes [1,7] a subdivision into digital and electronic stabilization methods may be encountered, based on the method for movement estimation, but that is not of primary importance from our point of view. Digital stabilization methods do not require additional equipment, because the estimation of vibrations is done through image analysis methods. This solution reduces the complexity of the system, so, at least potentially, also the costs, however a larger demand for computational power should be taken into consideration. A general diagram of the digital image stabilization algorithm [3,10] is shown in figure 2.

The system consists of two basic blocks: the movement estimation block and the movement compensation block. The purpose of the estimation block is to determine the global movement vector (GMV), which depending on the estimation

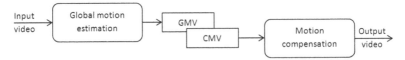

Fig. 2. A general diagram of digital image stabilization system

method can be determined by local movement vectors. The final result of estimation is strongly dependent on the local movement vector. In the compensation block based on the global movement vector the compesating movement vector (CMV) is determined, and the window located inside the image is translated by this vector. This block, other than performing simple vector translation, may also rotate the image. The details depend on the particular movement algorithm used.

Many digital image stabilization algorithms have been developed. They differ mostly by their local and global movement estimation method. A traditional solution is a block matching metod [3,5,6]. In order to decrease computational complexity several block are selected from the image and the movement vector is determined individually for these blocks. Based on those vectors, a global movement vector is calculated, which is then used as the basis for movement compensation. The advantage of these methods is a relatively low computational complexity, while there are considerable disadvantages related to the determination of local movement vectors in varying lighting conditions and in cases of multiple objects present in a single block. Amongst many available methods, plain matching and gray-coded bit plane matching were used as a starting points for the purpose of this paper. What follows is its short description, while the following chapter discusses a possible modification that increases its effectiveness and exactness.

2.1   Plain Matching Algorithm Overview

The straightforward solution leads to use of a discrete cross-correlation respectively which produces a matrix of elements. The elements with high value correspond to the locations where a chosen pattern and image match well (high correlation). It means that the value of element is a measure of similarity in relevant point and we can find locations in an image that are similar to the pattern.

On the input we have the pattern $P(x,y)$ and image $I(x,y)$. Because it is not necessary to search over the whole image I, a small square area M is defined (known as search block or window). This area specifies the maximal shift (p) in vertical and horizontal direction (fig. 3).

The discrete cross-correlation is defined as:

$$c(m,n) = \frac{1}{M^2} \sum_{x=0}^{M-1} \sum_{y=0}^{M-1} P(x,y) I(x+m,y+n) \tag{1}$$

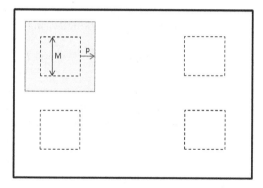

Fig. 3. Illustration of search window: M is a search window, p is the number of pixels to search over

where M is a square search window and $-\left(p + \frac{M}{2}\right) \leq m, n \leq \left(p + \frac{M}{2}\right)$, where p is the number of pixels to search over.

Matching according to this definition is problematic because correlation can also be high in locations where the image intensity is high, even if it doesn't match the pattern well. Better performance can be achieved by a normalized correlation [2,3]:

$$c(m,n) = \frac{\sum\limits_{x=0}^{M-1}\sum\limits_{y=0}^{M-1} P(x,y)\, I(x+m, y+n)}{\sqrt{\sum\limits_{x=0}^{M-1}\sum\limits_{y=0}^{M-1}(I(x+m, y+n))^2} \cdot \sqrt{\sum\limits_{x=0}^{M-1}\sum\limits_{y=0}^{M-1}(P(x,y))^2}} \qquad (2)$$

This approach is rarely used due to enormous time consumption (for every point of matrix c is necessary to perform $2N \times N$ multiplications and additions). The eq. 2 appears to be an ideal choice from the hardware processing point of view. However, in this paper we discuss only software methods, so this algorithm is not taken into account.

2.2   Gray-Coded Bit Plain Matching Algorithm Overview

Just like the prevoius method, the Gray-Coded Bit Plain Matching method belongs to the group of region matching methods. Its first description was in [4], but it was later modified many times and other method based on it have been introduced [3,8,9]. The basic assumption is the use of bit-plane images instead of 8-bit gray-coded images. Because of that Boolean binary operation may be used in algorithm implementations which greatly reduces computational complexity. Let's assume that $F(x,y)$ is the image and $G_i(x,y)$ are binary images representing it:

$$F(x,y) = a_{K-1}(x,y) \cdot 2^{K-1} + a_{K-2}(x,y) \cdot 2^{K-2} + \ldots + a_1(x,y) \cdot 2 + a_0 \quad (3)$$

where K is the number of binary images, which for 8-bit gray-coded images gives K=8. Binary images are determined based on the following relations:

$$G_i(x,y) = \begin{cases} a_i(x,y) \oplus a_{i+1}(x,y) & 0 \leq i \leq K-2 \\ a_i(x,y) & i = K-1 \end{cases} \tag{4}$$

In order to determine the movement vectors the correlation of binary images representing two successive video frames should be calculated. The correlation is defined as:

$$c(m,n) = \frac{1}{M^2} \sum_{x=0}^{M-1} \sum_{y=0}^{M-1} G_k^t(x,y) \oplus G_k^{t-1}(x+m,y+n) \tag{5}$$

where M is a square search window and:

$$-\left(p+\frac{M}{2}\right) \leq m,n \leq \left(p+\frac{M}{2}\right) \tag{6}$$

where p is the number of pixels to search over (fig. 3).

After minimizing the correlation c the resulting values for (m,n) are local movement vectors for all of the defined block (Fig. 3 illustrates four such blocks). All local motion vectors from each area along with the previous global motion vector are passed through a median operator to produce the current global motion vector estimate. In the method described in order to determine the values of local movement vectors a correlation for the whole search area should be calculated. The following chapter discusses a proposed modification which reduces the search area.

3   Using Data Aquired from Accelerometer

A solution, whereby a camera is integrated with an accelerometer providing data on acceleration along three axis, is proposed. Because it is assumed that the camera is located in a car, the accelerations resulting from the car's movements should be taken into consideration. The acceletometer is located so that the Z axis is perpendicular to the XY image plane. As in such cases vertical acceleration are dominant, only Y-axis accelerations were analyzed for the purpose of the experiment. Figure 4 shows a typical 26-second long record of Y-axis accelerations.

At the time of image acquisition the value of acceleration is also received and the variable h is calculated according to:

$$h_t = sign(a_t - a_{t-1}) \tag{7}$$

Because of the use of the *sign* function, h may take a value of -1, 0 or 1 (fig. 5).

Based on this value, the search area is narrowed accordingly. The value of -1 means that the acceleration was directed down, thus the search area should be

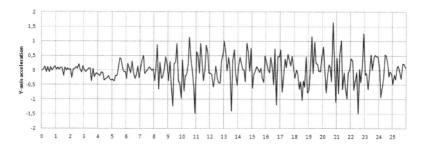

Fig. 4. A typical record of Y-axis accelerations $[m/s^2]$

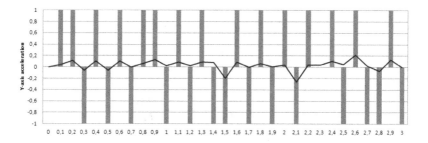

Fig. 5. Variable h calculated for sample accelerations record

narrowed from the bottom. The case for $h=1$ is analogous, while at 0 there is no movement along the vertical axis and the area is narrowed both from the bottom and from the top (Fig. 6). The value of h results in a change of coordinate pairs (m, n) according to eq. 6:

$$
\begin{aligned}
-\tfrac{M}{2} \le n \le \left(p + \tfrac{M}{2}\right) && when\ h = 1 \\
-\left(p + \tfrac{M}{2}\right) \le n \le \tfrac{M}{2} && when\ h = -1 \\
-\tfrac{M}{2} \le n \le \tfrac{M}{2} && when\ h = 0
\end{aligned}
\tag{8}
$$

4   Experiments

The algorithm described was implemented in the Octave environment with the Octave-video package. A test material were video sequences with a 720x480 pixels resolution and a framerate of 30 fps, synchronized with a set of accelerometric data. For the purposes of the experiment, 5 video sequences were selected (with a length of 60-120 frames) which varying vibration intensities (Table 1).

The experiments consisted of performing test for each video sequence ten times. On this basis it was determined the average processing time t_P for the proposed method. For purposes of comparison the same tests for classical method of Gray-Coded Bit Matching Plain were performed. The results are presented in Table 2.

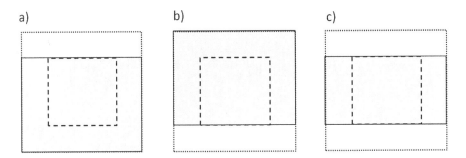

Fig. 6. An illustration of narrowed search areas: a) h=1; b) h=-1; c) h=0

Table 1. Test video sequence

Video sequence	Length [$frames$]	Average acceleration $\overline{|a_Y|}\left[\frac{m}{s^2}\right]$
1	60	0.13
2	90	0.36
3	120	0.52
4	60	0.73
5	90	0.95

Table 2. Experimental results for test video sequences

Video sequence	Gray-Coded Bit Plain Matching $\overline{t_P}$ [ms]	Proposed method $\overline{t_P}$ [ms]	Improvement [%]
1	65	50	23.1
2	62	53	14.5
3	65	51	21.5
4	65	50	23.1
5	68	55	19.1

The comparison to the results achieved by the use of the Gray-Coded Bit Plain Matching gave an average improvement of 20%.

During the testing period, however, errors in the selection of search area were noted, related to the imperfect synchronization of the video sequences to the acceletometric data. It should be stressed that the experiments were preliminary and were only designed to determine whether further tests and improvements to the algorithm should be implemented.

5   Conclusion

Due to the supplemention of the video sequences with accelerometric data it was possible to optimalize the algorithm for global movement vector determination and, furthermore, increase the effectiveness of the image stabilization method. So far only proof-of-concept experiments have been conducted. To determine, how far does the above-described method increase effectiveness in practical applications, it should be implemented and tested under real-life conditions. Currently, an attempt to integrate the acceletometric data with other movement estimation methods is underway.

References

1. Amanatiadis, A., Gasteratos, A., Papadakis, S., et al.: Image Stabilization in Active Robot Vision. In: Ude, A. (ed.) Robot Vision, pp. 261–274. InTech (2010)
2. Brooks, A.C.: Real-Time Digital Image Stabilization. International Journal of Autonomic Computing 1(2), 202–209 (2009)
3. Drahanský, M., Orság, F., Hanáček, P.: Accelerometer Based Digital Video Stabilization for General Security Surveillance Systems. International Journal of Security and Its Applications (1), 1–10 (2010)
4. Ko, S.J., Lee, S.H., Jeon, S.W.: Fast Digital Image Stabilizer Based on Gray-Coded Bit-Plane Matching, pp. 90–91. IEEE, USA (1999)
5. Morimoto, C., Chellappa, R.: Evaluation of Image Stabilization Algorithms. Electrical Engineering 5, 3–6 (2002)
6. Sachs, D., Nasiri, S., Goehl, D.: Image Stabilization Technology Overview (February 20, 2012), http://www.invensense.com/shared/pdf/ImageStabilizationWhitepaper_051606.pdf
7. Shih, F.Y., Stone, A.: A New Image Stabilization Model for Vehicle Navigation. Positioning 1, 8–17 (2010)
8. Vella, F., Castorina, A., Mancuso, M., Messina, G.: Robust Digital Image Stabilization Algorithm Using Block Motion Vectors, pp. 234–235. IEEE, USA (2002)
9. Yeh, Y.M., Wang, S.J., Chiang, H.C.: A Digital Camcorder Image Stabililzer Based on Gray Coded Bit-plane Matching. In: Proceedings of SPIE, vol. 4080, pp. 112–120 (2000)
10. Zhang, Y., Xie, M.: Robust Digital Image Stabilization Technique for Car Camera. Information Technology Journal 10(2), 335–347 (2011)

Redefining ITU-T P.912 Recommendation Requirements for Subjects of Quality Assessments in Recognition Tasks

Mikołaj I. Leszczuk[1], Artur Koń[1], Joel Dumke[2], and Lucjan Janowski[1]

[1] AGH University of Science and Technology, Department of Telecommunications,
al. Mickiewicza 30, Kraków, Poland,
{leszczuk,janowski}@kt.agh.edu.pl, arturkon87@gmail.com
[2] National Telecommunications and Information Administration (NTIA), Institute for
Telecommunication Sciences (ITS),
Boulder CO, USA
jdumke@its.bldrdoc.gov

Abstract. The transmission and analysis of video is often used for a variety of applications outside the entertainment sector, and generally this class of video is used to perform a specific task. Therefore it is crucial to measure, and ultimately, optimize task-based video quality. To develop accurate objective measurements and models for video quality assessment, subjective experiments must be performed. Problems of quality measurements for task-based video are partially addressed in a few preliminary standards and a Recommendation (ITU-T P.912, "Subjective Video Quality Assessment Methods for Recognition Tasks,") that mainly introduce basic definitions, methods of testing and requirements for subjects taking part in psychophysical experiments. Nevertheless, to the best of the authors' knowledge, the issue of requirements for subjects has been not verified in any specific academic research. Consequently, in this paper, we compare groups of subjects assessing video quality for task-based video. Once a comparison has been made for task-based video, specifications amendments for P.912 are developed. These will assist researchers of task-based video quality in identifying the subjects that will allow them to successfully perform the psychophysical experiment required.

Keywords: ITU-T, standards, systems, video, quality.

1 Introduction

The transmission and analysis of video is often used for a variety of applications outside the entertainment sector, and generally this class of (task-based) video is used to perform a specific recognition task. Examples of these applications include security, public safety, remote command and control, tele-medicine, and sign language. The Quality of Experience (QoE) concept for video content used for entertainment differs materially from the QoE of video used for recognition tasks because in the latter case, the subjective satisfaction of the user depends upon achieving the given task, e.g., event detection or object recognition. Additionally, the quality of video used by a human observer is

A. Dziech and A. Czyżewski (Eds.): MCSS 2012, CCIS 287, pp. 188–199, 2012.
© Springer-Verlag Berlin Heidelberg 2012

largely separate from the objective video quality useful in computer vision [8]. Therefore it is crucial to measure and ultimately optimize task-based video quality. This is discussed in more detail in [9].

There exist only a very limited set of quality standards for task-based video applications. Therefore, it is still necessary to define the requirements for such systems from the camera, to broadcast, to display. The nature of these requirements will depend on the task being performed.

Enormous work, mainly driven by the Video Quality Experts Group (VQEG) [12], has been carried out for the past several years in the area of consumer video quality. The VQEG is a group of experts from various backgrounds and affiliations, including participants from several internationally recognized organizations, working in the field of video quality assessment. The group was formed in October of 1997 at a meeting of video quality experts. The majority of participants are active in the International Telecommunication Union (ITU) and VQEG combines the expertise and resources found in several ITU Study Groups to work towards a common goal [12]. Unfortunately, many of the VQEG and ITU methods and recommendations (like ITU's Absolute Category Rating – ACR – described in ITU-T P.800 [2]) are not appropriate for the type of testing and research that task-based video, including closed-circuit television (CCTV), requires.

European Norm number 50132 [6] was created to ensure that CCTV systems are realized under the same rules and requirements in all European countries. The existence of a standard has opened an international market of video surveillance devices and technologies. By selecting components that are consistent with the standard, a user can achieve a properly working CCTV system. This technical regulation deals with different parts of a CCTV system including acquisition, transmission, storage, and playback of surveillance video. The standard consists of such sections as lenses, cameras, local and main control units, monitors, recording and hard copy equipment, video transmission, video motion detection equipment, and ancillary equipment. This norm is hardware-oriented as it is intended to unify European law in this field; thus, it does not define the quality of video from the point of view of recognition tasks.

To develop accurate objective measurements and models for video quality assessment, subjective tests (psychophysical experiments) must be performed. The ITU has recommendations that address the methodology for performing subjective tests in a rigorous manner [5], [3]. These methods are targeted at the entertainment application of video and were developed to assess a person's perceptual opinion of quality. They are not entirely appropriate for task-based applications, in which video is used to recognize objects, people or events.

Assessment principles for the maximization of task-based video quality are a relatively new field. Problems of quality measurements for task-based video are partially addressed in a few preliminary standards and a recommendation (ITU-T P.912, "Subjective Video Quality Assessment Methods for Recognition Tasks," 2008 [4,7]) that mainly introduce basic definitions, methods of testing and psycho-physical experiments. ITU-T P.912 describes multiple choice, single answer, and timed task subjective test methods, as well as the distinction between real-time and viewer-controlled viewing, and the concept of scenario groups to be used for these types of tests. Scenario groups are groups of very similar scenes with only small, controlled differences between them, which enable testing recognition ability while eliminating or greatly reducing the potential effect of scene memorization. While these concepts have been introduced specifically for task-based video applications in ITU-T P.912, more research is necessary to validate the methods and refine the data analysis.

Section 7.3 of ITU-T P.912 ("Subjects") says that, "Subjects who are experts in the application field of the target recognition video should be used. The number of subjects should follow the recommendations of ITU-T P.910 [3]." Expert subjects (police officers, doctors, etc.) are costly and difficult to hire compared to non-expert subjects (colleagues, friends, students, pensioners). Nevertheless, to the best of the authors' knowledge, this expert subject issue has not been verified in any specific academic research. There do exist some applicable ideas incorporated from industry. For example, large television companies hire expert subjects to monitor their quality [10]. However, there is no evidence that these companies have applied any serious effort to determine how these people compare to non-expert subjects. Almost all Mean Opinion Score (MOS) tests focus on non-experts. There is a belief that people who have more video knowledge or experience would give different results, but that has not ever been rigorously studied.

In this paper, we compare groups of subjects assessing video quality for task-based video. Once a comparison has been made for task-based video, specifications amendments for standards can be developed that will assist researchers of task-based video quality in identifying the subjects that will allow them to successfully perform the psychophysical experiment required.

Current efforts to remedy this lack of video quality standards and measurement methods for task-based video are described in this paper. Section 2 presents an experimental psychophysical experiment method for public safety video applications. Section 3 discusses results of a test using this method. Section 4 explains approaches to comparing data between tests. Standardization activities are described in Section 5. The paper is concluded in Section 6.

2   Experimental Method

This Section presents an experimental design including source video sequences (Subsection 2.1) and recognition test-plans (Subsection 2.2).

2.1   Source Video Sequences

The scene clips that were shown were produced using the following four categories of lighting condition scenarios:

1. outdoor, daytime light,
2. indoor, bright with flashing lights,
3. indoor, dim with flashing lights, and
4. indoor, dark with flashing lights.

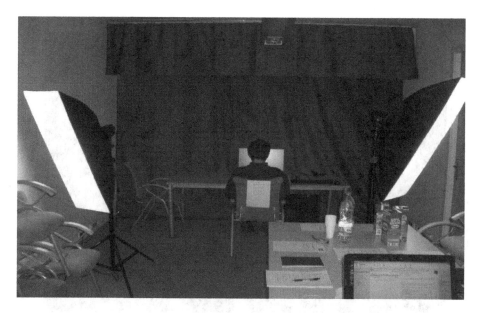

Fig. 1. A photo of the test environment used in this experiment

Three different distances were used to create the clips:

1. 5.2 meters for indoor scenarios,
2. 10.9 meters for outdoor scenarios, objects close, and
3. 14.6 meters for outdoor scenarios, objects far.

More information about the videos and scenarios is included in [13, 14].

2.2   Recognition Test-Plans

37 subjects were gathered for each of the tests. Each subject was asked to fill out a test form which included demographic and professional experience questions as well as the results of the vision tests. The form also informed subjects about the purpose of the experiment, what kind of videos would be presented, the subject's task, and what objects would be shown in the videos.

Viewing conditions in the room where the test took place followed Recommendation ITU-R BT.500-12 [5] and Recommendation ITU-T P.910 [3] (Fig. 1):

Subjects were asked to answer specific questions regarding content in the video. This was a multiple choice test; subjects had to identify the moving or stationary object from a constant list of seven objects. An example of the user interface is shown in Fig. 2.

Each subject saw 420 scenes. Each scene was approximately 7 seconds long. There were 960 different clips made from 96 original clips by compressing each source clip at two resolutions (CIF and VGA) and five bit rates for each resolution. Subjects were shown the scene, then asked to answer the question relating to the scene, as described above. Prior to the actual test, a practice session with four clips was presented to each

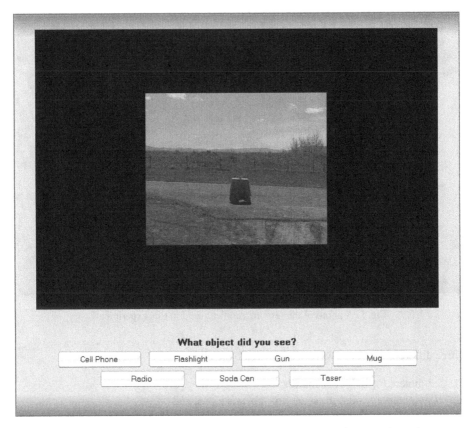

Fig. 2. User interface used for subjective target recognition task test performed

subject. There was no "don't know" option in the answers. Subjects could take break during the test, but this option was rarely used.

Experiments have been conducted at the Institute for Telecommunication Sciences (ITS) of the National Telecommunications and Information Administration (NTIA) in Boulder, CO, USA (Subsection 2.2) and at the Department of Telecommunications of AGH University of Science and Technology in Kraków, Poland (Subsection 2.2).

ITS-NTIA Test-Plan. NTIA-ITS performed the object recognition tests with two groups of subjects: the practitioner group [13, 14], and the non-practitioner group. Within the practitioner group, all subjects were volunteers and weren't paid for the test. Nevertheless, most of them received invitational travel to Boulder, CO, USA. Furthermore, all of them had experience in public safety, including: police, fire-fighter, Emergency Medical Services (EMS) and forensic video analysts. Very few were outside the range of 30–60 years old. Three of them had minor color vision problems–their results were not significantly different. Within the non-practitioner group, subjects had no experience in image recognition. All subjects were paid through a temporary work agency to take the test. None of them had experience in public safety. Subjects had a wide variety of ages, but skewed young. Two had minor color vision problems–their results were not significantly different.

AGH Test-Plan. Subjects had to have no experience in image recognition. All subjects were volunteers and they weren't paid for taking the test. None of them had experience in the public safety area.

Almost all subjects were 20-25 years old. One of the subjects had color vision problems which had no apparent effect on the subject's performance.

3   Results

This section presents experimental results. After an introduction (Subsection 3.1), the impact of memorization (Subsection 3.2), lighting (Subsection 3.3), motion (Subsection 3.4) and distance (Subsection 3.5) are discussed.

3.1   General Introduction

15540 total answers were collected from 14 scenario groups; 65% (10096) of the questions were answered correctly.

Subjects achieved the highest recognition rate with outdoor, stationary, close distance scenarios (89%) and the lowest rate with indoor, moving, dark light scenarios (25.4%). Because guessing was likely, each score was normalized using the following equation [1]:

$$R_A = R - \frac{W}{n-1}$$

where:

R_A – adjusted number of right answers
R – number of right answers
W – number of wrong answers,
n – number of answer choices

3.2   Impact of Memorization

There was some change in recognition rates over the course of the test. For the first half (210 clips), subjects achieved 62.4%, recognition. For the second half of the test, 67.5% of the items were correctly recognized. This growth can be seen in Fig. 3.

The trend line shows that the accuracy of answers given by subjects was growing during the test. It suggests that subjects were aided by memory effects as the test progressed.

Some subjects said that they remembered how objects were carried so in case of bad conditions they looked how the person was carrying the object. That aspect strongly depended on what object was presented on the clip.

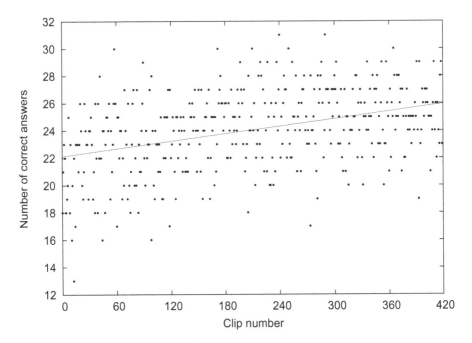

Fig. 3. Correlation between clip number and number of correct answers

3.3 Impact of lighting

Lighting conditions had significant influence on recognition. The best recognition rates were under daylight and bright lighting. Subjects achieved about 90% recognition for stationary objects with outdoor lighting as Fig. 6 shows and moving objects with bright lighting as Fig. 4 shows.

Fig. 4 shows that without proper lighting conditions correct recognition is very difficult. For indoor, moving objects, with dark lighting conditions recognition rates never exceed 40%, even for the 1536 kbit/s bit-rate. The small recognition rate difference between 256 kbit/s and 1536 kbit/s bit-rates shows that without proper light, increasing bit-rate does not improve recognition. The sole exception is between the 128 kbit/s and 256 kbit/s bit-rates, but for the such low video-stream bandwidth, the recognition rate is always much lower, regardless of other conditions. An interesting phenomena is that in both charts for some bit-rates there are better rates for Common Intermediate Format (CIF) resolution. For indoor, moving objects with dark lighting conditions recognition is 14% higher in the CIF resolution case. On the Fig. 4 it can be seen how better lighting improved the rate of correct answers. The recognition rate for 1536 kbit/s bit-rate parameter is 66%, which is 31% more than for worse lighting conditions.

There is a similar relation for indoor stationary objects with dim and dark (Fig. 5) lighting conditions. On both charts the same interesting phenomena as on the charts before can be seen. For some bit-rate parameter ranges (512–1536 kbit/s for dim

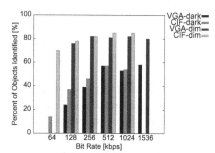

Fig. 4. Indoor, moving

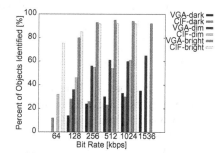

Fig. 5. Indoor, stationary

and 128–1536 kbit/s for dark lighting conditions) there is almost no recognition rate increase as bit-rate increases.

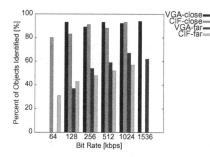

Fig. 6. Outdoor, stationary

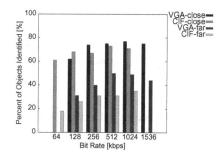

Fig. 7. Outdoor, moving

There is also a significant difference between indoor, moving object with bright and dim lighting conditions. In the second case for 1536 kbit/s bit-rate, 89% of objects were recognized correctly, which is a significant increase in comparison to 66% for dim light. It turns out that only in the cases of bright or daylight lighting conditions is there a high chance to recognize the object.

3.4   Impact of Motion

Results of the test show that motion has influence on recognition rate. Interestingly, motion's influence strongly depends on other conditions. For close, outdoor clips with stationary (Fig. 6) and moving (Fig. 7) objects, the recognition rates are 27% lower when objects are moving. Also, in the case of the indoor, dim lighting scenarios (Fig. 4 and Fig. 5 – respectively) motion causes high recognition rate decreases (29%).

On the other hand, for some scenario groups impact of motion isn't so significant. For outdoor, far, stationary (Fig. 6) and moving (Fig. 7) objects the difference is about 15% (The recognition rate drops from 23% for VGA with a 1536 kbit/s bit-rate to 6% for VGA with a 128 kbit/s bit-rate.), and 17% for close, dark lighting indoor clips. (The recognition rate drops from 32% for CIF with a 1024 kbit/s bit-rate to 1% for VGA with a 128 kbit/s bit-rate.)

3.5   Impact of Distance

Like other factors previously mentioned, distance (and projected size) has a great influence on recognition rate. This influence can be seen in a comparison of the outdoor, far, stationary objects scenario and the outdoor, close, stationary objects scenario (Fig. 6).

There is a significantly higher recognition rate for low bit-rates (39% for CIF resolution at 64 kbit/s), but for higher bandwidth values, the difference between recognition rates isn't so big as in case of smaller bit-rates. The reason for that is probably the fact that under some circumstances, conditions seem to be sufficient to make correct recognition. A similar dependence can be seen in the case of outdoor, moving objects for close and far distances (Fig. 7) .

4   Comparison

This section compares results among tests. After an introduction (Subsection 4.1), the impact of lighting, motion, and distance (Subsections 4.2, 4.3, and 4.4 – respectively) are discussed.

4.1   Introduction to Comparison

In the recognition psychophysical experiment performed by NTIA-ITS, two groups (non-paid professional and paid non-professional) achieved almost the same results, so both subject groups can be equally treated as "motivated subjects." Nevertheless, in contrast to the experiments with "motivated subjects," there is a great difference between results of "motivated subjects" and "unmotivated subjects."

"Motivated subjects" had more correct answers in all kinds of scenario groups–about 82%, which is 17% more than for "unmotivated subjects."[1] As in the "unmotivated subjects" experiment, the worst results were gathered for inside, dark, moving scenarios (42.4%), and the best for inside, bright light, moving scenarios (96.5%), which were well recognized also in "unmotivated subjects" experiment ("unmotivated subjects" did also perform well for those questions).

An interesting phenomena is that under good conditions (high bit-rate, enough lighting, etc.) the results of both groups were quite similar, but the difference grows fast while conditions are degrading, and in case of very bad conditions (like far, moving objects) the difference decreases again (see Fig. 8a):

- 43% difference for VGA, 64 kbit/s bit-rate, outside, far, stationary objects clips
- 6% difference for VGA, 1536 kbit/s bit-rate, outside, close, stationary objects clips
- 9% difference for CIF, 64 kbit/s bit-rate, outside, far, moving objects clips

---

[1] We are defining these subjects solely by their performance. There is no other reason they are categorized as unmotivated.

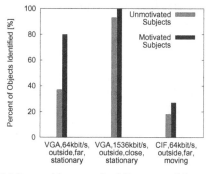

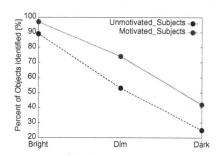

(a) Recognition rates in different conditions.

(b) Moving object scenarios in different lighting conditions.

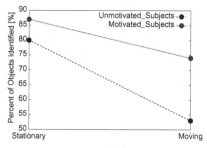

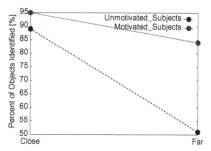

(c) Inside, dim lighting, differences between stationary and moving objects.

(d) Outdoor, stationary objects, differences between close and far distances.

Fig. 8. Comparison results

4.2 Lighting

Lighting played an important role in both experiments. For indoor scenarios with stationary object and dim lighting, the recognition level doesn't differ much among tests; "motivated subjects" achieved only about 7% greater ratio. For dark lighting conditions, the difference grows, achieving about 28% for VGA resolution with 1536 kbit/s bit-rate.

A similar dependence can be seen for indoor, moving object scenarios. With bright lighting conditions, "motivated subjects" achieved about 7% higher recognition only. For dim lighting conditions, the difference grows to 21%, and then falls to 17% for dark lighting (Fig. 8b). The decrease of difference between dim and dark lighting scenarios suggests that subjects crossed a "difficult point" and in the case of worse condition the difference will decrease. The difference between recognition ratio for dark lighting, VGA, 128 kbit/s bit-rate (only 6%) seems to support that theory.

4.3 Motion

The comparison of inside, dim lighting, moving and stationary objects also shows a significant difference between experiments. For stationary objects, "unmotivated subjects" achieved 7% lower recognition, for the same lighting conditions, but for moving objects, the difference changes to 21% (Fig. 8c).

4.4 Distance

For outdoor, stationary, close distance scenarios, "motivated subjects" achieved 95% recognition, which is 6% more than "unmotivated" ones, but for outdoor, stationary scenarios, for the greater distance, it was 83%, which is 32% more than "unmotivated" ones (Fig. 8d). For the CIF resolution, 64 kbit/s bit-rate, the result of "motivated subjects" was 43% greater.

5   Standardization Activities

Internationally, the number of people and organizations interested in this area continues to grow, and there was enough interest to motivate the creation of a task-based video project under VQEG. At one of the recent meetings of VQEG, a new project was formed for task-based video quality research. The Quality Assessment for Recognition Tasks (QART) project addresses precisely the problem of lack of quality standards for video monitoring. The initiative is co-chaired by NTIA-ITS, and AGH. Other members include research teams from Belgium, France, Germany, and South Korea. The purpose of QART is exactly the same as the other VQEG projects – to advance the field of quality assessment for task-based video through collaboration in the development of test methods, performance specifications and standards for task-based video, as well as predictive models based on network and other relevant parameters [11].

6   Conclusion and Further Research

We have extensively compared groups of subjects assessing task-based video quality. Clearly, in terms of the quality of psychophysical experiment (subjective test) results, it is more important to motivate the subjects than to acquire experts. They can either be paid or (for public safety scenarios) they can be police officers or practitioners of other public safety agencies.

Consequently the first sentence of Section 7.3 ("Subjects") of ITU-T P.912 could be rephrased as: "Subjects who are motivated should be used." Such an amendment has already been proposed at the recent VQEG meeting and is expected to be conveyed to the ITU-T. This minor amendment will significantly affect the methodology used by research community in conducting psychophysical experiments for video quality assessments in recognition tasks.

Further research and next steps in standardization will include more extensions to P.912 test methods, possibly covering prospective use of Computer Generated Imagery (CGI) and an introduction of a new concept of "video acuity" [15].

Acknowledgements. The authors would like to thank the European Community's Seventh Framework Program (FP7/2007-2013) for received funding for the research leading to these results under grant agreement № 218086 (INDECT) and the U.S. Department of Commerce for received funding for the research under the Public Safety Communications Research project.

References

1. ANSI S3.2, American National Standard Method for Measuring the Intelligibility of Speech over Communications Systems (1989)
2. ITU-T P.800, Methods for subjective determination of transmission quality (1996), http://www.itu.int/rec/T-REC-P.800-199608-I
3. ITU-T P.910, Subjective video quality assessment methods for multimedia applications (1999), http://www.itu.int/rec/T-REC-P.910-200804-I
4. ITU-T P.912, Subjective video quality assessment methods for recognition tasks (2008), http://www.itu.int/rec/T-REC-P.912-200808-I
5. ITU-R BT.500-12, Methodology for the subjective assessment of the quality of television pictures (2009), http://www.itu.int/rec/R-REC-BT.500-12-200909-I
6. CENELEC EN 50132, Alarm systems. CCTV surveillance systems for use in security applications (2011)
7. Ford, C.G., McFarland, M.A., Stange, I.W.: Subjective video quality assessment methods for recognition tasks. In: Proceedings of SPIE, vol. 7240, 72400Z–72400Z–11 (2009)
8. Leszczuk, M., Stange, I., Ford, C.: Determining image quality requirements for recognition tasks in generalized public safety video applications: Definitions, testing, standardization, and current trends. In: 2011 IEEE International Symposium on Broadband Multimedia Systems and Broadcasting (BMSB), pp. 1–5 (June 2011)
9. Leszczuk, M.: Assessing Task-Based Video Quality — A Journey from Subjective Psycho-Physical Experiments to Objective Quality Models. In: Dziech, A., Czyżewski, A. (eds.) MCSS 2011. CCIS, vol. 149, pp. 91–99. Springer, Heidelberg (2011), http://dx.doi.org/10.1007/978-3-642-21512-4_11
10. Spangler, T.: Golden eyes. Multichannel News (October 2009)
11. Szczuko, P., Romaniak, P., Leszczuk, M., Mirek, R., Pleva, M., Ondas, S., Szwoch, G., Korus, P., Kollmitzer, C., Dalka, P., Kotus, J., Ciarkowski, A., Dąbrowski, A., Pawłowski, P., Marciniak, T., Weychan, R., Misiorek, F.: D1.2, report on ns and cs hardware construction. Tech. rep., The INDECT Consortium: Intelligent Information System Supporting Observation, Searching and Detection for Security of Citizens in Urban Environment, European Seventh Framework Programme FP7-218086-collaborative project, Europa (2010), cop.
12. VQEG: The Video Quality Experts Group, http://www.vqeg.org/
13. VQiPS: Video quality tests for object recognition applications. Public Safety Communications DHS-TR-PSC-10-09, U.S. Department of Homeland Security's Office for Interoperability and Compatibility (June 2010)
14. VQiPS: Recorded-video quality tests for object recognition tasks. Public Safety Communications DHS-TR-PSC-11-01, U.S. Department of Homeland Security's Office for Interoperability and Compatibility (June 2011)
15. Watson, A.: Video acuity: A metric to quantify the effective performance of video systems. In: Imaging Systems Applications, p. IMD3. Optical Society of America (2011), http://www.opticsinfobase.org/abstract.cfm?URI=IS-2011-IMD3

Mobile Data Collection Networks for Wireless Sensors

Kai Li and Kien A. Hua

Department of Electrical Engineering and Computer Science, University of Central Florida
Orlando, Florida 32816, U.S.A.
kailee.cs@knights.ucf.edu, kienhua@eecs.ucf.edu

Abstract. Energy consumption is a major limitation of wireless sensor networks, with irreplaceable tiny sensors and multi-hop communication in sensor data aggregation as major contributors to this problem. While existing single hop data collection schemes reduce individual sensor energy consumption, they suffer from longer latency and low data delivery ratio. In face of these problems, we propose a *mobile data collection network* (MDCNet) as a new paradigm for wireless sensing applications. MDCNet is a fully self-deployed mesh network with virtual mesh nodes (mobile relay nodes), through which sensor data can be collected in a single hop and transmitted to the sinks. Our simulation results, based on NS-2, indicate that this new approach can achieve short latency and high data delivery ratio.

Keywords: Wireless sensor network, mobile relay node, mobile data collection network.

1 Introduction

Traditional wireless sensor networks (WSNs) are composed of densely-populated inexpensive tiny sensor nodes equipped with application-specific sensing units and communication components. The function of these sensors is to sample physical quantities (e.g. temperature, moisture, etc.) from the surroundings and route sensed data to a data processing center through base-stations (sinks). Since these tiny sensors have limited energy and they are often irreplaceable for many applications, energy consumption is of primary concern to ensure a long network operational time. Typically, data collection methods are based on sensors forming a connected network, through which data could be routed to the sink(s) through multiple hops. In this approach, a sensor is not only a data source, but also forwards data for other sensors. Such a data forwarding strategy may incur significant energy consumption. To reduce energy consumption and prolong the lifetime of WSNs, researchers have added mobility to WSN designs [7][8][9].

Mobility designs in WSNs falls into two major categories: **i) Using Mobile Sinks** [5][9][10]. Mobile sinks are used to replace static sinks as the endpoints of the data flow at the edge of the WSN. In general, sensors in the proximity of the stationary sinks will run out of energy faster because more data would pass through them to reach the sinks. This "energy hole problem" [6] results in premature cessation of net-

A. Dziech and A. Czyżewski (Eds.): MCSS 2012, CCIS 287, pp. 200–211, 2012.
© Springer-Verlag Berlin Heidelberg 2012

work operation. Mobile sinks address this problem by moving to different locations in the field to effectively balance the data-forwarding workload for their neighboring sensor nodes. This data collection paradigm can extend network lifetime for many applications. **ii) Using Messenger Nodes** [1][4][8]. Sensor nodes do not relay data in this approach. Instead, intermediate mobile nodes act as messengers between the sensors and the stationary sink. These messenger nodes move around in the sensor field to collect sensed data from the sensors and deliver the data to the sink in a manner similar to delay tolerant networks [12]. Since sensor nodes do not need to relay data, the Messenger Node approach also substantially reduces the energy consumption for sensor nodes to prolong their service time.

Although the two aforementioned data collection strategies alleviate the energy consumption problem, they have some limitations. Data delivery through messenger nodes may incur excessive data collection delay. Considering the limited buffer space in tiny sensors, this may result in data loss if some sensors have to wait for a long time before their next chance to transmit data. Although using mobile sinks does not have this issue, they are relatively more expensive mobile devices with direct access to Internet and sometimes powerful computing capability [11]. Deploying a large number of such mobile sinks to ensure short transmission delay for a large sensor field can be very expensive. This motivates us to consider a mobile data collection network (MDCNet) in this paper. An MDCNet consist of mobile relay nodes (MRN's) and one or more sink nodes. The MRNs form a network that allows electronic transmission of sensed data from the sensors to the sinks to minimize delay. MDCNet can be viewed as an advanced wireless mesh network [13] with the following two additional features: (1) unlike stationary mesh nodes which are deployed manually, our mobile relay nodes (MRNs) have the intelligence to survey the sensor field and automatically configure themselves into a mesh topology suitable for the given data collection task. (2) While mesh nodes are stationary, MRNs can move around in their designated region to achieve the effect of a virtual mesh node with a larger communication radius.

The advantage of the MDCNet approach is twofold. First, the degree of node mobility can be controlled (by deploying the proper number of MRN's) to achieve the desired tradeoff between transmission delay and overall system cost. This gives us a sensing system that features energy efficiency, shorter data collection latency, and higher data delivery ratio compared to existing mobile solutions. Second, since an MDCNet can be fully self-deployed, it is more suitable for inaccessible or hostile environments. In practice, the MRNs can be airborne (e.g., quadrocopter) or land-based (e.g., mini autonomous vehicle). This topic is beyond the scope of this paper. We simply refer to them as MRN's. To the best of our knowledge, MDCNet is the first data collection network designed for wireless sensing applications, and this paper is the first study of a fully self-deployed wireless mesh network with virtual mesh nodes.

The remainder of this paper is organized as follows. In Section 2, we briefly discuss some related work. Our MDCNet model is introduced and explained in detail in Section 3. We present our simulation results in Section 4. Finally, we conclude this paper in Section 5.

2   Related Work

Optimization of sink movement for energy minimization is first studied in [2], in which the authors used integer linear programming to determine the locations for each mobile sink to stop by periodically for data collection. Stefano et al. [5] addressed the same problem using a decentralized mixed linear programming model. Tang et al. [10] investigated this problem considering more practical situations when obstacles exist and mobile sinks can only move along certain paths in the field to collect data. The most recent work on network lifetime maximization using multiple cooperating mobile sinks is presented by Liang et al. [9]. He formulated the problem as optimizing h-hop-constrained multiple mobile sink movement and found the optimal trajectory for every sink using heuristics.

All of the abovementioned works formulates the data collection as an optimal path selection problem with different assumptions and constraints. A different method using intermediate mobile nodes is first proposed for WSNs by Shah et al. in [1]. He proposed to use randomly moving "Data Mules" for data gathering. Data Mules in the sensor field are used as forwarding agents. The idea is to save energy by using single hop routing (from the sensor to the "Mule") instead of more expensive multi-hop routing (from the sensor to the sink). The "Mule" eventually approaches the sink and delivers all collected data to the sink. In this architecture, energy is traded off for latency. Wang et al. [8] investigates the performance of a large dense network with one MRN and shows that the improvement in network lifetime over an all-static network is upper bounded by a factor of four. Communication involving relay nodes in those approaches only considers sensor-to-relay and relay-to-sink scenarios. Relay-to-relay communication, which could be potentially useful, is neglected or underutilized.

To the best of our knowledge, no work has considered the benefits of letting the data-collection-layer nodes communicate to form a network. One idea most similar to ours is presented in [3]. However, the relay nodes in [3] do not directly get data from sensors and they move to facilitate the communication between static data aggregation centers of different sensor clusters. We are dealing with a dynamic situation when all data collectors are moving and we jointly consider load assignment, data collection (from sensors), and data routing (among MRNs) problems, in addition to the initial fully automatic network deployment issues.

3   Proposed Models

Before presenting our data collection scheme, we would first like to specify the general assumptions about the WSN model we use. We are considering a WSN with sensors that have limited energy and buffer. The MRNs are rechargeable mobile devices with much larger buffer space. And they are also supposed to have much longer communication range than sensors, while not to the extent of accessing backbone network directly. Finally, we assume the MRNs are equipped with GPSs.

Based on those assumptions, we propose to use MDCNet to collect and route data for sensors. The MDCNet is a middle-layer network between the sensor network and the sink (see Fig. 1). It's composed of MRNs that move independently in the field to collect and forward data for a certain number of sensors. Sensors access the MDCNet to deliver their data by contacting MRN in a single hop. The data uploaded to a MRN will be routed through the MDCNet towards the static sink. The MDCNet is generally a partially and intermittently connected mobile ad hoc network. Data uploaded to a MRN will be temporarily buffered at each MRN on its path to the sink. MRNs only communicate with their neighboring peers when they need to send data.

In order to collect data effectively, the MDCNet has to satisfy several require-ments. Firstly, the number of sensors that every MRN serves should be balanced to average their utilization rate and reduce sensor contention. Secondly, most of the sensors' data should be collected in time to avoid data loss caused by sensor buffer overflows. Thirdly, a reliable data relay protocol among MRNs should be developed to make sure that data uploaded from sensors would arrive at the sink safely. To satis-fy the first requirements, we need to solve a load-balanced area partitioning problem, which constitute the first step of our data collection scheme. And we address the data collection and data transmission problems by developing corresponding communica-tion protocols.

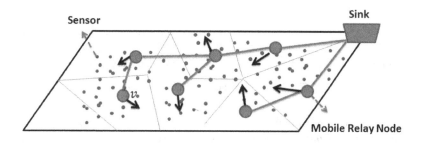

Fig. 1. Delay Tolerant Mobile Data Collection Network

3.1 Load Balanced Area Partitioning

Assuming different degree of global knowledge, we propose two algorithms to solve the load-balanced area partitioning problem.

A. Deterministic Area Partitioning

The deterministic area partitioning (DAP) solution is based on the assumption that the distribution of the sensors is known in advance and that MRN's initial position can be deployed by a central administrator. Consider, for example, the case when the sensors are evenly distributed over a square area. Load balancing could easily be achieved by equally partition the region and assign one MRN to each of the partitions (as illu-strated by Fig. 2(a)). In order for the sink in each square to provide full coverage of sensors in its assigned square, they are programmed to move in a snake-like manner as illustrated in Fig. 2. The distance between parallel paths are set to be $R/\sqrt{2}$, where R is the communication radius of the sensor. The $R/\sqrt{2}$ margin of the moving path from the border of the partition also makes sure that there is no overlapping of neighboring MRNs' services coverage (i.e. each sensor will only be able to communicate one MRN).

This simple solution to the area partitioning surely has some nice properties. It's easy to execute and would perfectly satisfy the load balance requirement. Latency is also controllable by varying the partition size. However, the assumption we made for the solution is not practical in most situations. The distribution of sensors is normally random and we generally do not know them in advance. Moreover, a centralized administration is not always feasible. Therefore, this scheme actually gives us an upper bound when all assumptions are valid or a lower bound otherwise.

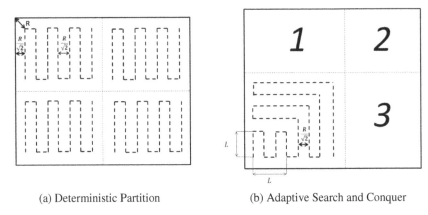

(a) Deterministic Partition   (b) Adaptive Search and Conquer

Fig. 2. Load Balanced Area Partitioning

B. Adaptive Search and Conquer

The *adaptive search and conquer* (ASC) strategy we present here assumes no prior knowledge of the sensor distribution and is totally distributed. We assume all the MRNs are located at the origin, say the bottom-left corner of the rectangular area at the beginning of network operation. MRNs incrementally enlarge their search space within a *target area* (TA) until they reach a predefined load factor (i.e. finding a certain number of sensors). Then they claim the area traversed so far as its *service area* (SA) and act as moving data collector for this SA. They will also notify other idle MRNs to search in the unexplored areas, if there are any. The search will not stop until the combination of all SAs cover the whole network area. After that, every MRN will move along certain paths within its SA to gather sensor data and relay them to the sink.

In detail, every MRN would set a random timer in the beginning. The MRN whose timer expires first would start out from the origin and take the whole rectangle as its TA to search for sensors. The searching is done by broadcasting *DISCOVER* messages periodically and keeping a counter of the number of different sensors that reply with *ACK* messages. The MRNs do exhaustive search in the following manner (see Fig. 2(b) and Fig. 3) in order to provide full coverage.

In the beginning, the moving path follows a snake-like pattern within an $L \times L$ square, where L is a preset parameter. If the MRN does not find enough sensors when it searched the initial square area, it begins to expand the searching square by $R/\sqrt{2}$ (recall that this step size is chosen to provide full coverage) every step and search along the border of the expanded square region. The reason for this searching strategy is that we want the conquered area for every MRN to approximate a compact square rather than a disproportioned long stripe.

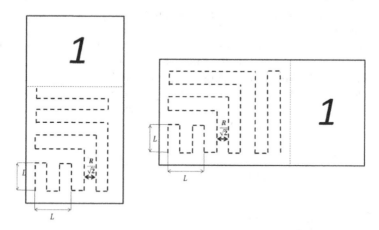

Fig. 3. Adaptive Area Partitioning Scenarios

Once the MRN has found a certain number of sensors, it will claim the rectangular area it has traversed so far, and at the same time, it will broadcast a *NOTICE* message to notify idle MRNs of the location information of the unexplored areas in its TA. The idle MRNs who hear the *NOTICE* message would make a list (in case they hear multiple messages) of the unconquered areas and set a random timer for the first item of its unconquered area list. If it doesn't hear any other MRNs set out before its timer expires, it will broadcast a *TIMEOUT* message and set this area as its TA and search for sensors in the same way as the first MRN. Other idle MRNs will cancel the timer they set for the first item in their unconquered area list, remove it from the list and set a new timer for the current first item in the updated unconquered area list, if it's not empty. Based on squareness of TA, every time a MRN claims a SA, its TA will be divided into 2 or 4 parts (see Fig. 3 for 2-part scenarios), and this partition process is done recursively and distributedly through communication among MRNs until the whole sensor field has been covered.

3.2   Data Collection Protocol

As soon as a MRN has conquered an area as a SA, it begins to move within its SA in a snake-like pattern (see Fig. 2(a)) to collect data from sensors it encounters along the way. Specifically, the MRNs broadcast *HELLO* messages periodically. Sensor that hears the *HELLO* message will reply an *ACK* message when it needs to upload its data. After receiving *ACK* messages from the sensor, the MRN will look up the service history record to make sure that sensors do not get repetitive service within a short interval. If the sensor satisfies the service requirement, the MRN will stop to receive data from the sensor until all its data has been uploaded.

3.3   Data Relay Protocol

The MRNs forms a delay tolerant ad hoc network through which gathered data will finally get to the sink. For the two different area partitioning schemes described earlier, the data relay hierarchy is derived in different manners.

For DAP, since the partition is pre-determined, the level of a MRN in the relay hierarchy is also pre-derived based on the partition it is assigned to. In detail, the parent of a MRN is set to the MRN that is closest to the sink among its eight-neighborhood. For example, consider the partition showed in Fig. 4(a), the corresponding relay hierarchy is showed in Fig. 4(b). During network initialization, the administrator would set the parent information for each MRN.

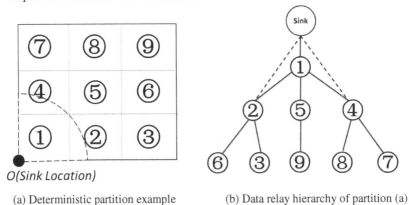

(a) Deterministic partition example   (b) Data relay hierarchy of partition (a)

Fig. 4. Deterministic Load-balanced Area Partition and Data Relay Hierarchy

While in the ASC strategy, the relay hierarchy is built automatically and dynamically when MRNs cooperatively communicate and search. The MRN that sets out first will set its parent to be the sink. Other MRNs will set their parent to be the MRN from whom they get their TA information. After the whole sensor field has been claimed, the hierarchical data flow structure will be completed. The data will flow in a bottom up fashion following the hierarchical structure. Consider the example shown in Fig. 5(a). MRN 1 is the first searching node. When it claims the left-bottom rectangle, it broadcasts a *NOTICE* message with the information of 3 unexplored areas. MRN 3, 4 and 2 won the competition of random timing, thus set their parent to MRN 1 and start their search in corresponding TAs. MRN 2, 3 and 4 further broadcast unexplored area information as they find enough sensors in their TA. In response to the new *NOTICE* messages, MRN 5 to 9 set out to search and finally claimed the remaining area. The corresponding data relay hierarchy is shown in Fig. 5(b).

In addition to the parent address information, each MRN will also keep a flag indicating its closeness to the sink. Specifically, if the part of a MRN's SA is within the communication radius of the sink, the flag will be set to true (corresponding to the dotted line in Fig. 4(b) and Fig. 5(b)), which indicates that this MRN will deliver its data directly to the sink instead of sending to its parent. However, if this MRN in no means could directly communicate with the sink, it will send *HELP* messages periodically to its parent node to seek for help. Upon receiving the *HELP* message, the parent node will reply a *READY* message as a signal of its readiness to accept data. Both the child and the parent will then stop to complete the data transmission until they finish.

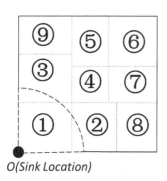
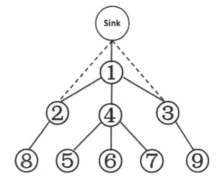

(a) Adaptive partition example   (b) Data relay hierarchy dynamically built for (a)

Fig. 5. Adaptive Load-balanced Area Partition and Data Relay Hierarchy

4   Experimental Evaluation

In this section, we will present the simulation results of our new data collection model. We implement the MDCNet in *ns-2.35* and carry out extensive experiments to evaluate the performance of our new model under various settings. Specifically, we set the size of the WSN to *100m* ×*100m*.The communication range of the sensors and the MRNs are set to *7m* and *40m* respectively. The moving speed of the MRNs is fixed to *2m/s*. The initial energy of sensors is set to 100J and energy consumption for transmitting and receiving data are both set to $10^{-3}J/bit$. Sensor generates a *10-bit* packet every *0.1* second and temporarily stores it in its *10KB* local buffer.

We evaluate the performance of our data collection strategy with two important WSN QoS metrics: data delivery ratio and latency. The metrics are acquired when the first sensor depletes its energy (generally considered in the community as the WSN lifetime). In the first set of simulations, we vary the number of uniformly distributed sensors N from *200* to *500* and compare the performance of deterministic and adaptive scheme. The load factor parameter L_f for adaptive MDCNet is set to *40* and we use *4* MRNs for the deterministic scheme. Each metric point was attained based on the average results of *20* simulation runs. The results of the two schemes along with the standard deviation are shown and compared in Fig. 6.

We note (from Fig. 6(a)) that in both schemes, there is an increase of data delivery ratio when we change N from *200* to *300*. This is because when N is below *300*, the number of sensors each MRN serves has not reached full load. Further increase the number of sensors will demonstrate the better load balancing characteristics of the adaptive scheme. In detail, when we further increase the number of sensors, the delivery ratio of deterministic scheme will drop more rapidly than the adaptive scheme. And the higher delivery ratio of the adaptive scheme can also be seen very clearly.

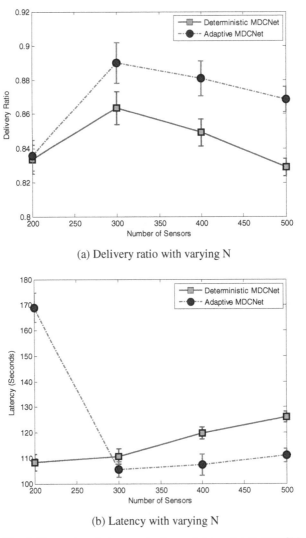

(a) Delivery ratio with varying N

(b) Latency with varying N

Fig. 6. Comparison between adaptive and deterministic MDCNet

The overload effect also accounts for the rapid increase in latency of the deterministic approach as shown in Fig 6(b). While in the adaptive scheme, since the load factor of each MRN is independent of the number of sensors, its actual load is less sensitive to the increase of sensor density. And the number of MRNs is determined dynamically by the total workload. The remarkably higher delay of the adaptive scheme when there are only *200* sensors is because the initial searching before MRNs begin to collect data takes a lot of time to finish (i.e. the initialization time is long) as a result of sparse sensor density. While the deterministic scheme does not have that initialization time since their SA is assigned in advance. As sensor density gets higher, the latency for the adaptive scheme decreases or changes very slowly, as demonstrates its advantages over the deterministic scheme in controlling the delay.

The second set of simulations is designed to evaluate the effect of load factor on the performance of the distributed MDCNet. We use a uniform grid-pattern distribution of sensors to rule out the effect of randomness. In this simulation, we fix the number of sensors to *300* and vary L_f (Load factor) from *20* to *100*. We evaluate the same metrics as we do in the first set of experiments. The results are shown in Fig. 7.

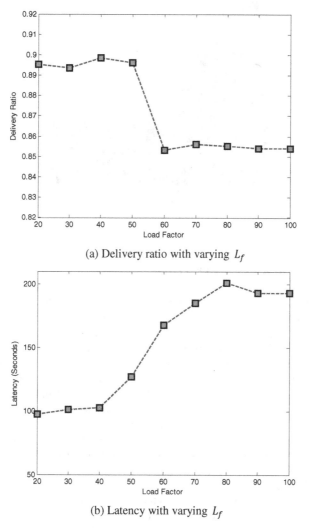

(a) Delivery ratio with varying L_f

(b) Latency with varying L_f

Fig. 7. Impact of load factor on delivery ratio, latency and number of MRNs

It can be noted that delivery ratio drops significantly when the load factor exceeds *50*. And the latency also begins to increase drastically around *50*. The behaviors of both metrics indicate that the maximum capacity of a MRN is around 50 sensors. Below the maximum capacity level, the latency and delivery ratio with respect to different load factor does not vary much. However, the comparable metrics are achieved at different costs (in terms of number of MRNs), which could be observed from Fig. 8. Considering all aspects, the optimal load factor in our settings should be between 40 and 50 and the choice should be a tradeoff between latency and cost.

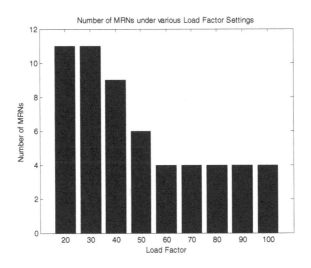

Fig. 8. Number of MRNs for different Load Factors

5   Conclusion and Future Work

In this paper, we propose a new paradigm—the MDCNet for effective data collection in WSNs. The major contribution of our work is twofold. Firstly, the new data collection paradigm shows promising QoS for WSNs in various ways: it saves sensor energy by limiting sensor's communication to single hop and achieves good latency and delivery ratio by letting the MRNs form a dynamic ad hoc network. Thus the new concept of data collection network will open a new window for further research on improving QoS for WSNs. Second, we jointly consider load balanced sensor assignment, effective data collection and reliable data transmission problems and give an implementation of the new data collection paradigm in *ns-2*.

On our basis, more advanced data collection solutions for the new paradigm can be motivated and developed under weaker assumptions. In particular, we are considering in our future work to further reduce assumptions about location-awareness (i.e. GPS devices) for MRNs. And the rectangularity of MRN's Service Area could also be generalized to irregular polygons such as the Voronoi diagrams considered in [14] More constraints (e.g. mobility of MRNs are constrained to certain paths due to the existence of obstacles) can also be considered to make the scheme more adaptive.

References

1. Shah, R., Roy, S., Jain, S., Brunette, W.: Data MULEs: Modeling a Three-tier Architecture for Sparse Sensor Networks. In: Proc. of the 1st IEEE SNPA (2003)
2. Gandham, S., Dawande, M., Prakash, R., Venkatesan, S.: Energy-efficient schemes for wireless sensor networks with multiple mobile base stations. In: Proceedings of IEEE GLOBECOM (December 2003)
3. Hou, Y.T., Shi, Y., Sherali, H.D., Midkiff, S.F.: Prolonging Sensor Network Lifetime with Energy Provisioning and Relay Node Placement. In: SECON 2005, pp. 295–304 (2005)
4. Jea, D., Somasundara, A., Srivastava, M.: Multiple Controlled Mobile Elements (Data Mules) for Data Collection in Sensor Networks. In: Proc. of the 1st IEEE/ACM DCOSS (2005)
5. Basagni, S., Carosi, A., Melachrinoudis, E., Petrioli, C., Wang, Z.M.: Controlled Sink Mobility for Prolonging Wireless Sensor Networks Lifetime. Wireless Networks 14(6), 831–858 (2007)
6. Li, J., Mohapatra: Analytical modeling and mitigation techniques for the energy hole problem in sensor networks. Pervasive Mobile Computing 3(3), 233–254 (2007)
7. Shi, Y., Hou, Y.T.: Theoretical results on base station movement problem for sensor network. In: IEEE INFOCOM (2008)
8. Wang, W., Srinivasan, V., Chua, K.C.: Extending the Lifetime of Wireless Sensor Networks through Mobile Relays. IEEE/ACM Transactions on Networking, 1108–1120 (2008)
9. Weifa, L., Jun, L.: Network lifetime maximization in sensor networks with multiple mobile sinks. In: IEEE Conference on Local Computer Networks, LCN (2011)
10. Tang, S., Yuan, J., Li, X., Liu, Y., Chen, G., Gu, M., Zhao, J., Dai, G.: DAWN: Energy Efficient Data Aggregation in WSN with Mobile Sinks. In: IEEE International Workshop on Quality of Service (2010)
11. Di Francesco, M., Das, S., Anastasi, G.: Data Collection in Wireless Sensor Networks with Mobile Elements: A Survey. ACM Transactions on Sensor Networks (2011)
12. Fall, K.: A delay-tolerant network architecture for challenged internets. In: Proceedings of the 2003 Conference on Applications, Technologies, Architectures, and Protocols for Computer Communications (SIGCOMM), pp. 27–34 (2003)
13. Hua, K.A., Xie, F.: A Dynamic Stream Merging Technique for Video-on-Demand Services over Wireless Mesh Access Networks. In: Proceedings of IEEE Conference on Sensor, Mesh and Ad Hoc Communications and Networks, SECON 2010 (2010)
14. Braun, T., Coulson, G., Staub, T.: Towards virtual mobility support in a federated testbed for wireless sensor networks. In: Proceedings of the Sixth Workshop on Wireless and Mobile Ad Hoc Networks, Kiel, Germany, March 10-11 (2011)

Intelligent Identification of Dangerous Situations Based on Fuzzy Classifiers

Aleksandra Maksimova

Institute of Applied Mathematics and Mechanics,
National Academy of Science of Ukraine, Donetsk, Ukraine
maximova.alexandra@mail.ru

Abstract. This paper deals with a method of dangerous situations identification based on fuzzy classifiers composition. In this article a formal hierarchical model of the situation presentation is offered. The security problem in a bank is given as an example. Dangerous situations can be recognized by analysis of pictures from security cameras. This applied problem is reduced to the problem of pattern recognition that can be efficiently solved by fuzzy classifiers. It is suggested to use conception of fuzzy portraits of pattern classes for creating inference rules. The scheme of fuzzy classifiers connection that allows to estimate information from security cameras is presented. The advantage of such method is the possibility of linguistic interpretation of results. The prototype of intelligent information systems for identification of dangerous situations is developed on the basis of suggested approach.

Keywords: pattern recognition, fuzzy classifier, fuzzy inference, data analysis.

1 Introduction

Information technologies are widely used in security and intelligent monitoring systems. In practice they are realized by combination of artificial and computational intelligent methods. Data presentation models determine methodology of such system realization. That is why the models of data presentation are important for information system creation. The author suggests the method of dangerous situation identification based on fuzzy classifiers where fuzzy portraits of pattern classes are used for data presentation.

Fuzzy classifiers are up-to-date trend of fuzzy inference systems for pattern recognition. Fuzzy inference systems have been proposed and practically applied to control problems of difficult formalizable processes. The fuzzy sets theory is the basis of fuzzy control but this theory was further developed in pattern recognition problems. [1].

The problem of dangerous situation or an object in the image identification in the common form can be reduced to the problem of pattern recognition [2, 3]. The use of fuzzy sets theory for intelligent monitoring has some unambiguous advantages. Fuzzy logic allows to create models based on human reasoning and avoids increasing complexity of the system. The fuzzy approach is very effective for multimedia data analysis. Methods of object identification on the picture are distinguished by great percent of false alarms. The quality rate can be estimated more effectively by multi-valued truth-space that is used in fuzzy logic theory [4].

A. Dziech and A. Czyżewski (Eds.): MCSS 2012, CCIS 287, pp. 212–219, 2012.

2   The Problem of Bank Security System Organization

Video surveillance systems to monitor situation on the objects of observation allow in on-line mode and are used in Bank Security Systems, for example. One of the subtasks of the system is the identification of critical situations that require urgent intervention. There is a common situation center, which receives video streams from all cameras installed in hazardous areas. An operator carries out the analysis of information delivering from scores of monitors. Depending on an organization of the situational center the number of monitors per operator varies. The automatic analysis of such videostreams is an actual problem of Bank Security System.

2.1   The General Formulation of the Problem

It is necessary to develop decision support system for monitoring and identification of dangerous situations. Let us consider the set of objects $X = \{X_i\}_{i=1}^n$, where n - is number of objects of observation. The object is defined as any room with observation cameras : $V_i = \{V_j\}_{j=1}^{k_i}$, where k_i - is number of cameras that are placed in X_i observation object.

Let us mark out three categories of situation for an object in a bank: «Dangerous», "Quite"   and   "Intensive".   Let   us   denote   the   linguistic   variable $L^* = \{"Situation", T, [0,100]\}$, $T \in \{"Damage", "Intensive", "Quiet"\}$ describing the situation.

Let us determine the local estimation of situation $S(V_{ij})$ called a fragment of situation, which can be received on the basis of analysis data from camera V_j. The situation on the object X_i is estimated on the analysis of situation fragments by $\{S(V_{ij}) \mid j = 1, k_i\}$.

The functionality of the system can be increased by observing the object "money collector", for example. The configuration of cameras location influences the quality of the system. In general, there are other types of sensors that can be used. For example, the microphones can transmit sounds from objects that can be analyzed.

2.2   The Model of Data Presentation

Let us present hierarchical model of data presentation by local situation fragments. For this purpose we carry out the decomposition of situation for bank security problem. Let us denote every local situation fragment as $\Phi_i = \{M_1, M_2, M_3\}$, where M_1 is internal-model of situation presentation, M_2 is external-model of situation presentation, M_3 is interpretation model and i is local fragment identifier.

The external model M_1 is a priori information about situation that is presented as a set of linguistic variables.

The internal model M_2 is used to describe current value of situation Φ_i in fuzzy set form.

The interpretation model M_3 defines the mechanism of transition from internal models of the situations fragments of lower level to internal model of current fragment.

The whole hierarchical model of data presentation is shown on Fig. 1. that is constructed as a result of empirical data analysis.

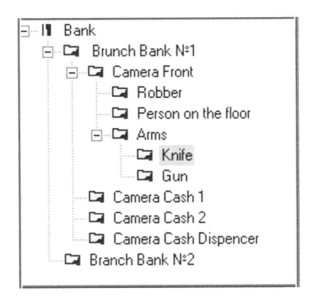

Fig. 1. The tree of the model of data presentation by local situation fragments for bank security problem

Each fragment of the situation for cameras V_{11}, V_{12} etc. has a complex structure $\Phi_{V_{11}} = \{\Phi_{V_{11\_1}}, ..., \Phi_{V_{11\_k_1}}\}$, $\Phi_{V_{12}} = \{\Phi_{V_{12\_1}}, ..., \Phi_{V_{12\_k_2}}\}$ etc. The local situation fragment for camera is estimated by the set of fragments of lower level: { "Man in the mask", "Weapon in the hand", "Man on the floor" }. Thus, to determine dangerous situation we have to determine dangerous objects on the lower level of the situation tree. The situation "Weapon in the hand" can be further decomposed like "Gun in the hand" and "Knife in the hand" situations. If we don't have any dangerous objects the situation is quite. If dangerous objects are detected with small degree of certainty the situation is intensive.

Let us consider two types of local situation fragments:

— *1 type*: the information is given from sensors;
— *2 type*: the information about situation is formed on the basis of local situation fragments from lower level..

The problem of local fragment identification at cameras level is solved in three stages:

1. Search of important object that can be indicated as dangerous.
2. Determine the semantic description of the scene on the basis of given objects.
3. Define the categories of local situation fragment.

On the first stage the method of pattern recognition ise used to detect unusual situation like "Man in the mask", "Knife in the hand", "Man with the gun" etc. As an example, they are considered in [5,6]. The results of this first stage algorithm is the semantic description of the scene in the form of information attributes vector. On the third stage the pattern recognition problem is solved as a problem in m-dimensional space. The formal mathematical statement for such problem is published in [7,8,9].

3   The General Scheme of Dangerous Situation Identification Method

This part of the paper deals with the principle of fuzzy classifier work. The scheme fuzzy classifiers communication is presented in the hierarchical structure of data presentation. For this reason it is proposed to calculate local situation fragment in the linguistic form. Every local situation fragment is described by the elementary fuzzy classifier.

The fuzzy classifiers use IF-THEN rules of inference with linguistic variables to solve the pattern recognition problem. The classifiers rules have linguistic variables as antecedents and labels of pattern classes as consequents, which can be represented as fuzzy sets or exact values. There are some rules of fuzzy inference for each class of images.

Let us consider the elementary classifier. The scheme is shown on the Fig. 2. The fuzzy classifier consists of knowledge base, data base and fuzzy inference method. The knowledge base is a set of fuzzy rules. The data base is a set of linguistic variables. The decision-making is carried out using knowledge base with selected inference method.

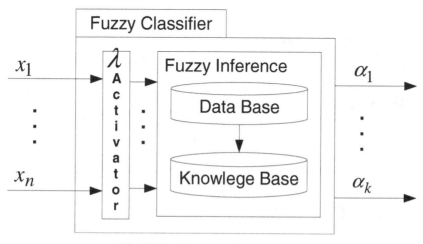

Fig. 2. The elementary fuzzy classifier

The input vector $\bar{x} = (x_1,...,x_n)$ comes to fuzzy classifier from sensors or from fuzzy classifiers of lower level. λ-activator carries out check input parameters and fuzzy classifier activation by the value of threshold λ. The fuzzy information vector $\tilde{\alpha} = (\alpha_1,...,\alpha_k)$, where k - is the number of classes of images and α_i - is degree of correspondence $\bar{x}$ to pattern class V_i, is received on the output.

The fuzzy rule is presented in the:

IF « x_1 is V_i » AND ... AND « x_n is V_i » THEN « V_i »| $\tilde{\alpha}_i^j = f(\bar{\mu})$,

where $\bar{\mu} = (\mu_1^i(x_1),...,\mu_m^i(x_m))$, f - is function that calculates the degree of correspondence given sample $\bar{x} = (x_1,...,x_m)$ to pattern class V_i, $\tilde{\alpha}_i^j$ -the degree of correspondence $\bar{x}$ to pattern class V_i by given rule, $\alpha_i = \max_{\forall j}(\tilde{\alpha}_i^j)$.

The linguistic variables that are used in IF-part of fuzzy rule corresponds to inputs of classifier. The elementary classifier is linear. It means that every fuzzy rule launches only ones.

3.1   The Scheme of Elementary Fuzzy Classifiers Cooperation

Each elementary fuzzy classifier inserts his match into the tree node. There is a set of information attributes $\bar{x}$ for each elementary fuzzy classifier. If $P = \{R_1,...,R_n\}$ - is a set of attributes names for whole problem and $\bar{c} = \{1,0\}^m$ is binary vector then $P_{pr}^{\bar{c}} = \{R_k \,|\, c_k = 1\}$ - is a set of attributes for current fuzzy classifier. Parameter $\bar{c}$, set of pattern classes V_{FC}, fuzzy information vector $\tilde{\alpha}$ for elementary fuzzy classifier are denoted.

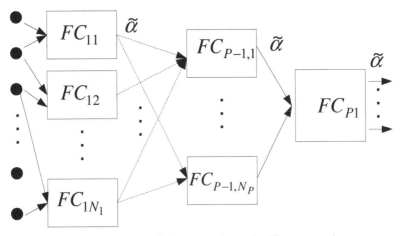

Fig. 3. The scheme of elementary fuzzy classifier cooperation

λ -activator for the second type of nodes activates the fuzzy classifier if input values are more than threshold λ. The hierarchical structure of fuzzy classifiers cooperation is created for pattern recognition problems with hierarchical model of data presentation. This approach allows to avoid conflict resolution because of linearity of elementary fuzzy classifiers and their structural configuration. Taking into account the fact, that classes of images can have intersections, this property is very important. The fuzzy inference mechanism is carried out from the leaves of the tree. Thus, more important conclusions on the basis of sensors data analysis are made.

3.2   XML-Format for Elementary Fuzzy Classifier Description

Fuzzy classifier description can be easily transformed into XML-format. Such a format has a lot of advantages especially for presentation of complex data structures. Here is an example how a part of linguistic variable "Knife" is described by xml-structure.

```xml
<?xml version="1.0" encoding="UTF-8" ?>
-<root>
-<variables>
<lingvo>
  <name>Knife</name>
  <universum>
    <begin>0</begin>
    <end>100</end> <step>1</step> <name>Percents</name>
    <type>2</type>
  </universum>
  <terms><term>
      <name>Knife is present</name>
      <function>
        <type>3</type><A>70</A><B>100</B>
      </function>
        ...
      </term>
    ...
  </terms>
</lingvo>
...
-<variables>
-</root>
```

4   The Knowledge Base Creation on the Basis of Fuzzy Portraits

In papers [8-10] the authors suggest the method of fuzzy portraits creation which are described by fuzzy rules and in fact are basic for fuzzy classifier.

The fuzzy portraits creation method consists of intra-class structure analysis and membership functions creation for linguistic variables on the basis of frequency analysis[10].

Let the fuzzy portrait S_j of class of images V_j be the fuzzy area in m-dimensional space that is described by the three objects $(L, Rule, F)$ where L is the set of linguistic variables $\{L_q \mid q = 1, m\}$, m is the number of attributes, $T_q = \{\mu_q^i(x) \mid i = 1, l\}$ is the set of terms, where l corresponds to the number of clusters R_j to current class V_j, R is the set of rules and $F = \{F_{AND}, F_{OR}\}$ are the pair of operations AND and OR correspondingly:

$$(\{L_q \mid q = 1,...,m\}, \{Rule_q \mid q = 1,...,l\}, \{F_{AND}, F_{OR}\}).$$

The inference rule $Rule_q$ is the following:

$$IF\ \mu_q^1(x)\ is\ A_q\ AND\ ...AND\ \mu_q^m(x)\ is\ A_q\ THAN\ V_j\ ,$$

where A_q is defined by $R(V_j)$ cluster. The number of such inference rules corresponds to the number of clusters detected by the cluster analysis procedure and can be different from different classes of images.

5   Conclusion

The prototype of intelligent information system for dangerous situation identification is developed on the basis of given approach. As an application of this method   the problem of dangerous situations recognition in the system of monitoring and bank security has been considered.

The developed prototype of system has been tested on the problem of weapon- in-hand identification from the video stream of monitoring camera.

There are some advantages for this approach. This system can be easily expanded by adding new nodes in the model of data presentation. The complexity of the system is only $O(n \cdot m)$, where n - is the number of fuzzy classifiers and m - is maximal number of rules in knowledge bases of classifiers. The linguistic interpretation of results can be carried out at each level of hierarchical model.

References

1. Ishibuchi, H., Nakashima, T., Nii, M.: Classification and modeling with linguistic information granules: Advanced approaches to linguistic data mining. Springer, Heidelberg (2004)
2. Bezdek, J.C., Keller, J.M., Krishnapuram, R., Pal, N.R.: Fuzzy Models and Algorithms for Pattern Recognition and Image Processing. Springer, New York (2005)

3. Zhuravlev, Y.I.: An algebraic approach to the solution of pattern recognition and identification problems. Probl. Kibernet. 33 (1978) (in Russian)
4. Konor, A.: Computational Intelligence: Principles, Techniques and Applications. Springer, Heidelberg (2005)
5. Żywicki, M., Matiolański, A., Orzechowski, T.M., Dziech, A.: Knife detection as a subset of object detection approach based on Haar cascades. In: Proceedings of 11th International Conference Pattern Recognition and Information Processing, Minsk, pp. 139–142 (2011)
6. Łukańko, T., Orzechowski, T.M., Dziech, A., Wassermann, J.: Testing fusion of LDA and PCA algorithms for face recognition with images preprocessed with Two-Dimensional Discrete Cosine Transform. In: Proceedings of the 15th WSEAS International Conference on Computers (2011)
7. Kozlovskii, V.A., Yu, A.: Maksimova Algorithm of Pattern Recognition with intra-class clustering. In: Proceedings of 11th International Conference Pattern Recognition and Information Processing, Minsk, pp. 54–57 (2011)
8. Kozlovskii, V.A., Maksimova, A.Y.: Solution of pattern recognition problem using fuzzy portraits of classes. Artificial Intelligence 4, 221–228 (2010) (in Russian)
9. Kozlovskii, V.A., Maksimova, A.Y.: Fuzzy system of pattern recognition for liquid petrochemical products classification. The Scientific Works of the DNTU: Series Informatics, Cybernetics and Computer Engineering 13(185), 200–205 (2011) (in Russian)
10. Kozlovskii, V.A., Maksimova, A.Y.: Pattern Recognition Algorithm based on fuzzy approach. Artifical Intelligentce 4, 594–599 (2008) (in Russian)

Face Recognition from Low Resolution Images

Tomasz Marciniak, Adam Dabrowski,
Agata Chmielewska, and Radosław Weychan

Poznań University of Technology, Chair of Control and System Engineering,
Division of Signal Processing and Electronic Systems,
ul. Piotrowo 3a, 60-965 Poznań, Poland
{tomasz.marciniak,adam.dabrowski,agata.chmielewska,
radoslaw.weychan}@put.poznan.pl

Abstract. This paper describes an analysis of the real-time system for face recognition from video monitoring images. First, we briefly describe main features of the standards for biometric face images. Available scientific databases have been checked for compliance with these biometric standards. Next, we concentrate on the analysis of the prepared face recognition application based on the eigenface approach. Finally, results of our face recognition experiments with images of reduced resolution are presented. It turned out that the proposed and tested algorithm is quite resistant to changing the resolution. The recognition results are acceptable even for low-resolution images (16×20 pixels).

Keywords: face recognition, biometric standards, low resolution images.

1 Introduction

Nowadays more and more automatic access systems are based on biometric techniques. Face recognition systems [1, 2] are characterized by low invasiveness of acquisition, and increasingly better reliability. The main problem that occurred in such systems is low resolution of details in the case of photos taken from long distances. The identification and recognition from this kind of pictures became an important scientific issue to solve [3, 4].

The authors of this paper have prepared a GUI (graphical user interface) application for experimental testing of face detection and recognition effectiveness. The Matlab environment was used, because of its widespread popularity and possibility for simulation of experiments. The developed application enables people identification using IP wireless camera and batch processing of databases with various quality of facial images. The application uses face recognition based on the eigenfaces approach.

The paper is organized as follows: Section 2 presents the regulations of biometric standards related to face identification, in Section 3 we shortly describe and compare face databases from various universities. Section 4 shows elements of our face recognition system. An influence of image resolution on the effectiveness of the

A. Dziech and A. Czyżewski (Eds.): MCSS 2012, CCIS 287, pp. 220–229, 2012.
© Springer-Verlag Berlin Heidelberg 2012

recognition process is discussed in Section 5 in a form of the FAR (*false acceptance rate*) / FRR (*false rejection rate*) graphs, while section 6 summarizes our work.

2   Biometric Face Recognition Standards Relevant to CCTV

Biometric standards describe general rules, directives, and features concerning biometric input data, such as e.g. face images.

Data interchange formats are one of four main kinds of the biometric standards and they specify contents and formats presentation for the exchange of biometric data [5]. The data presented below are based on two international standards: ISO/IEC 19794-5 (*Biometric data interchange formats - Part 5: Face image data,* 2005) [6] and ANSI/INCITS 385-2004 (*Face Recognition Format for Data Interchange,* 2004) [7].

Figure 1 shows an illustrative example of the face position in an image, according to these standards. It contains distances (in pixels) for a picture made with 320×240

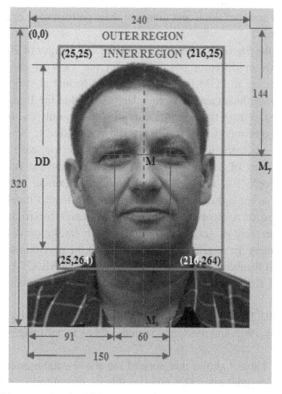

Fig. 1. Distances (in pixels) between main points in a front face picture

resolution. The most interesting regions have been separated (the inner region and the outer region). Dashed line M_X approximates horizontal midpoints of the mouth and of the bridge of the nose, M_Y line defines the line through the centre of the left eye and the center of the right eye.

At the intersection of these lines an M point is placed and it defines the center of the face. The x-coordinate M_X of M should be between 45 % and 55 % of the image width and the y-coordinate M_Y of M should be between 30 % and 50 % of the image height. The width of head should be between 50% and 75% of the image width, and the length of head – between 60% and 90% of the image height. Rotation of the head should be less than ±5° from frontal in every direction – roll, pitch, and yaw. This standard includes also a width-to-height ratio of the image, which should be between 1.25 to 1.34.

3   Scientific Databases of Faces in Relation to Biometric Standards

There are several face image databases prepared by academic institutions. Below we shortly described some of them and compare their performance against the standards for facial biometrics.

A database of the Sheffield University contains 564 images of 20 persons [8]. Each of them is shown in different positions, from the view of the profile to frontal position. The files are in PGM (*portable graymap*) format in various resolutions and 256-bit grayscale. A disadvantage of this database is that the frontal face images are not clearly separated from the others. From all pictures, only the frontal photos were selected for our experiments. The results are shown in Table 1.

One of the best-constructed database in terms of the use, ease of processing, and sorting files in terms of features is the "Yale Face Database B". It includes 5760 images of 10 people [9]. Each of the photographed person has 576 images in 9 positions and different lighting conditions. Every file in this database can be easily separated, because of a clear description of file names for frontal photos and the others. The pictures are of good quality and in high resolution (640×480 pixels) but still do not fully meet the requirements mentioned in the previous section (Table 1).

The next tested database for biometric standards is the MUCT Face Database from the University of Cape Town [10]. It consists of 3755 faces. All images are in the frontal position.

The Color FERET Database [11] from the George Mason University was collected between the years 1993 and 1996 and includes 2413 facial images, representing 856 individuals.

Summarizing, Table 1 shows that none of the above databases entirely respects the required biometric standards. The main problems in conforming the standards for these databases are: wrong proportions of image dimensions and too long distance of a person being photographed to the lens (particularly for older databases).

Table 1. Comparison of face databases in relation to the biometric standards

	Database name	Biometric norm	Sheffield	Yale Face Database B	MUCT	FERET
Information	Number of individuals	-	20	10	624	856
	Total number of files	-	564	5760	3755	2413
	Tested file	-	„1i012-.pgm"	„yaleB10_P00A+005E-10.pgm"	"i025ra-mn.jpg"	"00068_931230fb.ppm"
Results	The ratio of height to width of the image	1,25-1,34	**1,09**	**0,75**	1,31	**1,48**
	The rotation of the head	Smaller than 5°	0°	0°	2°	0°
	M_X [%]	45-55	52	53	51	50
	M_Y [%]	30-50	36	49	42	30
	The ratio of head width to image width	0,5-0,75	0,58	**0,39**	**0,47**	**0,34**
	The ratio of head height to image height	0,6-0,9	0,7	0,78	**0,53**	**0,33**
	A statement: whether the base complies with the biometric standards	-	No. The bad aspect ratio of picture	No. The picture is taken horizontally, so the ratio of the width of the head to the width of the image is too small.	No. The face is too far from the lens.	No. The face is too far from the lens and the bad aspect ratio of picture.

4   Elements of Face Recognition

4.1   Face Detection

Face detection in a picture is the preliminary step for face recognition. By detection we understand locating the face in the image or determination of its position.

There are many programs for face detection. An example of web application is *faint (The Face Annotation Interface)* [12]. Figure 2 shows an example of face detection realized with the use of this program. It can be noted that almost all faces were marked correctly.

Typically, the detection is realized in three stages. The first step is a reduction of impact of interfering factors with the use of histogram equalization and noise reduction. The next step is determination of areas with high probability where a face can be placed. The final stage is verification of the previously selected areas. Finally, the face is detected and marked [13].

There are several common approaches for face detection [2]. The first one consists in a face location based on the color of human skin. The human skin color is different in terms of lighting intensity (luminance) but has the same chroma. Thanks to that other elements in the image, which do not correspond to skin, can be effectively removed. Then, using mathematical morphology operations in the selected ROI (*Region of Interest*) further features can be isolated, which indicate the presence of a face in the picture [14].

Fig. 2. Examples of face detection with the *faint* program

Next technique is location of a face with the use of geometric models. This method is based on comparing the location of the selected models of the test face with the processed image. An advantage of such detection is an opportunity of working with static images in grayscale. This method is based on the knowledge of geometry of a typical human face and on dependencies between them – position, distance, etc.. This method is based on the use of the *Hausdorff distance* [15, 16]. The technique does not deal with changes in orientation or position of the face.

Face detection can be based also on the analysis of facial expressions (move of facial muscles and blinking). This method uses edge filters and morphological operations. Next, the statistical models were constructed to describe dependencies and then to compare the model with the face. Other moving objects in the background can quite seldom cause false recognitions.

4.2   Face Recognition

Initially, an experimental software *Face Recognition System 2.1* [17] was used. The program uses an algorithm based on *eigenfaces* [18, 19]. Face recognition gives a result based on a distance from the nearest class, according to the numbering assigned at the beginning of the individual photographs (indicating a person in the class). The program also provides an opportunity to work only with a photo taken earlier. Thus, there is no possibility for operation in real time. The software has an ability to add images to the database or facial recognition only after you select an image. In this case, we have to add images individually, manually entering the number of the class pictures.

The *Face Recognition System 2.1* does not meet the authors' needs, thus, we have decided to introduce the following modifications [20]:

- Batch processing; in separate folders we place a set of test images from the face database; the program checks the number of images in each folder, prepares a list, and then successively adds them to the program database. Each folder is assigned to another class (a person).
- Acquisition of the image from an IP wireless webcam (e.g. D-Link DSC-930L [21]) or a standard USB camera.
- Automatic color space conversion in case of the YUV standard cameras.
- Face detection in an image and also the background and noise removing; a skin color filter was used; we could not let the direct use of ROI, because images entering the base must be of the same size. For this purpose, facial images were scaled to the resolution of 100×120 pixels.

The application is equipped with the *graphical user interface* (GUI) as shown in Fig. 3. The program allows to work in two modes: real-time continuous mode and batch processing of images from the database.

Fig. 3. Face recognition program interface (Matlab environment) with example from *Yale Face Database B*

5   Influence of Image Resolution

In the presented experiment the *Yale Face Database B* has been used to distinguish the face positions. This database includes front rotated (pitch and yaw) face pictures of 10 individuals in various light conditions. For the tests only the front images have been chosen. The original resolution of 640×480 pixels has been downsampled 2, 4, 8 and even 16 times.

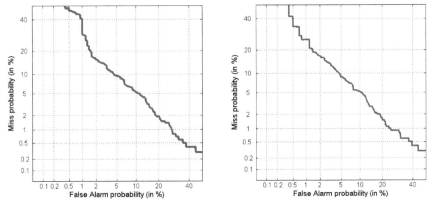

Fig. 4. FAR/FRR plot for 640×480 resolution   **Fig. 5.** FAR/FRR plot for 320×240 resolution

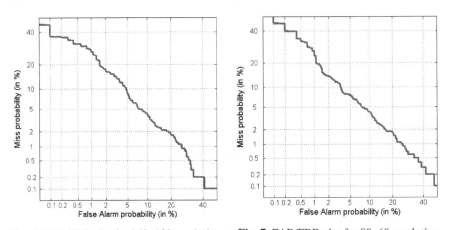

Fig. 6. FAR/FRR plot for 160×120 resolution   **Fig. 7.** FAR/FRR plot for 80×60 resolution

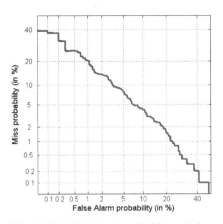

Fig. 8. FAR/FRR plot for 40×30 resolution

As reference images for the experiment, one front face picture of every 10th person (in the full light) has been chosen. Another 63 face images from the database were used for testing.

Figures 4-8 shows FAR / FRR plots for resolutions from 640×480 down to 40×30. As it can be seen, the computed curves are very similar to each other. Even a decimation by 16 times does not influence the recognition accuracy as much as for visual perception. EER (*equal error rate*) is in every case in the range of 6–8%. Differences are visible at the ends of intervals. Presented results show, that the front images with no rotated faces (which is the main requirement of the norms) can be very well distinguished between each other by using our approach.

To make this interesting results more visible, we have tried to use other databases with color images and greater number of persons: the MUCT database and the Color FERET database. However, the first of them does not meet all standards for face positions. The second database is very well structured but it has only two front pictures of every person, which is unacceptable in our experiment.

6   Conclusions

The proposed system for face recognition has an important advantage: the modified algorithm is prepared to be only weakly sensitive to low resolution images. Thus, correct face recognitions are achieved even for webcam images (included in the database). In result, the prepared algorithm may find widespread use in the video analysis in CCTV (*closed-circuit television*). Plots described in section 5 indicate a correct face recognition even if it has resolution of 16×20 pixels. It means, that persons can be recognized from a large distance (of several meters) by using basic monitoring systems. A confident image acquisition of the frontal face position can additionally be realized by placing the camera e.g. on the top of straight stairs.

The achieved EER in every case (even low resolution) was between 6–8%. An additional advantage is that the usual proportions of the training data to the testing data in typical biometric studies are 0.33. In our experiments this ratio was only 0.01.

Our analysis shows, that available face databases do not meet biometric standards. Among incorrect database features are: bad picture aspect ratios, too far distances between face and camera lens, and lack of proper labeling of file names of frontal photos and photos taken at an angle.

Acknowledgment. This paper was prepared within the INDECT project and partly with the DS funds.

References

[1] Davis, M., Popov, S., Surlea, C.: Real-Time Face Recognition from Surveillance Video. In: Zhang, J., Shao, L., Zhang, L., Jones, G.A. (eds.) Intelligent Video Event Analysis and Understanding. SCI, vol. 332, pp. 155–194. Springer, Heidelberg (2011)

[2] Marciniak, T., Drgas, S., Cetnarowicz, D.: Fast Face Localisation Using AdaBoost Algorithm and Identification with Matrix Decomposition Methods. In: Dziech, A., Czyżewski, A. (eds.) MCSS 2011. CCIS, vol. 149, pp. 242–250. Springer, Heidelberg (2011)

[3] Xu, Y., Jin, Z.: Down-sampling face images and low-resolution face recognition. In: The 3rd International Conference on Innovative Computing Information and Control, p. 392 (2008)

[4] Zou, W.W., Yuen, P.C.: Very Low Resolution Face Recognition Problem. IEEE Transactions on Image Processing 21(1), 327–340 (2012)

[5] Biometrics Technology Introduction, http://www.biometrics.gov/documents/biointro.pdf

[6] ISO/IEC 19794-5:2005, Information technology – Biometric data interchange formats – Part 5: Face image data (2005)

[7] ANSI/INCITS 385-2004, Information technology – Face Recognition Format for Data Interchange (2004)

[8] Database of the Sheffield University, http://www.sheffield.ac.uk/eee/research/iel/research/face

[9] Georghiades, A.S., Belhumeur, P.N., Kriegman, D.J.: From Few to Many: Illumination Cone Models for Face Recognition under Variable Lighting and Pose. IEEE Trans. Pattern Anal. Mach. Intelligence 23(6), 643–660 (2001)

[10] Milborrow, S., Morkel, J., Nicolls, F.: The MUCT Landmarked Face Database. Pattern Recognition Association of South Africa 2010 (2010), http://www.milbo.org/muct/

[11] Portion of the research in this paper use the FERET database of facial images collected under the FERET program, sponsored by the DOD Counterdrug Technology Development Program Office; Phillips, P.J., et al.: The FERET Evaluation Methodology for Face Recognition Algorithms. IEEE Trans. Pattern Analysis and Machine Intelligence 22, 1090–1104 (2000)

[12] The Face Annotation Interface, http://faint.sourceforge.net/

[13] Frischholz, R. W.: The Face Detection Homepage, http://www.facedetection.com/

[14] Kapur, J.P.: Face Detection in Color Images. EE499 Capstone Design Project, University of Washington Department of Electrical Engineering (1997), http://www.oocities.org/jaykapur/face.html

[15] Jesorsky, O., Kirchberg, K.J., Frischholz, R.W.: Robust Face Detection Using the Hausdorff Distance. BioID AG, Berlin, Germany (2001)

[16] Nilsson, M.: Face Detection algorithm for Matlab. Blekinge Institute of Technology School of Engineering Department of Signal Processing, Ronneby, Sweden (2006), http://www.mathworks.com/matlabcentral/fileexchange/13701-face-detection-in-matlab

[17] Rosa, L.: Face Recognition System 2.1 (2006), http://www.advancedsourcecode.com//face.asp

[18] Agarwal, M., et al.: Face Recognition using Principle Component Analysis, Eigenface and Neural Network. In: Conference on Signal Sensors, Sensors, Visualization, Imaging, Simulation and Materials (2010)

[19] Belhumeur, P.N., Hespanha, J.P., Kriegman, D.J.: Eigenfaces vs. Fisherfaces: Recognition Using Class Specific Linear Projection (1997)

[20] Rzepecki, S.: Real-time localization and identification of faces in video sequences, (M.Sc. Thesis), Supervisor: Marciniak, T., Poznan University of Technology (2011)

[21] Description of D-Link camera (2011), http://mydlink.dlink.com/products/DCS-930L, Data sheet, ftp://ftp10.dlink.com/pdfs/products/DCS-930L/DCS-930L_ds.pdf

Asynchronous Physical Unclonable Functions – AsyncPUF

Julian Murphy

Centre for Secure Information Technologies,
Queens University Belfast,
Belfast, BT3 9DT
United Kingdom
j.p.murphy@qub.ac.uk

Abstract. Physically Unclonable Functions (PUFs) exploit the physical characteristics of silicon and provide an alternative to storing digital encryption keys in non-volatile memory. A PUF maps a unique set of digital inputs to a corresponding set of digital outputs. In this paper, the use of asynchronous logic and design techniques to implement PUFs is advocated for Asynchronous Physically Unclonable Functions (APUFs). A new method of using asynchronous rings to implement PUFs is described called AsyncPUF which features inherent field programmability. It is both a novel and holistic PUF design compared to the existing state-of-the-art as it naturally addresses the two challenges facing PUFs to-date that prevent wide-spread adoption: robustness and entropy. Results of electrical simulation in a 90 nano-metre lithography process are presented and discussed.

Keywords: Cryptography, Physically Unclonable Functions, PUFs, Asynchronous Physically Unclonable Functions, Clockless Physically Unclonable Functions.

1 Introduction

Many security mechanisms are based upon the concept of a secret. Classic cryptography applications contain a secret key as input to encryption algorithms in order to scramble and decipher data. While they are secure against attack at the algorithm and mathematical level, it is commonly known that digitally-stored secret keys can be attacked or cloned relatively easily. In security tokens, such as smartcards, keys are stored on-chip in non-volatile memory. While field-programmable gate arrays (FPGAs) instead store keys in off-chip memory. This is because FPGA technology cannot easily integrate non-volatile memory, and besides read latency issues, it only acts to further increase vulnerability to attack.

Physical Unclonable Functions (PUFs) offer an efficient alternative to storing digital keys in on or off-chip memory. They exploit the physical lithography manufacturing variations of silicon integrated circuits (ICs). A PUF maps a unique set of digital inputs, known as challenges, to a corresponding set of digital outputs, known as

A. Dziech and A. Czyżewski (Eds.): MCSS 2012, CCIS 287, pp. 230–241, 2012.

responses, for use in challenge-response security protocols. Almost every year since 2000 there has been a new PUF design proposed as highlighted in Table 1.

Table 1. Different types of PUF

Year	PUF Type
2000-2004	Device mismatch [9], One-way function [10], Physical Random Function [11], Arbiter PUF [12]
2005-2008	Coating PUF [13], Ring Oscillator PUF [2], SRAM PUF [14], Butterfly PUF [15]
2009-2011	Power distribution PUF [16], Glitch PUF [17], Mecca PUF [18]
2012	**AsyncPUF (this paper)**

A typical challenge-response identity authentication scenario is illustrated in Fig. 1. Here, a challenge is given to an IC to authenticate its identity via the on-chip PUF. If the received response is not equal to the known challenge which has been recorded during manufacturing it is identified as fake.

Sadly, the unique benefits of silicon PUFs come with inherent stability design issues. In addition, in their basic configuration PUFs lack enough entropy to prevent modeling attacks [1]. However, it can be observed that PUFs are naturally asynchronous in nature. Insomuch as that they attempt to exploit asynchronous effects such as metastability, propagation delay or binary signal glitches. Therefore, it follows that asynchronous techniques may deliver much better PUF designs or provide an alternative to the existing state-of-the-art.

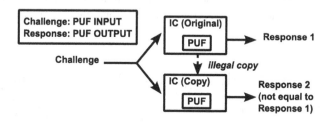

Fig. 1. Challenge-response authentication of chip identity using a PUF

In this paper, we present AsyncPUF which uses asynchronous rings for robust operation and to replace inverter chain ring oscillators used in ring oscillator PUFs [2] (RO-PUFs). It is fully digital and features inherent field programmability which naturally addresses the two challenges facing PUFs that prevents wide-spread adoption: robustness and entropy. Results of electrical simulation using a 90 nano-metre UMC lithography are discussed.

1.1   Contributions and Paper Organization

Our research, technical and scientific contributions are as follows:

- We propose Asynchronous Physically Unclonable Functions (APUFs).
- We advocate the use of asynchronous logic and techniques to implement PUFs.
- We propose ASYNCPUF which is inherently field-programmable to address robustness and entropy challenges. It uses asynchronous rings to replace inverter ring oscillators (IROs) used in ring oscillator PUFs [2] (RO-PUFs).

The remainder of the paper is organized as follows: Section 2 gives an overview of asynchronous logic. Section 3 discusses asynchronous rings. Section 4 describes ASYNCPUF. Section 5 presents results from electrical simulation. Section 6 draws conclusions.

2   Asynchronous Logic

The design of synchronous digital circuitry is based upon the discretisation of time, where a synchronous system changes from one state to the next at transitions of a system clock. The state is held in a set of registers and the next state outputs are derived from Boolean logic acting on the old state and present inputs. While the next state is copied through the registers on every rising and falling edge of a global clock signal. As such, the system exhibits deterministic behaviour as long as certain timing constraints on the inputs are met.

On the other hand, asynchronous designs do not follow this regime. In general, there is no global clock to govern the timing of state changes. Subsystems and components exchange information at mutually negotiated times. Therefore, certain parts of a design are always quiescent when they are not in use and hardware runs as faster as computational dependencies, input rate and the lithography device switching times.

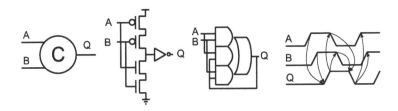

Fig. 2. The Muller C-element

As a field, it is historically seen as niche due to the profound understanding of concurrency, hardware, and semiconductors it takes to implement functionally correct designs. However, interest in the field has grown linearly in recent years in terms of applications as the fringes of Moore's Law have been reached and cyber security issues have become main-stream.

A plethora of design paradigms and techniques are known in literature. These range from high performance transistor level pipelines for processor design [3] and application to physical security [4]. The common denominator in all of which is the hysteresis capable Muller-C element [5] shown in Fig. 2. Both inputs must be equal to set or reset its output otherwise it holds its original state.

3   Asynchronous Rings

One of the most widely-used structures that use Muller-C elements are asynchronous rings (ARs) [7], which are purposely used here to implement ASYNCPUF. That is, as an alternative to inverter ring oscillators (IROs) in RO-PUFs for increased PUF stability and entropy.

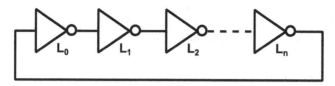

Fig. 3. L stage inverter ring oscillator

To illustrate how ARs operate an IRO structure is shown in Fig. 3. Here, L inverter stages are connected to form a ring. The oscillation time is the propagation delay of one logical transition all around the ring.

While an AR structure of L stages is shown in Fig. 4 and corresponds to the control path of a micro-pipeline [7]. Each stage is composed of a Muller C-element and an inverter, where for stage i: F_i is the forward input, R_i is the reverse input, and C_i is the output. The forward input value is written to the output if the forward and reverse input values are different otherwise the previous output is maintained.

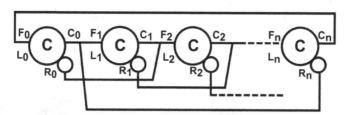

Fig. 4. Asynchronous Ring

3.1   Bubbles and Tokens

With reference to Fig. 4 the *bubbles and tokens* concept is as follows:

- Stage i contains a bubble if its output C_i is equal to the output of the previous stage C_{i-1}: $C_i = C_{i-1}$.

- Stage i contains a token if its output C_i is different from the output of the previous stage C_{i-1}: $C_i \neq C_{i-1}$.

Hence, for a 5 stage AR an initial state could be the token-bubble tuple:

$$\{Bubble_0, Token_0, Token_1, Bubble_1, Bubble_2\} \tag{1}$$

Which would correspond to the initial binary state:

$$\{S_0, S_1, S_2, S_3, S_4\} = \{1,0,1,1,1\} \tag{2}$$

As each stage i has a value of token or bubble determined by its output C_i and the output of the previous stage C_{i-1} the mapping from (1) to (2) is intuitive: $Token_0 = \{C_0, C_1\} = \{1,0\}$, $Token_1 = \{C_1, C_2\} = \{0,1\}$ $Bubble_1 = \{C_2, C_3\} = \{1,1\}$ etc.

Since it is possible to configure an AR with respect to bubbles and tokens, as explained above, they are naturally field-programmable and will increase the available entropy in an AR based PUF design i.e. ASYNCPUF.

3.2 Token and Bubble Propagation

Based on the token and bubbles concept, a token propagates from the stage i to the stage $i + 1$, if, and only if, the next stage $i + 1$ contains a bubble as shown in Fig. 5. In the same way, a bubble propagates from the stage i+1 to the previous stage i, if and only if, the previous stage i contains a token. Hence, ARs will have an oscillatory behaviour if the following conditions hold:

- $L \geq 3$ and $L = N_t + N_b$.
- $N_b > 1$, where N_b is the number of bubbles.
- N_t is a positive even number of tokens.

The oscillation depends on the stage timing parameters determined by process variability and the ratio N_t/N_b. It should be understood, while it is possible to maintain high frequencies in ARs, frequency decreases linearly with the number of stages in IROs. That is, different AR ring configurations will result in different frequencies for the same ring lengths.

Fig. 5. Token and bubble propagation

3.3 Noise

Both ARs and IROs exhibit thermal noise (known as jitter in the time-domain and phase noise in the frequency domain) such that the propagation delay will resemble a

Gaussian distribution. Fig. 6. illustrates the effect of jitter on an IRO in a 90 nano-metre SPICE transient noise analysis simulation using thermal noise with a bandwidth of 100KHz to 10GHz. A clear 71 pico-second variance is observable.

Where ARs and IROs differ is through how jitter accumulates. An IRO's period is defined by two loops of one token around the ring, and accumulates jitter from the number of crossed stages. But, in an AR, several tokens propagate in the ring simultaneously indicating the period is governed by the time between successive tokens. As such, each token crossing a stage experiences a variation in its propagation delay due to the jitter contribution of that particular stage. This is contrary to the IRO effect of jitter accumulation. This naturally provides improved robustness against noise instabilities caused by jitter in PUF designs, that is, by use of ARs instead of IROs.

In addition to Gaussian jitter, deterministic jitter occurs from non-random variations in propagation delays due to external global influences. The main difference is again in that in an AR several events propagate simultaneously, so deterministic jitter affects each event in the same way rather than the whole structure. This again leads to increased robustness in ARs versus IROs, and a more stable PUF design if ARs are used instead of IROs.

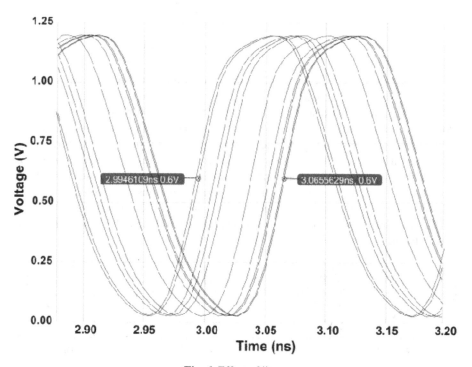

Fig. 6. Effect of jitter

4   AsyncPUF

We present in this section how to build AsyncPUF using asynchronous rings by replacing IROs in RO-PUFs.

A 1-bit RO-PUF is composed of 2 identically laid-out RO's, $R0_1$ and $R0_2$ with frequencies f_1 to f_2. They are selected using a pair of multiplexers that takes a bit of the PUF challenge as the select bit. Due to process variation, f_1 and f_2 will differ generating one response bit, R, of the PUF from comparison of the two frequencies measured by their respective counters. When enabled, R will be 1 if $f_1 > f_2$ otherwise 0, hence producing a single bit of a PUF response signature. The exemplary design in Fig. 7 produces a single PUF bit - n-bit PUF configurations are built by cascading these 1-bit RO-PUF structures.

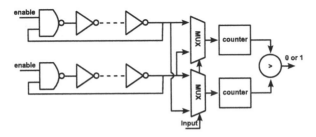

Fig. 7. Ring-oscillator based PUF design

Since IRO frequencies are closely matched, environmental effects can cause the oscillators to switch their outputs, for increasing temperature and/or decreasing voltage resulting in incorrect responses. It can be also observed large arrays of ring oscillators can cause a change in local chip temperature. These temperature stability issues are depicted on the left in Fig. 8. The ideal scenario is that the frequency difference should be sufficient to ensure consistent operation over temperature and voltage as shown on the right in Fig. 8. The approach to this problem in PUFs to-date has been to use error-correcting methods, which are expensive in terms of silicon area and add additional complexity to the challenge-response protocol. The other disadvantage of RO-PUFs is that they can be easily modelled to break the underlying security [1] to permit cloning.

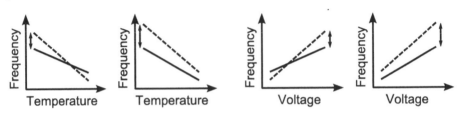

Fig. 8. Temperature and voltage effects on RO-PUFs

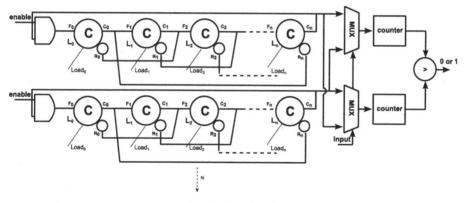

Fig. 9. AsyncPUF

AsyncPUF is an AR based PUF, as shown in Fig. 9, and gives the opportunity to address the above issues as well as noise. By configuring N_t and N_b, that is, by purposely controlling L, the number of stages, and their initial value by setting or resetting the Muller-C elements through the load inputs. By determining the configuration of the ARs with the maximum frequencies differences maximum reliability can be attained. This inherent configurable permits extremely low error rates by tending towards the ideal scenario. A further opportunity is to calibrate the AsyncPUF configurability according to different operating conditions. For example, the entire operating range of temperature and voltage could be divided into regions and have different AR load bit patterns.

Indeed AsyncPUF offers the opportunity to not only increase robustness through tolerance to environmental effects, but also the fact they can be re-configured increases entropy to address modelling attacks. This is because, as discussed, ARs can be easily configured to change their frequency by controlling N_t and N_b. Thus varying N_t and the load bit patterns in-field will result in completely new PUF designs, therefore thwarting modelling as no two PUFs are the same. Another alternative is to allocate different values randomly during manufacture and store in on-chip non-volatile memory.

It should be noted, for correct operation the AR run-time has to be low enough so that the counters do not overflow. Hence, care has to be taken to ensure the counters are matched to the estimated frequencies. It is worth noting also, other methods are perfectly plausible to convert the varying AR frequencies to a binary bit, rather than using a pure multiplexer approach. How RO-PUFs are cascaded for n-bit PUFs may also differ e.g. AR re-use.

5   Results

Experiments were performed using Monte Carlo SPICE analysis on the highest accuracy setting with a 90 nano-metre UMC lithography process and thermal noise with a bandwidth of 100KHz to 10GHz. Firstly, ARs were characterized to quantify how their oscillation frequency is affected by intra-die and inter-die process variation i.e. to understand their response to the lithography effects PUFs exploit. Simulations were conducted for a 6-stage AR using a 20 nano-second window and 1000 iterations for the two types of process variation (die-to-die and within-die). They took approximately 8 hours to complete on a high-end multi-core Linux server under the Cadence Design Framework. Fig. 10 shows the results from each of the 1000 simulations.

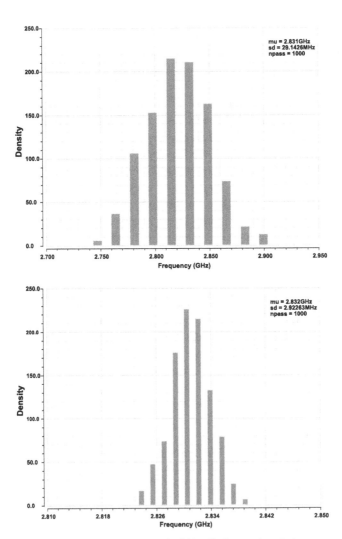

Fig. 10. Die-to-die (top) and within-die (bottom) variation

The ARs exhibit clear frequency deviations confirming their suitability for use as PUFs. For die-to-die variation an average frequency of 2.83GHz is obtained and a standard deviation of 29.14MHz, which indicates a die-to-die variation of 1.03%. And for within-die variation an average frequency of 2.83GHz is obtained and a standard deviation of 2.92MHz, which indicates a within-die variation of 0.10%. Clearly, the variation in AR frequency is greater between silicon wafers than on the same wafer for this particular lithography process; while both results exhibit a bell-curve Gaussian distribution.

Next 20 ASYNCPUFs of length 6, 12 and 18 each able to generate 32-bits of a response (i.e. 64 rings) were constructed, which was found in the setup phase to allow practical SPICE simulation. Note, using four different AR configurations a 128-bit response output can be generated, which highlights the trade-offs that are possible with ASYNCPUF due to its inherent field-programmability.

This time both die-to-die and within-die process variation SPICE simulation switches were activated together for analogous electrical simulation of 20 ASYNCPUF silicon chips. Matlab was used to parse and process the simulation data obtained and to generate random input challenges. Using two well-known PUF metrics, uniqueness and reliability (defined below), ASYNCPUF was evaluated. Both uniqueness and reliability results were captured at supply voltages between 0.4 V and 1.1 V, and temperatures ranging from -30C to 100C. Note, these result graphs were produced by Matlab rather than exported directly from Cadence as in Fig. 11. And to fit within the paper length, the presented results highlight the effect of temperature effects only. This is also because temperature affects PUFs silicon chips more than regulated voltage that can be viewed as a constant variable.

- Uniqueness is a measure of how easily an individual PUF can be differentiated; and quantifies the hamming distance between the responses of different ICs implementing the same PUF design that have been challenged with the same input. It is characterized by the probability density distribution (PDF) of the hamming distances, where PUFs with PDF curves centred at half the number of response bits and tall are more easily identifiable (unique) than PUFs with flatter curves.
- Reliability is a measure of how easily a given PUF can reproduce the same output response for the same input challenge. This is measured by the bits that remain unchanged under varying environmental conditions with the same input challenge. The PDF representing hamming distance of the response characterizes reliability of the same PUF subject to different environmental conditions i.e. changes in temperature and supply voltage. PUFs with PDF curves centred at 0 and tall are more stable than PUFs with flatter curves.

It was observed with increasing length of the ring, uniqueness is consistent, with a slight tendency for a stronger PDF the longer the length, shown on the left in **Fig. 11.** This result was consistent across all ASYNCPUF lengths initialized with arbitrary token patterns that satisfy the requirements in Section 2.

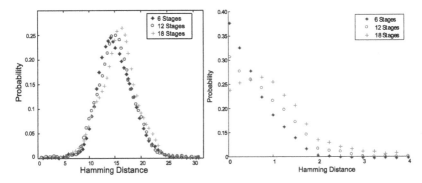

Fig. 11. Uniqueness and reliability of ASYNCPUF with respect to temperature

Fig. 11 on the right shows the effect of the stage length for ASYNCPUF for reliability. It was observed for ASYNCPUF that 6 stages are most stable followed by 12 and 18 stages. Therefore it can be concluded that shorter stages leads to better stability. This can be exploited for area efficient PUF implementations.

6   Conclusions

We have proposed using asynchronous logic to address the inherent issues with physically unclonable functions. We have presented and described a method of using asynchronous rings to implement a novel APUF architecture design, ASYNCPUF to enable increased robustness and entropy. We presented Monte Carlo SPICE analysis results of uniqueness and reliability. The results represent as close as possible to physical silicon chip results. It is common practice to rely on statistical SPICE transistor level simulation based on foundry process information before actual physical implementation. Due to the requirements of asynchronous circuits to be correct by construction (hysteresis from feedback) FPGAs were not used.

As cryptographic primitives, PUFs have several useful applications in security but are most frequently used for device authentication (i.e. Fig. 1). However, a new level of robustness and entropy for APUFs allows increased resistance to modelling attacks and makes it feasible for real-life efficient design. And also enables new applications: secret key generation, Intellectual Property (IP) protection, and to prevent product counterfeiting – or even to use in a software and hardware scenario.

Our future work is to consider application of asynchronous techniques to further PUF technologies and tape-out of a silicon chip. For instance, it would be possible to build PUF designs using elements from asynchronous elastic controllers [4] or eager monotonic logic [5]. Or alternative structures could be used instead of C-elements to implement ASYNCPUF ring stages that are widely published in literature e.g. GasP.

References

1. Ührmair, U.R., Sehnke, F., Ölter, J.S., Dror, G.: Modeling attacks on physical unclonable functions. In: Proceedings of 17th ACM Conference on Computer and Communications Security, pp. 237–249 (2010)

2. Suh, G.E., Devadas, S.: Physical unclonable functions for device authentication and secret key generation. In: Proceedings of the 44th Annual Design Automation Conference, DAC 2007, New York, USA, pp. 9–14 (2007)

3. Sutherland, I., Fairbanks, S.: GasP: a minimal FIFO control. In: Seventh International Symposium on Asynchronous Circuits and Systems, pp. 46–53 (2001)

4. Murphy, J., Yakovlev, A.: An Alternating Spacer AES Crypto-processor. In: Proceedings of the 32nd European Solid-State Circuits Conference, pp. 126–129 (September 2006)

5. Muller, D.E., Bartky, W.S.: A Theory of Asynchronous Circuits. In: Proc. Int'l Symp. Theory of Switching, Part 1, pp. 204–243. Harvard Univ. Press (1959)

6. Williams, T.E., Horowitz, M.A.: A Zero-Overhead Self-Timed 160-ns 54-b CMOS Divider. IEEE Journal of Solid-State Circuits 26(11), 1651–1661 (1991)

7. Sutherland, I.E.: Micropipelines. Communications of ACM 32(6), 720–738 (1998)

8. Ebergen, J.C., Fairbanks, S., Sutherland, I.E.: Predicting performance of micropipelines using charlie diagrams. In: Proceedings of Fourth International Conference on Asynchronous Circuits and Systems, pp. 238–246 (1998)

9. Lofstrom, K., Daasch, W., Taylor, D.: Ic identication circuit using device mismatch. Digest of Technical Papers, IEEE International Conference in Solid-State Circuits (ISSCC), pp. 372–373 (2000)

10. Pappu, R.S., Recht, B., Taylor, J., Gershenfeld, N.: Physical one-way functions. Science 297, 2026–2030 (2002)

11. Gassend, B., Clarke, D., van Dijk, M., Devadas, S.: Silicon physical random functions. In: Proceedings of the 9th ACM Conference on Computer and Communications Security (CCS), New York, USA, pp. 148–160 (2002)

12. Lim, D., Lee, J., Gassend, B., Suh, G., van Dijk, M., Devadas, S.: Extracting secret keys from integrated circuits. IEEE Transactions on Very Large Scale Integration Systems 13(10), 1200–1205 (2005)

13. Tuyls, P., Schrijen, G.-J., Škorić, B., van Geloven, J., Verhaegh, N., Wolters, R.: Read-Proof Hardware from Protective Coatings. In: Goubin, L., Matsui, M. (eds.) CHES 2006. LNCS, vol. 4249, pp. 369–383. Springer, Heidelberg (2006)

14. Guajardo, J., Kumar, S.S., Schrijen, G.-J., Tuyls, P.: FPGA Intrinsic PUFs and Their Use for IP Protection. In: Paillier, P., Verbauwhede, I. (eds.) CHES 2007. LNCS, vol. 4727, pp. 63–80. Springer, Heidelberg (2007)

15. Kumar, S., Guajardo, J., Maes, R., Schrijen, G.-J., Tuyls, P.: Extended abstract: The butterfly puf protecting ip on every fpga. In: IEEE International Workshop on Hardware-Oriented Security and Trust (HOST), pp. 67–70 (2008)

16. Helinski, R., Acharyya, D., Plusquellic, J.: A physical unclonable function defined using power distribution system equivalent resistance variations. In: Proceedings of the 46th Annual Design Automation Conference (DAC), New York, USA, pp. 676–681 (2009)

17. Suzuki, D., Shimizu, K.: The Glitch PUF: A New Delay-PUF Architecture Exploiting Glitch Shapes. In: Mangard, S., Standaert, F.-X. (eds.) CHES 2010. LNCS, vol. 6225, pp. 366–382. Springer, Heidelberg (2010)

18. Krishna, A.R., Narasimhan, S., Wang, X., Bhunia, S.: MECCA: A Robust Low-Overhead PUF Using Embedded Memory Array. In: Preneel, B., Takagi, T. (eds.) CHES 2011. LNCS, vol. 6917, pp. 407–420. Springer, Heidelberg (2011)

Scenario-Driven System for Open Source Intelligence

Edward Nawarecki, Grzegorz Dobrowolski, Jacek Dajda,
Aleksander Byrski, and Marek Kisiel-Dorohinicki

Department of Computer Science, AGH University of Science and Technology,
Al. Mickiewicza 30, 30-059 Kraków, Poland
{nawar,grzela,dajda,olekb,doroh}@agh.edu.pl

Abstract. The system that performs information integration based on formalized scenarios is presented. Heterogeneous data are collected from different sources, transformed and integrated in the course of a process supervised by a user (here a police analyst). The data comes from generally accessible sources, thus the presented system follows an open-source intelligence paradigm. The core of the paper is introduction to the system idea and functionality. All considerations are illustrated with an exemplary pedophilia content localization scenario.

1 Introduction

Development of the Internet and other so-called electronic media creates a new situation in which the offered services very often help in planning and committing crimes and more over some of them become direct aims of illegal activities.

Fortunately, taking advantage of these services as sources of information (often publicly accessible) creates possibilities of penetration of crime organizations and realization of certain preventive actions or investigations. Fulfillment of all functionalities necessary for acquiring and analysis of such sources requires development of dedicated software tools using the newest technologies and advanced methods of artificial intelligence. It is crucial, especially, when supporting and visualization functions of the system are taking into account.

The information system realized in the framework of the INDECT project (called INDECT-MAS [3]) is a solution of this class.

In section 2 of the paper the overall idea of INDECT-MAS is presented with special emphasis on the supporting role of the system in various questions of criminal analysis. Next, the formal description of a scenario as the driving force of all activities of the system and how external information sources are integrated according to the given scenario is presented. Section 4 shows how the information sources are integrated leading to the analyst's expertise. A scenario is a base for putting together several narrowly-oriented tools that allow to obtain information from the sources in appropriate forms and to carry out analysis entirely. At the end of the paper a real-life, but a bit simplified, scenario of pedophilia content localization is presented together with its orchestration by the system and how the analyst can get to the conclusion being supported in the proposed way.

A. Dziech and A. Czyżewski (Eds.): MCSS 2012, CCIS 287, pp. 242–251, 2012.
© Springer-Verlag Berlin Heidelberg 2012

2   INDECT-MAS Concept

INDECT-MAS system [3] is dedicated to support activities of public security agencies, in particular Police departments of criminal analysis. Generally speaking, the support is based on acquiring information from heterogeneous sources, and transforming them to the form that is easy for interpretation and processing by criminal analysts.

The considerations presented here focused on the open sources of information (Internet), however using other sources (according to the law constraints) is also acceptable. The schema of INDECT-MAS is shown in Fig. 1

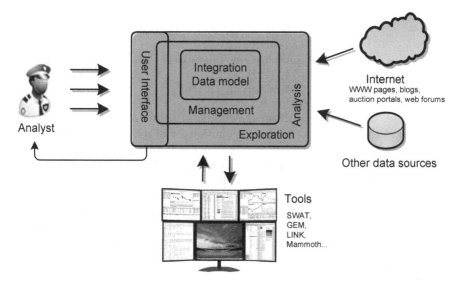

Fig. 1. The idea of INDECT-MAS

The main agent holding the initiative and supervising the work of the system is *analyst*, who prepares a *scenario* of the system activities and realizes it by using the dedicated interface. The system explores the information and data sources pointed out in the scenario, along with performing their analyses, using originally constructed software tools. The results of the analyses are integrated using predefined procedures in the *management layer*. Its effect is a *criminal analysis data model* that may be transformed to the form ready for interpretation. Independently on the integration process, the analyst can interact with achieved partial results, adjusting the realized scenario.

Internal structure of INDECT-MAS is designed using agent-oriented methodology, that improves its modularity and allows easy interpretation of functioning of its subsystems. It is easy to observe, that the crucial role in the work of the system plays the analyst, defining the scenarios and supervising their realizations.

3   Scenarios Construction

Constructing of a scenario allows realization of a certain task that depends strongly on the intuition and experience of the analyst. Because the structure of the scenario should assure its unambiguous interpretation by the system, its format should be appropriately constructed and standardized. The standardization affects the actions of the system supporting the work of the analyst [6].

In order to standardize the scenarios, the following notions and definitions are proposed.

Scenario is a sequence of actions performed in order to solve a certain task, that may be written as:

$$SCEN = [A_1, A_2, \ldots, A_n].$$ (1)

Action A_i is an aggregate characterizing operations realized in the system:

$$A_i = (a_{i1}, a_{i2}, \ldots, a_{im})$$ (2)

It is to note, that elements of the description of action $a_{ik} \in A_i$ do not have a sequential nature (as in the scenario), but consist of defined procedures (often complex) realized by the system.

As a typical example, the following description of the action may be given: D_i =(specification of data stream, specification of data source, tools used for exploration or transformation, result specification).

In the case of typical tasks (e.g. searching for criminal), it is possible to predefine some task elements, in order to help the analyst to construct the scenario. Moreover, preparing the list of predefined scenarios, adjusted to the realization of typical tasks may be considered.

Preparing a following set:

$$SCEN(T_k) = [A_1^k, A_2^k, \ldots, A_n^k]$$ (3)

where: T_k is k-th type of task, A_i^k is i-th action in the scenario Z_k, makes possible immediate beginning of its realization, because particular actions A_i^k are already defined.

It has to be mentioned, that in certain cases, the scenario elements (parameters) may be personal data, characteristics of certain people, time or place of events.

A complete example illustrating the construction of a scenario and its results will be given later in this contribution.

The Internet-based scenarios rely on data coming from such Internet sources as web pages, forums, blogs, and on-line auctions. These scenarios require the Internet connection and therefore a need for some security mechanisms may raise.

On the contrary, the scenarios that rely on phone bills, bank transfers, and text documents operate on data that can be provided as a file and therefore the Internet connection may not be necessary (though it may be possible to obtain the data from Internet services as well). These scenarios usually describe such operations as: searching, matching similar connections or transfers, pattern discovery and visualization.

Finally, there are also scenarios based on the use of external databases and services. More details on these databases can be found in the Deliverable D5.1 entitled *Preliminary report on police and prosecutor repositories and access procedures.*

Needless to say that the identified so far scenarios do not form an exhaustive and closed list. One can imagine adding new data source to extend the list of possible scenarios. A detailed specification of the already identified scenarios is under development and will be provided in the future.

4   A Sketch of the System Logic

In order to logically structurize the work of the police analyst, it is proposed to take advantage of the object-based analysis. Applying the analysis in the system is illustrated in Figure 2 which shows a part of the identified notions and interrelations among them.

It is assumed that every consistent portion of information forms an object. Like objects in the software object-oriented analysis, every object can be described by a set of attributes and operations (which can be performed on the data represented by the object). Obviously, similar objects form a class of objects called a *type*. For example, the analyst can operate on objects of type *Person*. Every person is described by the same attributes such as a name, a surname, a phone number, an address, etc. On the contrary, another object of type *Car* can be described by a different set of attributes such as a license number, a type of car, a manufacturer, a make, model, and year of production.

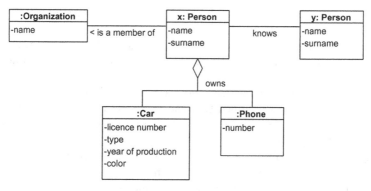

Fig. 2. The object-oriented structuralization of data (an example)

Moreover, an object can relate to other objects. As an example, the ownership relation between the object of type *Person* and an object of type *Car* can be given. Obviously, depending on objects, different relations can be identified and specified. Based on these relations, an automated analysis can be more thorough as it can use the knowledge of available relations to search deeper for additional information and clues. For example, with a defined relation between a suspect and a specific person the system can analyze both of them, instead of acquiring information on the suspect only.

Therefore, the user interaction with the system can be described by the following set of general actions:

- placing (adding) an object in the workspace,
- choosing an operation (transformation) offered (also suggested) by the system,
- transforming an object according to user preferences,
- performing automatic system reasoning based on objects modifications and user suggestions.

The way of interacting with user may be described as realization of one of the following modes.

- The first mode, autonomous work of the analyst is assured. The main menu of the system proposes a list of data sources available to the analyst user in a given case, as well as the list of all core system functionalities corresponding to available data sources.
- Second mode provides user with limited consultancy. The analyst, apart from the choice of the core functionality and importing corresponding data, enters a short description of a given case compliant with a chosen template-ontology.
- Third mode supports the user in a fully autonomous way. The beginning of the workflow is the same as for second mode. The system operates in an automated and autonomous way without the analyst participation. The analyst only receives the results of the analyzed scenarios.

Depending on the mode of the system work the system may support the user in her decisions, refrain from interacting and take initiative in full by performing transformations of objects on its own. Such changes (introduced automatically) will be annotated and the user will be able to modify or reject them according to his/her will.

The interaction between the user and the system is carried out according to predefined scenarios. It is assumed that the scenarios are realized in the form of workflows. In order to explain the idea, it is presented as a process consisting of several distinctive phases.

Firstly, verbal description of a particular scenario is conditioned so as each its element could be linked to the adequate mean of its realization. The means are: software components of the system – some of them providing access to external information sources and others – representing in the system by the user. It is worth to mention that the latter forms a channel introducing important information coming from sources that cannot be automatically integrated as well as a channel that allows for controlling of the system work including workflows formation also.

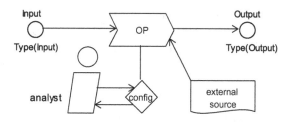

Fig. 3. Workflow operator

Let a basic element of a workflow (also representation of the scenario element) be described by notion of *operator*. An operator is responsible then for transforming the input data to a specific output. For example an operator can be responsible for filtering the specific data. To be more specific, imagine an operator that filters out all phone numbers from a specific country from a phone bill of a large size. The alternative way of the operator operation is wrapping an external source of information. Then input data become a query to the source, output – query result.

Formally every operator has an input and an output of specific predefined types that define the context of the operator. Additionally, an operator can have a configuration input used to tune the operator to current needs by a user. A wrapping operator can communicate with its external source. Fig. 3 illustrates this concept.

Obviously, the number of possible inputs or outputs can be greater than one. The exact number can differ depending on a given operator. However, when we assume that the input or output types (we can call them composite) may consist of other types (we can call them basic), the operator illustrated in Fig. 3 also refers to these cases.

Now we can pass to the second phase of the discussed process – assembling of a workflow. Having defined the operator as a core workflow element, the question is how to join two of them to form the base of the scenario realization finally.

Two operators are combined in such a way that the output of a given operator becomes the input of another. The operator can be join when they are compatible in the following sense if one of the following conditions is fulfilled:

- Type1(Output) equals Type2(Input)
- Type2(Input) is a generalization of Type1(Output)

If the above condition is true the components can cooperate in one of two modes:

- fully automatic, immediately after the first processing is ready or the specific condition with respect to the data is fulfilled;
- controlled by a user – the data are passed between the operators as the consequence of the user's decision.

Both modes do not exclude the situation when the same operators are joined forming a kind of loop that can be obviously: automatic or controlled.

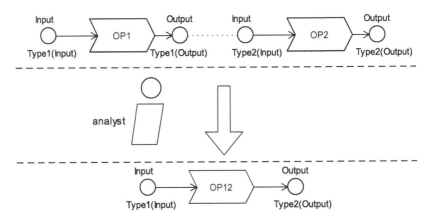

Fig. 4. The rule of effective workflow assembling

The controlled mode opens the second way (after possibility of operators configuration)in which a user can influence a particular workflow for the sake of the analysis (information integration) he/she carried out.

Figure 4 shows how a workflow of two operators is built. Obviously, a combination of operators can be considered as a composite operator with the same input type as for the first operator and the output type of the second operator. This observation is especially important when automatic joining is in operation. The whole workflow can be considered as a composite operator also.

Having defined how to combine the operators we are ready to construct whole workflows. Apart from operators, an important element of a workflow is a *dataset*. While operators are responsible in general for processing and form functionality of a workflow, datasets store information that operators work on. Thus datasets are all the time under the user's consideration. Some datasets directly contain description of the objects defined as crucial for the criminal analysis.

The third phase is realization of the assembled workflow and thus the scenario under consideration. It must be observed that, in usual cases, the realization passes paths marked out by the workflow several times but with different data.

One can say that while the workflow represents the general schema of how analysis ought to be carried out, the effective path shows how the analysis has run in the particular case. It shows all variants that have been checked prior to the successful one has been accomplished. Thus the realization fully depends on the user's ideas of how to work with the system that, in turn, depends on what information is disclosed along the workflow run.

The same causes may lead to reconfiguration (re-assembling) of a workflow. The presented system is opened to such possibility which makes it flexible when it is perceived as a set of cooperating modules or components. An example of such component architecture which can be easily adopted to the workflow concept is presented in [5].

The above presentation can be closed with some general architectural remark. The workflow can be perceived at 2 levels (layers): on GUI layer (Interface Layer) and model layer (Module Layer). While Module Layer is responsible for the workflow logic, the Interface Layer provides a user with all its consequences – ability to support his/her analysis with as reach as it is possible visualization and ergonomic manipulation.

5   An Exemplary Real-Life Scenario Realization

According to the description given in Section 3 a simplified exemplary scenario is given below. Necessary for the scenario realization specialized software tools that process the information from the open sources are also briefly presented.

Let us take the scenario in the narrative form firstly.

1. The Police seizes a number of illegal multimedia materials (pedophile content): images, movies, etc. Based on the assumption, that the pedophile society often exchanges the same material, the hash-codes for the multimedia content are computed.
2. The hash-codes are submitted to IBIS crawler hoping to find the desired content on the web. As an outcome, number of IP numbers are given.
3. The hash-codes are submitted to GEM crawler to search the content in P2P network.
4. Outcome of IBIS and GEM-based search is a list of IP numbers containing illicit material.
5. Using abcLog tool, address data of the Internet providers is found.
6. Police starts operational work in order to retrieve the personality of the users from the Internet providers.
7. After seizing the hardware, PyFlag/FLUSH [4,1] is used to localize illicit material on the data storage.

This scenario is illustrated in Fig. 5. The software tools mentioned in the scenario (WWW index, GEM, abcLog and PyFlag/FLUSH) are advanced applications prepared for certain types of data processing.

The realized scenario consists of actions A_1, A_2, A_3. Each action contains elements describing its character, data source, used software tool and the structure of result. In the discussed case, these actions are defined as follows:

$$A_1 = [Internet, fileshash - codes, WWWindex, IPnumbers]$$
$$A_2 = [P2PNetwork, fileshash - codes, GEM, IPnumbers]$$
$$A_3 = [resultsofA_1andA_2, IPnumbers, abcLog, Internetproviderdata]$$

The external, operational actions lead to acquiring the personal data of suspects from their Internet providers, allowing to fulfill the final action:

$$A_4 = [Suspectdata, filesystem, PyFlag/FLUSH, illicitmultimedia]$$

More exemplary scenarios with realization proposals can be found in [2].

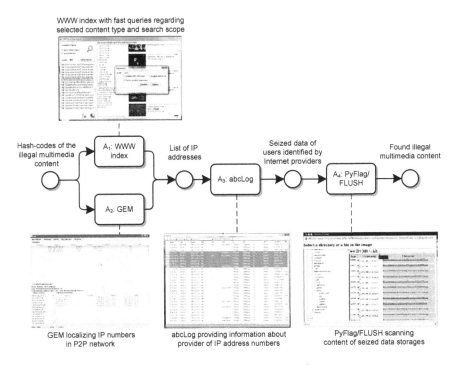

Fig. 5. Example scenario and its realization

6   Conclusion

The general concept of INDECT-MAS system has been presented, along with the principles of scenario orchestration which is the main element of internal organization of the system. A workflow which is generated and performed allows integration of indispensable information that comes from several sources. Moreover the information is presented to the analyst in the especially processed form that magnify possibility of inference drawn from it.

In this way the analyst obtains especially tailored for him as well as, each time, suitable for the assessment he carries out computer support.

Several more detailed but also important aspects of the system could not be presented in such a short publication. Among them the crucial question of a versatile graphical user interface that, in fact, determines success (usefulness) of the whole project. Solutions to the mentioned aspects are at final levels of maturity now and will be the subject of publication in the near future.

It is worthwhile to underline at the end that the system is designed to be open in the sense of introducing new sources (tools which wrapped them) into it. The used in the presented example tools do not complete a possible list. Internal architecture and mechanisms of the INDECT-MAS are ready to adapt other useful also third-party tools.

Acknowledgment. The research leading to these results has received funding from the European Community's Seventh Framework Program – Project INDECT (FP7/2007-2013, grant agreement no. 218086).

The work described in this paper was partially supported by The European Union by means of European Social Fund, PO KL Priority IV: Higher Education and Research, Activity 4.1: Improvement and Development of Didactic Potential of the University and Increasing Number of Students of the Faculties Crucial for the National Economy Based on Knowledge, Subactivity 4.1.1: Improvement of the Didactic Potential of the AGH University of Science and Technology "Human Assets", No. UDA-POKL.04.01.01-00-367/08-00.

References

1. Byrski, A.: Effective analysis of overtaken data storage. In: Artificial Intelligence Methods Supporting Public Security. AGH University of Science and Technology Press (2009) (in polish)
2. Byrski, A., Dajda, J., Dobrowolski, G., Kisiel-Dorohinicki, M., Koźlak, J., Nawarecki, E., Śnieżyński, B., Turek, W., Zygmunt, A., Uruena, M., Berger-Sabbatel, G., Duda, A., Korczyński, M.: Specification of scenarios, software agents and components for operational and investigation activities support. Technical report, University of Science and Technology, Krakow, Poland (2011)
3. Byrski, A., Kisiel-Dorohinicki, M., Dajda, J., Dobrowolski, G., Nawarecki, E.: Hierarchical Multi-Agent System for Heterogeneous Data Integration. In: Bouvry, P., González-Vélez, H., Kołodziej, J. (eds.) Intelligent Decision Systems. SCI, vol. 362, pp. 165–186. Springer, Heidelberg (2011)
4. Byrski, A., Stryjewski, W., Czechowicz, B.: Adaptation of pyflag to efficient analysis of seized computer data storage. The Journal of Digital Forensics, Security and Law 5 (2010)
5. Dajda, J., Debski, R., Byrski, A., Kisiel-Dorohinicki, M.: Component-based architecture for systems, services and data integration in support for criminal analysis. Journal of Telecommunications and Information Technology (2012)
6. Nawarecki, E., Dobrowolski, G.: Information – knowledge – intelligence in systems supporting public security. In: Practical Elements of Organized Crime and Terrorism Fighting: New Technologies and Operational Actions. Wolters Kluwer Poland (2009) (in polish)

Towards Hardware Implementation of INDECT Block Cipher

Marcin Niemiec, Jakub Dudek, Łukasz Romański, and Marcin Święty

AGH University of Science and Technology
Al. Mickiewicza 30, 30-059 Krakow
`niemiec@kt.agh.edu.pl`, `jakub@dudek.in`,
`{lukasz.romanski,marcin.swiety}@wp.eu`

Abstract. This paper presents the first steps towards hardware implementation of INDECT Block Cipher (IBC) — a new symmetric block cipher invented in INDECT project. Currently, end-users can encrypt or decrypt single files by the software implementation of IBC but migration to hardware allows to speed up the encryption and decryption processes. In the paper, the authors describe software and hardware environments (Xilinx Spartan platform and System Generator environment) where the implementation is performed. Also, the models of IBC encryptor and decryptor developed in System Generator environment are presented in detail. Additionally, some considerations and propositions of Concurrent Error Detection in the hardware IBC architecture are described. Beside the descriptions of the main achievements, future development and next steps towards final hardware implementation are also considered.

Keywords: security, hardware implementation, symmetric cryptography, block ciphers, INDECT Block Cipher.

1 Introduction

INDECT Block Cipher (IBC) is a new symmetric cryptography algorithm based on substitution - permutation network proposed and described in detail in [2]. Its software implementation was widely presented in [3]. In these documents we mentioned that software implementation of IBC, was first of all created to show functionality of new cipher, but it is possible to improve the implementation performance in the future. Usually, performance is a critical issue in a real deployment of cryptographic algorithms. Therefore we decided to create the hardware implementations of encryption/decryption modules.

This paper presents some key achievements towards hardware implementation of INDECT Block Cipher. At the beginning, development environment and target device are presented. After that, the models of IBC encryptor and decryptor are shown. At the end, the authors describes some ideas and solutions of Concurrent Error Detection in hardware implementation of IBC.

A. Dziech and A. Czyżewski (Eds.): MCSS 2012, CCIS 287, pp. 252–261, 2012.

2 Development Environment and Target Device

In order to create ready and operational device that is both efficient and fast, the FPGA solutions were taken into account. Perspective of iterative development and requirement of easy code generation were also an major objective in selection process of targeted device. The need of fast implementation and well known architecture pointed the Spartan family by Xilinx Inc. Ability to integrate MATLAB/Simulink models with synthesis process has proven to be very accurate and helpful in case of IBC deployment on FPGA device. For that purpose, the System Generator software has been used. It allows to take full advantage of already implemented classes and modules as well as create highly optimized code. In this section both environments, software and hardware, are presented.

2.1 Hardware

From wide spectrum of Spartan family devices the Spartan-3AN boards, platform based on 90 nm technology were chosen. Reason for that was low cost and more than enough efficiency provided by this solution. Support of both 32-bit MicroBlaze Embedded Processor and 8-bit PicoBlaze Embedded Controller makes these hardware an excellent choice when considering creation of cipher device that communicates by standardized interfaces like Ethernet.

Moreover, this board allows to implement security measures not only as an software security methods, but as well can provide protection on hardware level. In the light of cryptography application and general security, the unique Device DNA allows to implement limited-functionality, time-bomb, and self-destruct methods preventing reverse-engineering, cloning and overbuilding. Unique serial numbers also provide flexibility in deploying the device with authentication mechanisms. Spartan-3AN board is presented in Fig. 1.

Whole Spartan-3A family devices, and Spartan-3AN is fully compatible with the 3A family, consists of five fundamental programmable function elements:

- Configurable Logic Blocs,
- Input/Output Blocks,
- Block RAM,
- Multiplier Blocks, and
- Digital Clock Manager (DCM) Blocks.

First of enlisted, the Configurable Logic Blocks, contains Look-up Tables (LUTs) that build logic and storage elements (flip-flops or latches). Input/Output Blocks control data flow between I/O ports and device internal logic. It should be noted, that data flows on this level can be bidirectional and the 3-state operations are supported. Block RAM is data storage for the design. As for Multiplier Blocks (MB), the length of binary word accepted by MB is 18-bits. The last type of fundamental element is DCM Block, that allows flexible and fully digital clock control. It is self-calibrating solution for distributing, delaying, multiplying, dividing and phase-shifting of internal clock signals.

Fig. 1. Spartan-3AN board

Each of Spartan-3AN board consist of various ports and input/output controls like LCD, LEDs or switches. List of most important available controls and ports is as follows [6]:

- Two lines, 16 characters LCD screen
- PS/2 port (supporting compatible mouse and keyboard)
- VGA display port, 12-bit color,
- 10/100 Ethernet PHY
- Two nine-pin RS-232 ports, both DCE and DTE style,
- USB-based programming interface,
- LEDs, Buttons, Switches, Rotary knob.

In our work we use the XC3S700AN version of Spartan-3AN board, that can be characterized by memory's parameters and I/O specifications.

- System Gates **700K**
- Logic Cells **13,248**
- Dedicated Multipliers **20**
- Block RAM Blocks **20**

- Block RAM Bits **360K**
- Distributed RAM Bits **92K**
- Flash Size Bits **8M**
- User Flash Bits **5M**
- DCMs (Digital Clock Managers) **8**
- I/O Standards **26**
- Max Differential I/O **165**
- Max Single Ended I/O **372**

2.2 Software

System Generator was designed mainly for Digital Signal Processing and takes full advantage of MATLAB/Simulink environment. It allows to use model-based designs along with solid and revised implemented building blocks. That functionality and ability to create custom blocks and modules opens an excellent opportunity for building complete and hierarchical design that can be debugged and tested on each level of detail depth. Also, there is no need to know every little technicality of hardware description language used or possess an extensive experience in FPGA field. All underlying FPGA implementation processes are automatically performed including synthesis and placing and routing.

Xilinx provides an excellent and complete DSP Block set, that can be adapted to cipher implementation purposes. An great number of 90 of these building blocks consist of both, common functions like delaying, multiplexing, and complex modules implementing algorithms like Fast Fourier Transform (FFT) or various filters. Due to the fact that System Generator is highly integrated with MATLAB environment, designer is able to provide MATLAB algorithmic models that can be incorporated into hardware design.

3   Model of Encryptor/Decryptor

For the purpose of the hardware implementation of IBC cipher, its low-level hardware model was implemented in Xilinx System Generator. In particular, encryption and decryption blocks of the 8 round/128-bit key length version of the algorithm were developed. General view of the whole model is presented in Fig. 2. Algorithm works on 256-bit data blocks, hence, encryption block has 32 1-byte inputs.

Each round of the algorithm consists of substitution performed with use of substitution S-box and permutation using permutation S-box. S-boxes are implemented as RAM System Generator blocks as the memory read is an exact representation of substitution process as long as RAM block was initialized with proper S-box data earlier. Such an initialization is called S-box generation in IBC and this is the moment when the key is passed to the algorithm. The secret (private) key data is used to generate highly nonlinear AES-like S-boxes as described in [2]. Fig. 3 presents the S-box generation process implementation in hardware.

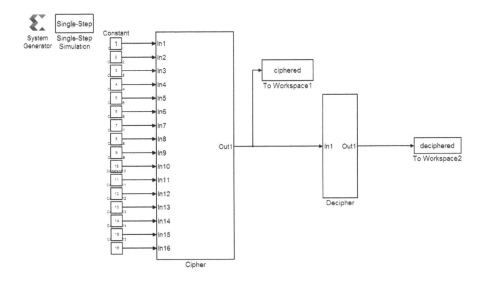

Fig. 2. General view of IBC System Generator model

Eight bytes of the key are required to initialize one IBC S-box. One byte of such S-box is constructed as follows: logical AND operation is performed on standard AES S-box and each key byte, then, bits of each of resulting 8 bytes are XORed together and resulting 8 bits, concatenated, form one byte of IBC S-box. After the process of memories initialization data can be passed for the processing in algorithm rounds. Consecutive rounds of the algorithm process the output data from the preceding rounds and transfer results to the input of the following round and each round consists of the same set of operations. That is why, only operations of first round of encryption process are going to be described in detail (decryption process consists of the same set of operations

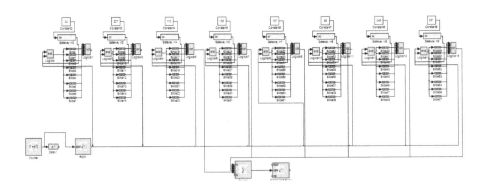

Fig. 3. System Generator implementation of IBC S-box generation process

performed in reverse order). Part of encryption device responsible for the one round operations is presented in Fig. 4.

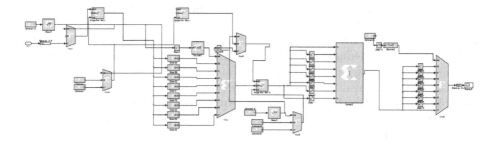

Fig. 4. Operations of one round of IBC encryption process

Input data is passed through the multiplexer to the address input of substitution RAM block (S-box). As a result, substituted byte can be obtained on the RAM output. This byte is sliced into single bits on which permutation operation is performed. They are saved in the auxiliary RAM block, which address input is driven by the output from permutation RAM. Permuted bits are then read, one by one, from auxiliary RAM block and converted to byte form again which is suitable for the next round operations. While the algorithm proceeds some invalid data may appear on the output. To avoid that, some additional blocks are used to store the proper bytes on the output line before the next proper byte arrives.

Implementation of the model in System Generator enables designer to easily check via the simulation process what data appears at any time at any point of the circuit. Therefore, verification of the correctness of the algorithm operation could be performed before the implementation on the FPGA board. Several test input vectors were applied to the encryption block and the same data was observed on the output of decryption block after some time which confirms proper implementation of both circuits. Conformity with IBC software application was also positively verified. These results allow to predict successful implementation of the model on FPGA board during the further research.

4   CED Implementations

For the hardware-implemented IBC cipher was decided to add the functionality of the Concurrent Error Detection (CED). This feature allows to detect permanent and transient errors in whole circuit. In this section both the CED and its implementation were described.

4.1   Concurrent Error Detection

Concurrent Error Detection is a class of digital electronic circuits self-checking methods which, as its name suggests, aims at detecting faults in circuit concurrently to its normal operation. The main aim of CED techniques is to assure that electronic device posses data integrity property, i.e. produces correct output results or if not, an error is reported. Thus, CED implementation always requires some kind of redundancy, namely: information, hardware or time redundancy [4].

Information redundancy occurs when we add some additional data to original information with use of coding techniques well known from data communications applications such as Error Detecting Codes (EDC), Forward Error Correction (FEC) or Cycling Redundancy Check (CRC). Hardware redundancy involves adding or duplicating some parts of hardware structure in order to be able to compare results from different hardware blocks and draw a conclusion about possible errors. Finally, time redundancy requires resending the same or specifically modified input data again through the same circuit and comparing if the output results change in time.

Worth to mention, that although hardware redundancy requires significant hardware area overhead, also information and time redundancy CED always requires adding some additional circuits/blocks to the original scheme.

4.2   CED in Cryptography

Concurrent Error Detection technique is going to be used in encryption/decryption device. Thus, it is very important to analyze CED capabilities for cryptographic applications. Fortunately, CED, especially information redundancy type, may bring many advantages if taken into consideration while developing new crypto device, namely:

- In cryptographic applications it is very important to have properly working encryption/decryption devices as some kind of faults in their internal structure may cause specific errors in encrypted cipher text available to potential attacker which may give him some information about the key used for encryption process or the message encrypted itself. Thus, cryptographic devices are those for which use of CED should be seriously considered.
- In cryptographic devices capable of performing both encryption and decryption process we have comfortable situation in which we expect the same data on the input of encryption process as on the output of decryption process, namely a message. That is why we can use information redundancy CED techniques which, as mentioned earlier, do not require much hardware or performance overhead and can be applied to a device project in a relatively easy way.
- Less hardware area overhead in CED techniques using Error Detection Codes gives possibility to use it with devices where it is very important to make the circuit as small as possible, such as, e.g. mobile phones or IP cameras.
- Finally, cryptographic algorithms are usually constructed in such way that single internal error may cause multiple errors in the output data, thus techniques which are capable of single-faults detection are not applicable in this area. CED using information redundancy provides better properties as far as multiple-faults detections is considered.

4.3   CED in IBC Crypto Device

As mentioned earlier, CED technique is going to be applied in hardware implementation of INDECT Block Cipher. As far as this application is considered, every type of CED: hardware, time and information redundancy were considered. Information redundancy scheme with Error Detection Codes is going to be used in the future because of its advantages in cryptographic implementations mentioned in the earlier section. For now some combination of hardware and time redundancy scheme with CED was applied. The high-level diagram of the encryption device with algorithm-level CED implemented in System Generator is shown in Fig. 5.

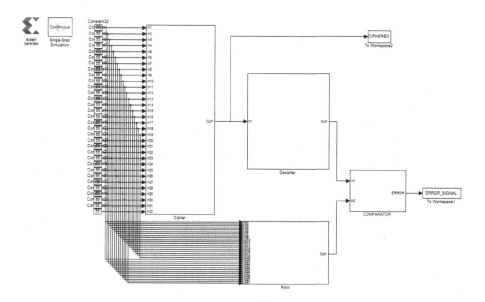

Fig. 5. Xilinx System Generator high-level scheme of algorithm-level CED applied to IBC encryption circuit

In this implementation, hardware redundancy in encryption device involves adding decryption circuit in order to be able to compare original input data stored in additional RAM memory with decrypted message in Comparator block. Of course, such solution imposes also serious performance overhead characteristic to time redundancy CED techniques because in order to be able to send an encrypted message we have to be sure that it is error-free and thus decrypt it what takes about the same amount of time as encryption process. Thus both time and hardware overhead reach 100%.

Because of these disadvantages information redundancy CED with use of Convolutional Coder and Viterbi Decoder blocks should be applied in the next stage of the hardware implementation. Such a scheme is presented in Fig. 6.

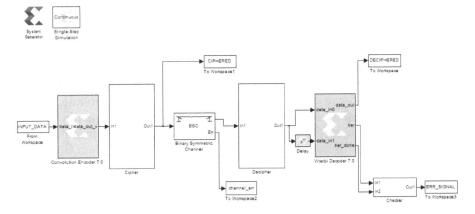

Fig. 6. Xilinx System Generator high-level scheme of information redundancy CED based on convolutional encoding applied to IBC encryption and decryption circuits

For this CED scheme, not only the encryption device but the whole encryption/decryption system should be considered. Plain text (INPUT_DATA) is coded with convolutional encoder and ciphered in IBC encryption block. Than it goes through the communication channel (modeled here as Binary Symmetric Channel) and arrives at the input of decipher device. Output from the decipher block is passed to Viterbi decoder in order to receive plain text back on its output. If Viterbi decoder detects some errors they are reported by Checker module. This scheme has several advantages over the algorithm-level one considered earlier. First of all, hardware area overhead is significantly lower as one has to add only Convolutional Encoder and Viterbi Decoder blocks to the whole system instead of doubling its hardware structure (installing redundant decipher block in encryption device and cipher block in decryption device). In addition, such scheme has almost no impact on performance of whole encryption/decryption process as its execution time is increased only by convolutional coding/decoding time. This is much better than 100% performance overhead in the case of algorithm-level CED. That is why, after all, information redundancy CED scheme based on convolutional codes will be applied to hardware implementation of IBC in the nearest future.

5   Conclusions

This paper presents first achievements towards hardware implementation of IN-DECT Block Cipher. The crucial milestone of this work was the construction of IBC encryptor and decryptor. It is worth to mention that tests confirmed proper construction and implementation of both circuits. The big part of this work focused on the ideas and solutions of Concurrent Error Detection in hardware implementation of IBC. Deployment of these methods allow to detect permanent and transient errors in whole encryption/decryption process.

The future work will be focused on the implementation of IBC encryptor/decryptor model on FPGA board as well as the verification of the correctness of the algorithms operations. Also, the definition of input/output data format as well as construction of interface between the device and end-user's PC will be important steps. Additionally, information redundancy CED with use of convolutional codes and Viterbi decoder will be applied as the next stage of the hardware implementation.

Acknowledgments. This work has been performed in the framework of the EU Project INDECT (*Intelligent information system supporting observation, searching and detection for security of citizens in urban environment*)–grant agreement number: 218086.

Development of application's functionality and graphical interface have been co-financed by the European Regional Development Fund under the Innovative Economy Operational Programme, INSIGMA project no. POIG.01.01.02-00-062/09.

References

1. INDECT Project, `http://www.indect-project.eu`
2. Niemiec, M., Machowski, Ł., Święty, M., Dudek, J., Romański, Ł., Stoianov, N.: D8.3 Specification of new constructed block cipher and evaluation of its vulnerability to errors. INDECT Project Deliverable (2010)
3. Dudek, J., Machowski, Ł., Romański, Ł., Święty, M.: Software Implementation of New Symmetric Block Cipher. In: Dziech, A., Czyżewski, A. (eds.) MCSS 2011. CCIS, vol. 149, pp. 216–224. Springer, Heidelberg (2011)
4. De, K., Natarajan, C., Nair, D., Banerjee, P.: RSYN: A System for Automated Synthesis of Reliable Multilevel Circuits. IEEE Trans. VLSI 2, 186–195 (1994)
5. Fernandez-Gomez, S., Rodriguez-Andina, J.J., Mandado, E.: Concurrent Error Detection in Block Ciphers. In: Proc. IEEE Int. Test Conf., Atlantic City, NJ, pp. 979–984 (2000)
6. Xilinx webpage. Spartan-3A/3AN FPGA Starter Kit: Board User Guide, version 1.1 (2008)

A Novel Keystream Generator with Cryptographic Purpose

Nikolai Nikolov and Nikolai Stoianov

Technical University of Sofia, INDECT Project Team,
8, Kliment Ohridski St., 1000 Sofia, Bulgaria
nkl_stnv@tu-sofia.bg

Abstract. In this paper a novel key stream generator is presented. Linear Feedback Shift Registers are widely used for creating pseudo random number generators with good statistic characteristics. Proposed generator is based on LFSR. For creating algorithm five new irreducible polynomials are used. Algorithm is based on four groups of polynomials and on additional polynomial used for feedback management. Proposed algorithm and generator have good linear complexity. Repetition period of generator is calculated and results proof that it meets requirements for PRNGs. Block scheme of PRNG is presented. This algorithm is tested whit NIST's suite for statistical testing of Random Number Generators. Received results are shown and they show that proposed and tested generator can be used for generating of cryptography keys and for using it as basis in stream ciphers.

Keywords: stream ciphers, key stream generator, linear feedback shift register, LFSR, irreducible polynomials.

1 Introduction

The steam cipher development as an effective protective tool for the information transmitted in the modern communication systems increases the requirements towards them. The reliable data protection from the cryptographic point of view imposes the use of modern highly productive generators of cryptographic keys in combination with sophisticated traditional algorithms. An alternative for satisfying the increased requirements is the introduction of non-linear in the algorithms, use of irreducible polynomials of higher level and dynamic change of the feedbacks in the generation process of the streams, used for cryptographic keys.

2 LFSRs and Polynomials

Linear Feedback Shift Register (LFSR) is main component of sets of cryptographic keys. The main reasons for their use are [1], [2]:

- LFSR can easily hardware implemented;
- Sequences with a long repetition periods can be produced;

A. Dziech and A. Czyżewski (Eds.): MCSS 2012, pp. 262–269, 2012.

- Sequences with good statistic characteristics can be generated;
- Because of their structure they can be analyzed via algebraic methods.

Definition [1]: LFSR with a length L consists of L elements numbered from 0, 1,..., L-1, each of those elements can be stored only in one bit, has only one input and output and clock, that defines the movement of data. In each moment of time the following operation can be conducted:

- The content of the element 0 is delivered to the exit and a part of the output sequence is formed;
- The content of the element i is moved to element i-1 for each i, where $1 \leq i \leq L-1$;
- The new content of the element L-1 is a bit for feedback s_j, which is summed by module 2 with the previous content of the elements 0,1,..., L-1.

LFSR defined in this way is marked with $s\langle L, C(D)\rangle$, where $C(D) = 1 + c_1 D + c_2 D^2 + ... + c_L D^L \in Z_2(D)$ is connection polynomial.

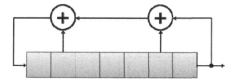

Fig. 1. General representation of LFSR

Sequences generated via the use of LFSR /Linear Feed Shift Register/ have very good statistic parameters, but they have very low linear complexity, that makes them vulnerable for different cryptographic attacks, especially for the attacks of Berlekamp-Massey. One of the ways to improve the cryptographic characteristics of the pseudo random sequences is the introduction of non-linarite during the generation and use of incorporated generators of pseudo random sequences.

Irreducible polynomials used for the generation are given in Table 1.

Table 1. Used irreducible polynomials

1	$x^{257}+x^{138}+x^8+x^7+x^6+x^5+x^4+x^3+x^2+x^1+x^0$	GF(2)
2	$x^{247}+x^{24}+x^8+x^7+x^6+x^5+x^4+x^3+x^2+x^1+x^0$	GF(2)
3	$x^{251}+x^{64}+x^8+x^7+x^6+x^5+x^4+x^3+x^2+x^1+x^0$	GF(2)
4	$x^{241}+x^{14}+x^8+x^7+x^6+x^5+x^4+x^3+x^2+x^1+x^0$	GF(2)
5	$x^{17}+3x^2+2.x^1+3.x^0$	GF(5)

The selection of periods for individual linear shifting registers is maximum mutually simple lengths L1,L2,...,Lj, where the equivalent period is calculated via Formula 1.:

$$S_0 = \prod L_j \qquad L_j = P_j^{N_j} - 1 \qquad L_1, L_2 \ldots L_j \qquad (1)$$

Replacing the values in Formula 1, we receive as a result for the equivalent period:

$$S_0 = (2^{257} - 1).(2^{247} - 1).(2^{251} - 1).(2^{241} - 1).(5^{17} - 1) \approx 5.10935e + 311 \qquad (2)$$

The repetition period of the resultant generator meets the requirements towards the Pseudo Random Number Generator.

The block-scheme of the Pseudo Random Number Generator, which is proposed and tested, is shown on figure 2.

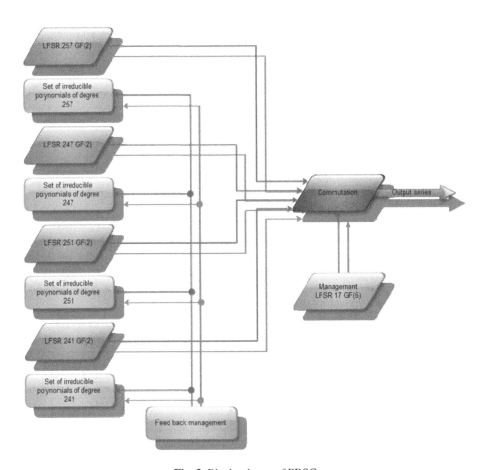

Fig. 2. Block scheme of PRSG

The polynomials used in the software for random sequences generation is given in Table 2.

Table 2. Used polynomials in algorithm

	Polynomials of level 257 GF(2) used
1	$x^{257}+x^{138}+x^8+x^7+x^6+x^5+x^4+x^3+x^2+x^1+x^0$
2	$x^{257}+x^{103}+x^9+x^7+x^6+x^5+x^4+x^3+x^2+x^1+x^0$
3	$x^{257}+x^{119}+x^9+x^7+x^6+x^5+x^4+x^3+x^2+x^1+x^0$
4	$x^{257}+x^{17}+x^{10}+x^7+x^6+x^5+x^4+x^3+x^2+x^1+x^0$
	Polynomials of level 247 GF(2) used
1	$x^{247}+x^{24}+x^8+x^7+x^6+x^5+x^4+x^3+x^2+x^1+x^0$
2	$x^{247}+x^{36}+x^8+x^7+x^6+x^5+x^4+x^3+x^2+x^1+x^0$
3	$x^{247}+x^{98}+x^8+x^7+x^6+x^5+x^4+x^3+x^2+x^1+x^0$
4	$x^{247}+x^{38}+x^9+x^7+x^6+x^5+x^4+x^3+x^2+x^1+x^0$
	Polynomials of level 251 GF(2) used
1	$x^{251}+x^{64}+x^8+x^7+x^6+x^5+x^4+x^3+x^2+x^1+x^0$
2	$x^{251}+x^{72}+x^8+x^7+x^6+x^5+x^4+x^3+x^2+x^1+x^0$
3	$x^{251}+x^{105}+x^8+x^7+x^6+x^5+x^4+x^3+x^2+x^1+x^0$
4	$x^{251}+x^{143}+x^9+x^7+x^6+x^5+x^4+x^3+x^2+x^1+x^0$
	Polynomials of level 241 GF(2) used
1	$x^{241}+x^{14}+x^8+x^7+x^6+x^5+x^4+x^3+x^2+x^1+x^0$
2	$x^{241}+x^{60}+x^8+x^7+x^6+x^5+x^4+x^3+x^2+x^1+x^0$
3	$x^{241}+x^{224}+x^8+x^7+x^6+x^5+x^4+x^3+x^2+x^1+x^0$
4	$x^{241}+x^{54}+x^9+x^7+x^6+x^5+x^4+x^3+x^2+x^1+x^0$
	Polynomial of level 17 GF(5) used
1	$x^{17}+3x^2+2.x^1+3.x^0$

Via introduction of dynamic change of the feedbacks the priory relativity is additionally increased.

3 Test Results

For the test of statistic characteristics of the generated sequences by the proposed algorithm the NIST's test suite is used [3], [4]. Input values for tested bit streams are:

```
Block Frequency:           block length     = 128
Overlapping Templates:     template length  = 9
Non-overlapping Templates: template length  = 9
Serial:                    block length     = 16
Approximate Entropy:       block length     = 10
Linear Complexity:         block length     = 500
Universal:                 number of blocks = 7
                           block length     = 1280
Number of bit streams generated             = 1000
Length of bit streams                       = 1000000
```

The statistic results for the incorporated generator are given in Table 3

Table 3. Received statistical results

	P-VALUE	PROPORTION	Y/N	STATISTICAL TEST
1.	0.699313	0.989	pass	frequency
2.	0.476911	0.983	pass	block-frequency
3.	0.881662	0.993	pass	cumulative-sums
4.	0.779188	0.989	pass	runs
5.	0.076187	0.994	pass	longest-run
6.	0.676615	0.992	pass	rank
7.	0.32985	0.994	pass	fft
8.	0.872425	0.989	pass	nonperiodic-templates
9.	0.440975	0.987	pass	overlapping-templates
10.	0.894918	0.99	pass	universal
11.	0.008266	0.991	pass	apen
12.	0.276873	0.9882	pass	random-excursions
13.	0.470757	0.9831	pass	random-excursions-variant
14.	0.15991	0.988	pass	serial
15.	0.000276	0.989	pass	lempel-ziv
16.	0.286836	0.992	pass	linear-complexity

The results in table 3 show that all the analyzed sequences have successfully passed the tests and meet the necessary norms.

Additional results of testing Keystream Generator are shown in figure 3 to 8.

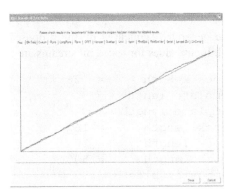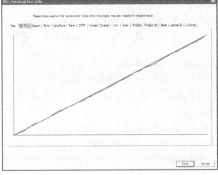

Fig. 3. Keystream Generator Frequency Test Results

Fig. 4. Keystream Generator Block-Frequency Test Results

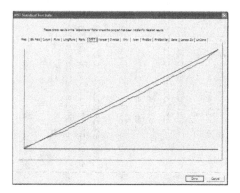

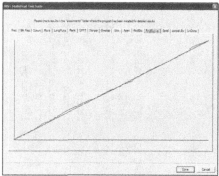

Fig. 5. Keystream Generator DFFT Test Results

Fig. 6. Keystream Generator Random-Excursions-Variant Test Results

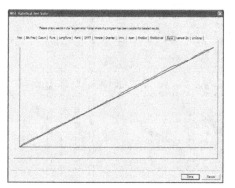

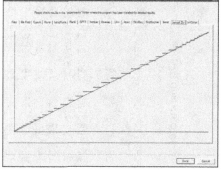

Fig. 7. Keystream Generator Serial Test Results

Fig. 8. Keystream Generator Lempel-Ziv Test Results

Results of testing the keystream generator.

RESULTS FOR THE UNIFORMITY OF P-VALUES AND THE PROPORTION OF PASSING SEQUENCES

C1	C2	C3	C4	C5	C6	C7	C8	C9	C10	P-VALUE	PROPORTION	STATISTICAL TEST
94	107	85	106	111	94	96	82	120	105	0.177628	0.9920	frequency
94	104	107	99	109	91	114	87	105	90	0.581082	0.9840	block-frequency
96	94	111	93	111	107	87	102	94	105	0.713641	0.9920	cumulative-sums
97	93	101	87	97	98	116	129	94	88	0.086109	0.9930	runs
109	100	81	101	108	115	87	95	98	106	0.378705	0.9880	longest-run
103	94	99	89	89	95	96	110	123	102	0.382115	0.9910	rank
103	129	97	113	95	109	104	89	98	63	0.001730	0.9980	fft
91	96	99	113	93	105	104	95	98	106	0.896345	0.9910	nonperiodic-templates
112	104	101	104	82	92	117	93	95	100	0.411840	0.9890	universal
94	100	96	102	102	103	89	96	102	116	0.862883	0.9880	apen
60	59	76	67	64	63	57	65	58	58	0.841281	0.9936	random-excursions
54	60	73	60	69	62	53	65	67	64	0.757536	0.9920	random-excursions-variant
92	99	102	101	105	81	107	91	118	104	0.413628	0.9920	serial
100	76	103	126	88	105	94	98	114	96	0.051611	0.9920	lempel-ziv
110	93	76	119	111	104	100	106	96	85	0.090936	0.9910	linear-complexity

The minimum pass rate for each statistical test with the exception of the random excursion (variant) test is approximately = 0.980561 for a sample size = 1000 binary sequences. The minimum pass rate for the random excursion (variant) test is approximately 0.978079 for a sample size = 627 binary sequences.

4 Conclusion

Key stream generators or so called Pseudo Random Number Generators are widely used for creating bit streams with good statistical characteristics. LFSR are mostly used for creating of Pseudo Random Number Generator (PRNG). Most of the used generators don't pass all statistical tests of NIST's suite. Proposed in this paper generator based on manageable feedback with usage of four groups of irreducible polynomials of degree 257, 251, 247 and 241 have very good statistic characteristics. This is proof by received results from NIST's suite. Also this way to create bit stream generator is novel in other works generators have only one polynomial in each group and feedback is nod based on polynomial. Have in mind received results it is our contention that this generator can be used for key stream generation in various applications and cryptographic algorithms and devices. Also this PRNG can be used in existing or novel stream ciphers.

Acknowledgments. This work has been funded by the EU Project INDECT (Intelligent information system supporting observation, searching and detection for security of citizens in urban environment) — grant agreement number: 218086.

References

1. Menezes, A., van Oorschot, P., Vanstone, S.: Handbook of Applied Cryptography. CRC Press (1996)
2. van Tilborg, H.C.A.: Encyclopaedia of Cryptography and Security. Springer Science+Business Media, Inc. (2005) ISBN-10: (eBook) 0-387-23483-7
3. Rukhin, A., Soto, J., Nechvatal, J., Smid, M., Barker, E., Leigh, S., Levenson, M., Vangel, M., Banks, D., Heckert, A., Dray, J., Vo, S.: A Statistical Test Suite for Random and Pseudo-Random Number Generators for Cryptographic Application, p. 162. NIST Special Publication 800-22 (with revision May 15, 2001)
4. Soto, J.: Statistical Testing of Random Number Generators, p. 12. NIST Special Publication

Detecting Predatory Behaviour from Online Textual Chats

Suraj Jung Pandey, Ioannis Klapaftis, and Suresh Manandhar

University of York,
Heslington, York, YO10 5GH, UK
{suraj,giannis,suresh}@cs.york.ac.uk

Abstract. This paper presents a novel methodology for learning the behavioural profiles of sexual predators by using state-of-the-art machine learning and computational linguistics methods. The presented methodology targets at distinguishing between predatory and non-predatory conversations and is evaluated in real-world data. All the text fragments within a malicious chat is not of predatory nature. Thus it is necessary to distinguish the predatory fragments from non-predatory ones. This distinction is made by implementing the notion of n-grams which captures predatory sequences from conversations. The paper uses as features both content words and stylistic features within conversations. The content words are weighed using tf-idf measure. Experiments show that content words alone are not enough to make distinction between predatory and non-predatory chats. The implementation of various stylistic features however improves the performance of the system.

Keywords: natural language processing, svm, text classification, offensive chats.

1 Introduction

Communicating online is becoming easier, effective and cheaper every passing day. Whether it's instant messaging, emails or the variety of social networking sites, we now have a plethora of ways to communicate and share information with each other. Such online, usually textual based conversation serves as efficient and instantaneous platform for communication. Unfortunately, these advantages of the online chats can also be used for criminal offences – a sexual predator can easily use these mediums to gain access and exploit their victims. This treacherous crime gets worse when the online chats are used by adult sexual predators for exploiting children. These predators use the dark shadows of the Internet and its ease to communicate with unsuspecting children with the motive of engaging in sexual activity.

Crime statistics have been collected since 1857, hence the amount of collected data is the most important resource for discovering trends in crime and analysing crime patterns [1]. Chat logs of convicted sexual predators are recorded. Such chat logs can be used as a queue to identify any malicious ongoing chat. This

A. Dziech and A. Czyżewski (Eds.): MCSS 2012, CCIS 287, pp. 270–281, 2012.
© Springer-Verlag Berlin Heidelberg 2012

can be achieved by measuring the similarity between the recorded chats and live chats.

The manual analysis of collected data is a time-consuming, error-prone and labour-intensive process that is unable to deal with the vast amounts of data that exists in the Internet. Behavioural profiling adds computational support to the task of crime data analysis in order to overcome the strain put on human resources. This computational support focuses on learning the behavioural patterns of known offenders based on solved crimes and then using these patterns to predict the location or the time crimes are likely to occur, identify their perpetrators [2] etc. Note that behavioural profiling does not refer to racial profiling, which is unethical, illegal and ineffective [2]. In contrast, it refers to capturing and modeling certain characteristics of offenders that are expressed during an illegal activity and using these in a proactive setting.

This paper presents a method for behavioural profiling; especially for detecting sexual predators by using state-of-the-art machine learning and computational linguistics methods. The model is created by training on both predatory chat logs and non-predatory chat logs. The chat logs are collected from various online forums, blogs and social networks. The resulting software is able to detect and raise an alarm whenever it senses a predator type activity in any monitored live chats. For building a classifier we use Support Vector Machine (SVM). The choice of SVM is due to its high performance in other text classification task [3].

The major contribution of this paper is that it presents a model which can learn from archived chats whether the chats are of predatory nature or not. The paper concludes that using words as features do not provide sufficient evidence in distinguishing predatory chats from the chats of sexual nature between two adults. The other contribution of this paper is the analysis we present about the predatory chats. We conclude that there is a certain pattern followed by predators to lure unsuspecting children. For example, they do not rush directly into conversation of sexual nature. We capture such pattern by using the concept of n-grams. We convert each chat logs into series of n-gram feature vectors. The paper also explains the use of stylistic feature present in chat logs. We then conclude that these features are essential in distinguishing between predatory chats and the chats of sexual nature between two adults.

2   Related Work

In this section, we provide a brief survey of the field of behavioural or offender profiling and show how different methods and features can be used to detect and locate offenders.

It has been shown from data analysis that an offender shows more tendency to target certain location and commit offence in shorter duration. Same hypothesis can be applied to sexual predators. Using significance testing [4] and clustering method [5] it has been shown that the repeat offence is done mostly in same surrounding and in shorter span of time. The clustering of the offence was done using SOM [6]. Although not tackled in this paper, such data can be used to get informations about the chat servers we need to monitor regularly.

Bache et al. [7,8] presented a language modeling method for linking behavioural features with characteristics of offenders. The main assumption behind such model is that offender committing a crime generates a document that contains offenders characteristics (gender, age, occupation etc.) and the type of crime they committed (burglary, shoplifting, assault etc.). The vocabulary in such documents can be used to create models that link the characteristics of offenders to the different type of crimes. Such models use likelihood function (crime given certain characteristics) and prior knowledge to infer possible suspects characteristics. We also use the terms in the documents to create models for detecting sexual predators and also use likelihood function to weight these terms.

Another field of behavioural profiling is authorship identification. Authorship identification is the task of associating a document with its original author. Emails can be used not only for legitimate activities, but also for illegal activities like threats, terrorism promotion etc. By learning the writing style of any person, authorship identification can be used to detect (1) if a document was not authored by a certain person (misuse of others identity) and (2) if a document was authored by a certain suspect.

Given a reasonable amount of data written by a certain author, different patterns can be learned in order to build a profile for that author. The profile can later be used to verify if the owner of a document is the person in concern. A simple supervised approach to learn patterns like proportion of blank lines, average sentence length, presence of greeting, signature etc. show highly encouraging results to detect legitimate authors [9]. We term these features as stylistic feature and use such features to detect pattern which identifies sexual predators.

3   A Method for Detecting Online Sexual Predators

This section presents a novel methodology for exploiting unstructured textual data for the task of behavioural profiling, and specifically for identifying predatory conversations, in which one of the persons involved in the conversation is a sexual predator. The pipeline of the proposed methodology is shown in Figure 1.

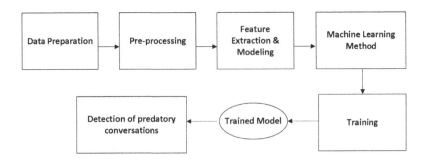

Fig. 1. Behavioural Profiling: Pipeline of our method

In the next section, we describe the main components of each stage of the unified methodology along with examples that highlight the main tasks performed.

3.1 Data Preparation

Data preparation is the stage of collecting, aggregating and storing out training and testing data under a common unified format. The task of data preparation is divided into two parts: (1) download conversations from forums, blogs and other social networks, and (2) classify each conversation as *predatory* or *non-predatory*, our conversations are stored in standard textual format an example of which is shown below:

Person_A (12/23/06 7:16:53 AM): hi
Person_B (12/23/06 7:16:54 AM): pfffttttttt.....loser
Person_B (12/23/06 8:34:29 AM): sup
Person_A (12/23/06 8:34:42 AM): location?
Person_A (12/30/06 2:21:54 PM): hello

In this example, each line corresponds to a different person, date and time are eclosed in parenthesis, while the text after a semicolon corresponds to a conversation. Such a format, allows us to reduce size of data compared to using a XML based representation, it allows us to easily apply the Natural Language Processing tools for feature extraction, and finally it is easy to read and inspect.

3.2 Pre-processing

The aim of this stage is to apply a set of pre-compiled NLP tools to the unstructured text produced by the *Data Processing* stage, so as to extract a set of contextual and stylistic features that might be useful for the task of behavioural profiling. As can be observed in Figure 2, the pre-processing pipeline consists of six stages analysed below:

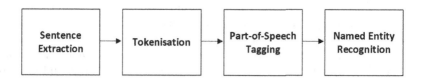

Fig. 2. Text processing

- **Sentence Extraction:**
  In this stage, the text given as an input is segmented into sentences. For instance given the text: *Hello! How are you! George you like to chat with me?*
  the following three sentences will be the output of this stage
  1. *Hello!*
  2. *How are you!*
  3. *Would you like to chat with me ?*

- **Tokenisation:**
  In this stage, the tokens forming one coherent sentence are separate, in order to perform the next stages of the pre-processing pipeline. Given the third sentence of our example, the outcome of tokenisation is shown in first column of Table 1.

Table 1. Pre-processing pipeline example

Token	Lemma	POS	NER
George	would	proper noun	Person
you	you	pronoun	N/A
like	like	verb	N/A
to	to	to	N/A
chat	chat	verb	N/A
with	with	preposition	N/A
me	me	pronoun	N/A

- **Part-of-speech Tagging:**
  Part-Of-Speech (POS) tagging refers to tagging the words in a given sentence according to their corresponding to a particular part of speech, based on both its dictionary definition and the context in which it appears. Note that in our pre-processing, the output of POS tagging also includes an additional sub-stage , i.e. *lemmatisation*, whose aim is to group together the different inflected forms of a given word and represent them by a common string, i.e. the lemma. Columns 2 and 3 of Table 1 provide the lemmas and POS tags of all the words of our example. We use Stanford's POS tagger [10].
- **Named Entity Recognition (NER):**
  NER is the process of automatically assigning words or phrases to a set of predefined categories, such *PERSON, ORGANISATION, LOCATION* and others. NER is an important and significant stage of the pre-processing stage, since it allows the extraction of semantic features compared to contextual ones. For instance, in our example (Table 1) the output of lemmatisation and POS-tagging for the token *George* is *George* and *proper noun*. In this setting, *George* is considered as a noun in the sentence and no other information can be inferred. However, by applying NER we can infer that *George* is an *Person*, a semantic feature that can be exploited in the next stages of our methodology. We use Stanford's NER [11].

3.3   Feature Extraction and Modeling

The aim of feature extraction is to extract the features considered to be important, so as to associate a given criminal offense with the corresponding feature set. The majority of supervised methods in behavioural profiling cast the problem as a classification task, in which a set of annotated training training documents are used to learn a classification model. The learned model is then used to classify new unseen documents.

In our setting, a naive formulation would consider as a document the whole chat between two persons. The chat could then be formulated as a vector of features (word, named entities, etc.) weighted according to their contextual importance within the chat. In contrast to this naive formulation followed by traditional supervised methods, in our work we explicitly acknowledge the fact that certain parts of a dialogue between a sexual predator and an under-age person are normal and cannot be considered as illegal. For example, given the following chat taken by a conversation between a sexual predator and an under-age person, we observe that none of the phrases used indicate any illegal activity.

Person_A (12/23/06 7:16:53 AM): hi
Person_B (12/23/06 7:16:54 AM): pfffttttttt.....loser
Person_B (12/23/06 8:34:29 AM): sup
Person_A (12/23/06 8:34:42 AM): location?
Person_A (12/30/06 2:21:54 PM): hello

Only some of the fragments in the chat indicate predatory behaviour as shown in the example below:

Person 1 [11:44 A.M.]: how old ru?
Person 2 [11:44 A.M.]: 45
Person 1 [11:44 A.M.]: —13
Person 2 [11:44 A.M.]: you look very hot
Person 2 [11:45 A.M.]: and sexy

Therefore, our target in a given conversation is to identify such fragments, which can be considered as predatory. Next, given the collected fragments we could classify a chat as predatory or normal based on the proportion of detected predatory fragments. To formulate the solution to our problem we use the notion of a n-gram. An n-gram is a contiguous sequence of n items from a given sequence of text or speech. Hence, given a chat between two people, we extract all line 3-grams of the given chat where each (line) gram is a line entered by a user. The following (line) 3-grams were extracted from the chat of our second example:

1. 3-gram, ID:1
   Person 1 [11:44 A.M.]: how old ru?
   Person 2 [11:44 A.M.]: 45
   Person 1 [11:44 A.M.]: —13

2. 3-gram, ID:2
   Person 2 [11:44 A.M.]: 45
   Person 1 [11:44 A.M.]: —13
   Person 2 [11:44 A.M.]: you look very hot

3. 3-gram, ID:3
    Person 1 [11:44 A.M.]: —13
    Person 2 [11:44 A.M.]: you look very hot
    Person 2 [11:45 A.M.]: and sexy

Once, we have extracted all the 3-grams of a given chat we need to model computationally each one of them. This is achieved by associating each extracted 3-gram with a vector of features. Traditional approaches to Natural Language Processing use as features words, or bigrams weighted by different statistical measures such as *pointwise mutual information* or *TF.IDF*. Such features aim at capturing the semantics of a given document and computationally model its meaning. In our work we do not only follow this approach, but we also introduce a set of stylistic features, whose aims is to capture the writing style of different people. Such features include the number of upper-case letters, the use of pronouns, the average sentence length etc. The inclusion of such features aims to distinguish between people that differ in age, habits, mental state etc. Table 2 shows the contextual and stylistic features used in our work.

Table 2. Contextual & Stylistic Features of our Method

Feature	Type	Weighting
Nouns	Contextual	TF.IDF
Verbs	Contextual	TF.IDF
Named Entities	Contextual	TF.IDF
Bigrams	Contextual	TF.IDF
Average word length	Stylistic	COUNT
Number of function words	Stylistic	COUNT
Average number of characters	Stylistic	COUNT
Average number of uppercase characters	Stylistic	COUNT
Average number of alphabetic characters	Stylistic	COUNT
Average number of digits	Stylistic	COUNT
Total number of punctuations	Stylistic	COUNT

As can be observed in Table 2 we use four kinds of contextual features, i.e. nouns, verbs, named entities and bigrams weighted by TF.IDF. Let a chat $C = \{X_1, X_2, \ldots, X_n\}$ consist of n 3-grams. The TF.IDF weight of feature f_i in 3-gram X_j, is defined in Equation 1, where $F(f_i, X_j)$ is the frequency of feature f_i in X_j, $|X_j|$ is the size of X_j, $|C|$ is the total number of extracted 3-grams of a chat and $1 \leq k \leq |C|$.

$$TF.IDF(f_i, X_j) = \frac{F(f_i, X_j)}{|X_j|} \times \log \frac{|C|}{|X_k : f_i \in X_k|} \qquad (1)$$

3.4   Machine Learning Method

The training of our model and the classification of extracted 3-grams is based on Support Vector Machines (SVMs). SVMs are one of the most popular and reliable supervised learning methods that have been used in different tasks including text classification [3], question classification [12], word sense disambiguation [13,14] and relation extraction [15].

Let us assume that we have a set of input vectors X_i (one vector for each 3-gram), while the target is to predict their labels, i.e. $Y_i = +/-1$ (predatory or non-predatory). The theoretical bounds on the generalization error specify the following:

- The upper bound on the generalization error does not depend on the dimensionality of the space.
- The bound is minimized by maximizing a quantity called the margin (Figure 3), i.e. the minimal distance between the hyperplane separating the two classes (1, -1) and the closest data points (support vectors) of each class.

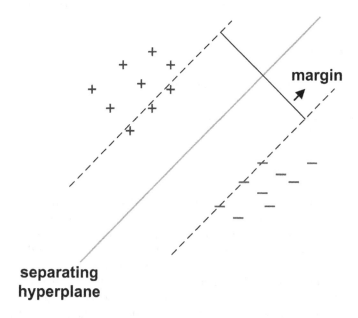

Fig. 3. Hyperplane separating 2 classes

A hyperplane can be written as a set of points x satisfying $w \cdot x - b = 0$, where w is a normal vector of weights perpendicular to the hyperplane and b is a constant. Note that $\cdot$ denotes the dot product. Maximizing the margin, i.e. the minimal distance between the hyperplane and the two classes (1,-1), is equivalent to minimizing $\frac{\|w\|}{2}$, subject to the constraint that $y_i \times (w \cdot x) - b \geq 1$.

$$W(\alpha) = \sum_{i=1}^{m} \alpha_i - \frac{1}{2} \sum_{i,j=1}^{m} a_i a_j y_i (x_i \cdot x_j) \tag{2}$$

This constraint optimization problem can formulated as finding the optimum of a quadratic function 2, where α_i are Lagrange multipliers and m is the number of training pairs (x_i, y_i). Finally for binary classification the decision function is shown in Equation 3, where q is a new test instance.

$$D(q) = sign(\sum_{j=1}^{m} a_j y_j (x_j \cdot q) + b)) \tag{3}$$

Using Equation 3, we can classify a given 3-gram as illegal or not. In the next step, we decide whether the whole chat or conversation should be considered as predatory or not, so as to raise an alarm. Let C be a chat and $PD(C)$ the set of 3-grams classified as illegal. Then chat C is considered as predatory if $\frac{PD(C)}{|C|} \geq h$, where h is a user-defined threshold. In other words, a chat C is considered as predatory, if the proportion of illegal 3-grams is greater than h.

4   Evaluation

In this section we present the results of our evaluation. In the first section, we describe the evaluation setting and datasets, next we evaluate our classification accuracy on extracted trigrams and finally we present the complete evaluation on new chats and conversations.

4.1   Evaluation Setting

The annotated data used for training the SVM classifier come from two different sources. The predatory or illegal conversations come from the website *http://www.pervertedjustice.com*. This organisation consists of volunteers, whose pose themselves as under-age persons and engage into chats with different people. Whenever a chat is considered as illegal from a criminal justice court and the corresponding adult person is convicted by that court, his/her chats aer made publicly available. The predatory conversations we collected from that website are 548. Similarly, for our non-predatory conversions we made use of the website *http://www.omegle.com*, a web place, where people can chat about a variety of topics and issues.

Our method is evaluated in two modes that are described in the next two sections. In the first one, we train our Support Vector Machine on a set of positive and negative 3-grams and evaluate the classification accuracy by applying 10-fold cross validation. In the second setting, we evaluate the classification accuracy in whole chats by varying the parameter h of our method (Section 3.4). For all experiments we used LIBSVM [16].

4.2   Evaluating Classification Accuracy on 3-Grams

Table 3 shows the results of our evaluation on classifying 3-grams. As can be observed, our method achieves a classification accuracy of 73.71%. The *Random* baseline that randomly classifies a 3-gram as predatory or not achieves an accuracy of 50%, which is significantly lower than the performance of our method.

Table 3. Results on 3-grams

SVM-Feature	Classification Accuracy (%)
Unigram + Bigram + Stylistic	72.57%
Unigram + Stylistic	70.71
Unigram (words)	50.0
Bigrams (words)	50.5
Stylistic (words)	68.66
Random	50

Furthermore, we observe that content features such as unigrams and bigrams perform as well as the *Random* baseline and do not clearly improve upon that. In contrast, our choice of stylistic features performs significantly better than content ones (68.66%), which shows that the adopted features are able to capture the writing style of predators as well as of non-predators. Our future work aims at focusing and extending the stylistic feature set used in this work.

Finally, the combination of content and stylistic features appears to perform best and shows that although content features are generally poor discriminators, they capture a small amount of information that stylistic features do not, in effect increasing the classification accuracy.

4.3   Evaluating Classification Accuracy on Chats

In this evaluation setting we used our two best forming models to classify chats as predatory and non-predatory.

Table 4. Results on chats

SVM-Feature	Classification Accuracy (%)
Unigram + Bigram + Stylistic	74.8%
Unigram + Stylistic	76.23%

Table 4 shows the results of our evaluation on the chat data. For the training, we fist randomly separated 25% of chat files from our dataset. Then we converted remaining 75% into 3-grams features and trained the model. Then from the 25% data we converted each chat separately into 3-grams. If for each chat file (containing its collection of 3-grams) more than 50% of 3-grams in that file are classified correctly than we considered the current chat file to be classified correctly. Thus, we can see that within this setting we can classify the whole chats as either predatory and non-predatory. This is important because although we trained in 3-grams we need an mechanism to classify the whole chat data. In future we aim to focus more on the stylistic feature and develop more sophisticated method to convert 3-grams results to an overall chat results.

5   Conclusion

This paper has presented a novel method for detecting online sexual predators. Our background survey has shown that most behavioural profiling methods focus on geographical profiling, whose aim is to predict locations considered to be likely for an offence to take place. Standard language modeling methods have also been applied to infer the characteristics of a given offender. Additionally, standard machine learning methods such as SVMs have been applied to tasks related to behavioural profiling such as authorship identification.

In this paper, we have presented a methodology that aims to identify sexual predators from chats or conversations in forums, blogs and other social networks. We presented a method of detecting pattern in conversation using the notion of n-grams. Our method can classify a single chat in less than 30 seconds, the processing time including both pre-processing and classification. We also showed that words alone are not enough to detect predatory chats. Accuracy of detection of predatory chats increases once we use the stylistic features.

References

1. Grover, V., Adderley, R., Bramer, M.: Review of current crime prediction techniques. In: Ellis, A.T.R., Allen, T. (eds.) Applications and Innovations in Intelligent Systems XIV, pp. 233–247 (2007)
2. Mena, J.: Investigative Data Mining for Security and Criminal Detection. Academic Pr. Inc. (April 2003)
3. Joachims, T.: Text Categorization with Support Vector Machines: Learning with Many Relevant Features. In: Nédellec, C., Rouveirol, C. (eds.) ECML 1998. LNCS, vol. 1398, pp. 137–142. Springer, Heidelberg (1998), http://citeseer.ist.psu.edu/joachims97text.html
4. Johnson, S.D., Bowers, K.J.: The burglary as clue to the future: The beginnings of prospective hot-spotting. European Journal of Criminology 1(2), 237–255 (2004)
5. Adderley, R.: The Use of Data Mining Techniques in Operational Crime Fighting. In: Chen, H., Moore, R., Zeng, D.D., Leavitt, J. (eds.) ISI 2004. LNCS, vol. 3073, pp. 418–425. Springer, Heidelberg (2004)
6. Kohonen, T.: Self-organized formation of topologically correct feature maps, pp. 509–521 (1988)
7. Bache, R., Crestani, F.: Estimating real-valued characteristics of criminals from their recorded crimes. In: CIKM 2008: Proceeding of the 17th ACM Conference on Information and Knowledge Management, pp. 1385–1386. ACM, New York (2008)

8. Bache, R., Crestani, F., Canter, D., Youngs, D.: A language modelling approach to linking criminal styles with offender characteristics. Data & Knowledge Engineering 69(3), 303–315 (2010)
9. de Vel, O., Anderson, A., Corney, M., Mohay, G.: Mining e-mail content for author identification forensics. SIGMOD Rec. 30(4), 55–64 (2001)
10. Toutanova, K., Klein, D., Manning, C.D., Singer, Y.: Feature-rich part-of-speech tagging with a cyclic dependency network. In: Proceedings of HLT-NAACL, pp. 252–259 (2003)
11. Finkel, J.R., Grenager, T., Manning, C.: Incorporating non-local information into information extraction systems by gibbs sampling. In: Proceedings of the 43rd Annual Meeting on Association for Computational Linguistics, ACL 2005, pp. 363–370. Association for Computational Linguistics, Stroudsburg (2005), http://dx.doi.org/10.3115/1219840.1219885
12. Moschitti, A., Quarteroni, S., Basili, R., Manandhar, S.: Exploiting syntactic and shallow semantic kernels for question answer classification. In: Proceedings of the 45th Annual Meeting of the Association for Computational Linguistics. Association for Computational Linguistics (2007)
13. Joshi, M., Pedersen, T., Maclin, R., Pakhomov, S.: Kernel methods for word sense disambiguation and acronym expansion. In: Proceedings of the 21st National Conference on Artificial Intelligence, vol. 2, pp. 1879–1880. AAAI Press (2006), http://portal.acm.org/citation.cfm?id=1597348.1597488
14. Lee, Y.K., Ng, H.T., Chia, T.K.: Supervised word sense disambiguation with support vector machines and multiple knowledge sources. In: Mihalcea, R., Edmonds, P. (eds.) Senseval-3: Third International Workshop on the Evaluation of Systems for the Semantic Analysis of Text, pp. 137–140. Association for Computational Linguistics, Barcelona (2004)
15. Zelenko, D., Aone, C., Richardella, A.: Kernel methods for relation extraction. J. Mach. Learn. Res. 3, 1083–1106 (2003), http://portal.acm.org/citation.cfm?id=944919.944964
16. Chang, C.-C., Lin, C.-J.: LIBSVM: A library for support vector machines. ACM Transactions on Intelligent Systems and Technology 2, 27:1–27:27 (2011) software, http://www.csie.ntu.edu.tw/~cjlin/libsvm

Voice Trust in Public Switched Telephone Networks

Zbigniew Piotrowski and Waldemar Grabiec

Military University of Technology, Faculty of Electronics,
gen. S. Kaliskiego 2 str., 00-908 Warsaw, Poland
{zpiotrowski,wgrabiec}@wat.edu.pl

Abstract. This paper discusses the results of research work performed to change the speaker's voice profile and edit the voice message in an unauthorized way. The results of conducted research work confirm that the intentional voice modeling in telephone lines can result in gaining or losing trust in the subscriber, thus gaining or losing access to confidential messages in an unauthorized way. The analysis of security regarding the trust in voice identity leads to developing new, unconventional protection methods: using hidden identification in the form of watermarking tokens and the tests of susceptibility to impersonation of the key corporation personnel and government officials.

Keywords: voice trust, impersonation, voice morphing, voice profile, voice message, watermarking token, hidden identification, data hiding.

1 Introduction

As a result of technology development and its widespread accessibility, the techniques used by intruders in telephone lines and radio networks become more and more refined and effective. Using commercial, advanced speech processing algorithms combined with proven socio-technical mechanisms in order to steal (impersonate) someone's voice identity in open, unsecured lines becomes more and more common. Here we should mention the methods of swindling especially elderly persons out of information, using the so-called "grandson impersonation method" [1] or obtaining unauthorized voice access to the call-centers of commercial banks.

Another problem is "injecting" unauthorized voice messages into already preset voice tracts or modifying key speech phrases in real-time both in unilateral and bilateral mode. The research work performed shows that it is hard for a subscriber to tell the difference between a well-prepared false voice message and the original one. It is even more difficult because in the established connections the subscriber encounters the so-called communication quality of the signal i.e. a considerable decrease in speech quality and intelligibility. The quality and intelligibility worsens as a result of limiting signals to the frequency passband of a telephone line (PSTN, GSM) and radio band (HF/VHF), including by using

A. Dziech and A. Czyżewski (Eds.): MCSS 2012, CCIS 287, pp. 282–291, 2012.

lossy compression codecs (GSM, WiFi) and the occurrence of noise and artifacts in the connection established.

An important aspect here is the matter of trust in the subscriber. The confidence, as shown by research work performed, is "allocated" subjectively. A verification is conducted during the conversation by allocating the voice heard to the voice profile of the person known to the subscriber and by building a knowledge profile about this subscriber based on the speech context. This profile, as the telephone or radio conversation continues, is usually effectively verified by checking the subscriber for shared knowledge (the so-called shared secret), however in the case of short conversations, the key point is the recognition of the subscriber's voice profile at the very beginning of the call. One of the more interesting and effective techniques developed is the change of subscriber's voice profile during the conversation. In spite of the shared knowledge profile of the subscribers, the voice profile verification turns out to be negative. In this case the telephone talk is usually intentionally interrupted by the subscriber who has verified their interlocutor's voice profile negatively. The impersonation engineering can also adopt another option: known and approved voice profile at negatively verified knowledge profile about the subscriber.

2   Definition of Trust in Voice Profile and the Voice Profile Delegation

There are a lot of definitions for trust, for example, [2] shows that the trust is an assured reliance on the character, ability, strength or truth of someone or something, one in which confidence is placed. In [3] the trust definition is as follows: a relationship created at the direction of an individual, in which one or more persons hold the individual's property subject to certain duties to use and protect it for the benefit of others. Furthermore in [3] there is the basic concept: the person who creates the trust is the settlor. The person who holds the property for another's benefit is the trustee. The person who is benefited by the trust is the beneficiary, or cestui que trust. The property that comprises the trust is the trust res, corpus, principal, or subject matter. For example, a parent signs over certain stock to a bank to manage for a child, with instructions to give the dividend checks to him each year until he becomes 21 years of age, at which time he is to receive all the stock. The parent is the settlor, the bank is the trustee, the stock is the trust res, and the child is the beneficiary [3].

In [4] we find out that the trust is confidence in the honesty or integrity of a person or thing, and moreover trust means to have confidence, faith or hope in someone or something. It is possible to formulate the following definition for the trust in voice profile. The trust in the voice profile is the certainty of a given voice service subscriber that the voice heard belongs to the specified subscriber. The specific subscriber can be particularly a person they know or do not know. The subscriber's voice properly allocated to a given person via the subjective allocation process by the human auditory system (HAS) authorizes access to a selected range of subscriber's information. Other range of information is available to the subscriber being a member of a family, commune, design team, co-workers from a company etc. while other range is available for a person the subscriber does not know.

The introduction of the hidden identification of the subscriber in the form of embedding digital watermark signature in a speech signal during an established communication session [5][6] protects against the subscriber's impersonation and against an unauthorized message edition and, consequently, may lead to an intentional voice identity delegation. The "A" subscriber with the digital watermark binary signature "α" can delegate his signature to other voice profile (the "B" subscriber) to represent the „A" subscriber and "act" on behalf of the settlor (person who establishes trust). In this case the "B" subscriber represents the "A" subscriber by his profile, as he proves his identity with the digital watermark binary signature "α". In particular, the artificially generated virtual voice synthesized via text-to-speech (TTS) techniques in telephone systems can be allocated to a specified subscriber. The impersonation, i.e. stealing the voice identity of such a "virtual" subscriber is far more difficult because of the necessity to gain access to the binary watermarking signature defining the subscriber's voice identity. Without the mechanism of hidden watermark signature embedding in the speech signal, it is not possible to delegate authorization resulting from a given voice profile.

3   Experimental Results - Voice Trust in PSTN

Public Switched Telephone Network (PSTN) is the oldest and as a result, the most popular information transfer system. The communication between PSTN network subscribers is performed via voice – being the most convenient medium for people, allowing users, at the same time, to recognize a mood, emotions etc. of the interlocutor. One of the basic risks for telephone communication (apart from bugging) is stealing subscriber's identity (impersonation or spoofing). Practically, all subscribers using unencrypted telephone lines are exposed to impersonation/spoofing. However, the following entities are especially exposed to such attacks: government officials, public order officers (e.g. police, paramilitary organizations, army), bank personnel (medium and higher level). To perform research work regarding the voice profile change a test stand for three playback (listening) tests has been prepared.

3.1   Test Bed Configuration

A schematic diagram of the test bed using the PSTN is presented in Fig.1 As a commutation component of the PSTN the Ascotel bcs64 private exchange manufactured by ASCOM [7] from Switzerland is used. This is a Private Automatic Branch Exchange (PABX) intended for small and medium companies from all industries. It allows for connecting from 20 to 190 subscribers and is designed to provide telecommunications in offices. Fitting the PABX with appropriate cards makes it possible to connect analog telephones with pulse or tone dialing Dual-Tone Multi Frequency (DTMF), group 4 faxes meeting the ETSI requirements

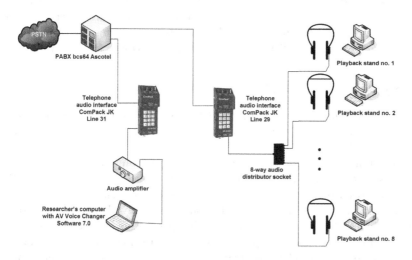

Fig. 1. Schematic diagram of the prepared test bed within a PSTN

or via the S contact – devices equipped with an ISDN card for PCs. A universal telephone audio interface ComPack [8] co-operated with a PSTN making it possible, i.a., to connect audio input from a PC. ComPack JK is equipped with a keypad to select the phone numbers of subscribers connected to the network and a simulation switch for picking up and hanging off the receiver (circuit switch). The device is fitted with a microphone headset and can be used as a regular telephone.

On the researcher's computer AV Voice Changer Software 7.0 Diamond was installed [9]. This software was used to provide voice (speech) morphing in real-time. The program made it possible to change man's, woman's or child's voice into a different one with any timbre or tone. It made it possible to record (the Recorder option) and play (the Player option) audio files with the morphing effect or without it. The program offers a wide set of functions, i.a. Voice Comparator, allowing for comparing and changing any voices.

The ComPack telephone interfaces were connected to the analog lines of the exchange. These lines have pre-defined numbers: 29 and 31. One ComPack interface (No. 29) through its Line Input port is connected with the researcher's computer containing voice samples for particular tests. To the Head phones port of the second ComPack (No. 31) eight playback stands were connected via an 8-way audio distributor socket. Each stand includes a set of stereo headphones (type: HT600 manufactured by Technics) and a PC. To automate the operations for the needs of the research work special web-based electronic forms were developed.

40 listeners participated in each experiment. The listeners taking part in the experiment were asked to answer, based on the listened recordings, questions on

a current basis by completing a form. After completing the experiments the forms were saved to the computer hard drive. Three tests were prepared with selected voice identity changes and voice message edition techniques for the need of the research. In each of the tests listeners played ten voice tracks (voice samples).

3.2   Processing Voice Samples

The experiments included: voice morphing (Test No. 1), voice message modification (Test No. 2) and composing the voice message transmitted via a telephone connection (Test No. 3).

Voice morphing - voice or speech morphing is a technique that involves converting one speaker's voice into the form sounding as other speaker's voice [10-15]. This transformation changes the distinctive speaker's features at the same time maintaining the voice message intelligibility. The key element of the voice morphing is changing the laryngeal tone and the spectral envelope corresponding to the vocal tract. The voice morphing is the basic technology used to change subscriber's voice identity. The voice transformation in test No. 1 was performed using the AV Voice Changer Software 7.0 Diamond.

Editing the voice message - a voice message can be edited by inserting or cutting specified voice phrases from the recorded voice message. By editing a voice message it is possible to change the speech context or weave in the voices for different speakers in particular speech phrases e.g. a fragment of the voice sample belongs to the speaker A, the next one belongs to the speaker B. Cutting or inserting certain key words e.g. no or yes can lead to changes in the context of the original message.

Voice message integrity - the damage to the integrity consists in composing a voice message using a few previously recorded voice materials (voice samples coming from the same speaker). The so-called speech generator (presented in Tab. 1) was used for the need of research work. One of the typical features of such speech generators is the fact that phrases and words read at any sequence make sentences that, in spite of making little sense, maintain their consistency. Speeches can be generated by connecting any phrase from the first column cell with the phrase read from any cell of the second column. The reader each time reads other fragments to avoid sequence repetition. Interestingly, the speech generator used in the research work allows for generating 10 thousand sentences, making it possible to make a "speech" lasting 40 hours.

3.3   Research Work Results

Test No. 1 - Voice Morphing. In this test listeners were asked to assess whether the voice material played (voice sample) was the voice of the person stealing the voice identity (attacker), the host voice, unmodified (target voice) or it was impossible to find it unambiguously who the voice sample belonged to. Each of the listeners before playing back the samples, familiarized themselves

Table 1. Speech generator used in test No. 3

1	2	3	4
Dear Colleagues	fulfilling the program tasks presented	forces us to analyze	the existing administration and financial conditions.
On the other hand	the scope and place of training the staff	plays an important role in forming	further development directions.
Similarly	the constant increase in the number and scope of our operations	requires precising and specifying	the system of common participation.
However, we should not forget that	the current organization structure	helps to prepare and complete	the attitudes of participants towards the tasks put by the organization.
This is the way	the new model of organizational operation	provides the participation for a wide group in forming	new proposals.
Everyday practice proves that	the further development of various operation forms	plays important roles in creating	the direction in progressive education.
There is no need for further justification of the problem importance because	the constant information and material protection of our operations	makes it possible to create to a arger extent	the system of personnel training to meet the proper needs.
Differentiated and extensive experience	enforcing and developing structures	makes us appreciate the importance of	the proper conditions for activity stimulation.
The care of the organization and particularly	consulting a wide group of staff	presents an interesting approach to check	the development model.
Higher conceptual assumptions, and	starting a widespread action of creating attitudes	results in the process of implementation and modernization	of interaction forms.

with an unmodified voice sample example and with the example of attacker's voice sample. Therefore, the target voice was the voice the attacker wanted to steal (impersonate). All the samples played in the PSTN were said by the attacker. The results of test No. 1 are shown in Fig. 2.

On the basis of results from test No. 1 it turns out that there is a non-zero probability of error in speaker's voice profile identification (for voice sample No.4,5,7,8,9 and 10 more than 10% of cases provided an error assessment showing that the voice belongs to the target), while for samples No. 1,3,5,7,8,9 and 10 listeners in 10% did not indicate the person the voice belonged to.

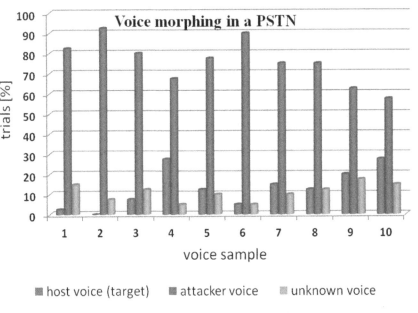

Fig. 2. Result of test No. 1 – voice morphing in a PSTN

Test No. 2 – Editing the Voice Message. The test consisted in specifying whether the voice message was modified (edited). The modifications involved cutting and inserting a voice phrase to a previously recorded signal. In voice samples (Fig.3) No.: 1-4 and 6-9 there were four such modifications, e.g. two cuts and two insertions or any other combination of four changes. The test listeners were informed in advance that only one out of five voice samples in the group was free from modifications. The test covered two groups containing five voice samples each. In each group the voice samples sounded similarly, but the last sentence within the group was the original, unmodified one. The first group of voice samples includes phrases from 1 to 5, whereas the second group of voice samples includes the phrases from 6 to 10 (Fig.3). All voice samples were performed by one reader. As a result of performing test No. 2 it was found that in 3 out of 10 cases the majority of listeners made a mistake indicating that the voice sample was unmodified (samples No. 1,2,6). Samples No. 5 and 10 were not subject to edition. The result of the experiment unambiguously shows that here is a non-zero probability of incorrect, subjective assessment made by a PSTN subscriber that a specified voice message should be recognized as original even if is intentionally edited.

Test No. 3. – Voice Composition. In this test the group of listeners (PSTN subscribers) was asked to assess whether a specified voice message is consistent and was recorded during one conversation or was composed of a few separately recorded voice messages. In each of the 10 cases the reader was the same (Fig.4). The voice samples no. 3 and 7 were free from any modifications (original and

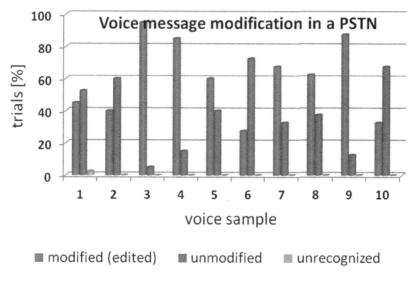

Fig. 3. Result of test No. 2 – voice message modification in a PSTN

consistent speeches). As a result of performing the test No. 3 it was found that in the case of consistent voice samples (samples No. 3 and 7) the majority of listeners (PSTN subscribers) assessed them as composed of separate recordings. Moreover, the listeners made a mistake while assessing that the sample No. 1 was consistent and maintained integrity between particular phrases. For samples No. 1,2,4,5,6,8,9 and 10 more than 10% assessments were erroneous and the listeners assessed the samples composed of separate phrases as consistent and maintaining integrity.

4   Methods of Defense against the Change of Voice Identity

As shown in [5] and in [6], using an information hiding technique can assist considerably in indentifying a subscriber, particularly in open, unencrypted connections, and allows for assessing the consistency of voice message sent. Watermark embedding in the host voice signal, representing the subscriber's Personal Identification Number, makes it possible to extract the binary watermark signature on the recipient's side of the telephone line. The message consistency is provided by the abbreviation function calculated on the basis of voice signal that is then compared with the one decoded at the recipient's side from the watermark. Another defense method is apparently prevention and providing trainings on the voice profile changes, the unauthorized edition of voice messages.

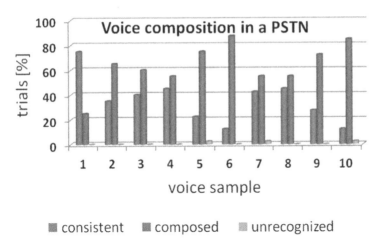

Fig. 4. Result of test No. 3 – damage to the integrity of voice message in a PSTN

5   Summary

The problem of unauthorized voice message identity in PSTNs using unencrypted connections plays an important role for the telecommunications security. Third parties have always been willing to acquire confidential information and such situation will still continue. Therefore, it should be taken into account that in addition to other attacks towards the classic PSTN subscribers (e.g. bugging) stealing the subscribers identity (impersonation) using voice impersonation methods and message edition will be more and more common. The attackers will often take advantage of psychological techniques, basing on mental weaknesses or a mere naivety and the complete negligence of subscribers attacked. The above-mentioned "grandson impersonation method" [1] is a perfect example for that. There are also examples from everyday life when strangers acquire information, sometimes very important, from their interlocutors during a telephone talk impersonating somebody else and even not changing their voice (e.g. the cases of acquiring important information by journalists from governmental officials). The state-of-the-art technology, including mainly the progress in digital signal processing results in developing a number of software and hardware implemented methods, allowing for effective voice impersonation.

Currently, many important state institutions use unencrypted telephone lines. This is mainly for financial reasons. Therefore, these are probably the places where voice impersonation attacks should be taken into account.

Apart from improving the security mechanisms (e.g. identification) an important element will be the training operations covering the potential victims of the attacks.

If the attacker uses a top quality voice morphing method it will be almost impossible to tell the difference between the original and processed voice of the interlocutor. However, paying attention, while talking on the phone, to certain distinctive interlocutor's voice features (e.g. voice breaking as a result of being nervous, unnatural rate of articulating words, etc.) makes it possible in a way to avoid the attack by e.g. interrupting the phone call or changing the subject of conversation into a more neutral one. The voice (speech) signal generated as a result of synthesis in an artificial way sounds to some extent unnatural, which can be detected by the human auditory system (HAS). Paying special attention to the certain above-listed articulation details can protect us, the telephony subscribers, against an unauthorized disclosure of important information to third parties. Summing up, the test results performed prove the attacks towards the voice identity of PSTN subscribers using open, unencrypted connections to be effective. The tests have been performed under conditions close to reality.

References

1. http://gorzow.gazeta.pl/gorzow/Na_wnuczka
2. http://www.merriam-webster.com/dictionary/trust
3. http://legal-dictionary.thefreedictionary.com/trust
4. http://www.yourdictionary.com/trust
5. Piotrowski, Z., Gajewski, P.: Identity management In VHF radio systems, Computational Methods and Experimental Measurements XV. In: CMEM XV, pp. 473–481. WIT Press, Southampton (2011) ISBN: 978-1-84564-540-3, ISSN: 1746-4064, ISSN: 1743-355X
6. Piotrowski, Z.: Drift Correction Modulation scheme for digital signal processing. Mathematical and Computer Modelling, September 16 (2011), doi: 10.1016/j.mcm
7. http://www.ascom.com/en/index/
8. http://www.jkaudio.com
9. http://www.audio4fun.com/
10. Abe, M.: Speech morphing by gradually changing spectrum parameter and fundamental frequency. In: Proceedings of International Conference on Spoken Language Processing, ICSLP, vol. 4, pp. 2235–2238. NTT Human Interface Laboratories, Japan (1996)
11. http://www.scribd.com/doc/17605831/Voice-Morphing
12. Yanagisawa, K., Huckvale, M.: Accent morphing as a technique to improve the intelligibility of foreign-accented speech. ICPhS XVII ID1486 Sarbriicken, August 6-10 (2007)
13. Furuya, K., Moriyama, T., Ozawa, S.: Generation of speaker mixture voice using spectrum morphing. IEEE (2007)
14. Slaney, M., Covell, M., Lassiter, B.: Automatic Audio Morphing. In: International Conference on Acoustics, Speech and Signal Processing, Atlanta, GA, May 7-10 (1996)
15. Mousa, A.: Voice conversion using pitch shifting algorithm by time stretching with PSOLA and re-sampling. Journal of Electrical Engineering 61(1), 57–61 (2010)
16. http://www.mm.pl/~rados/tresc/gener.html

Fundamental Frequency Extraction in Speech Emotion Recognition

Bartłomiej Stasiak and Krzysztof Rychlicki-Kicior

Institute of Information Technology, Technical University of Łódź
ul. Wólczańska 215, 90-924 Łódź, Poland
basta@ics.p.lodz.pl, krzysztof.rychlicki-kicior@makimo.pl

Abstract. Emotion recognition in a speech signal has received much attention recently, due to its usefulness in many applications associated with human – computer interaction. Fundamental frequency recognition in a speech signal is one of the most crucial factors in successful emotion recognition. In this work, parameters of an autocorrelation – based algorithm for fundamental frequency detection are analysed on the example of Berlin emotion speech database (EMO-DB). The obtained results show that lower-than-standard values of the upper limit of the analysed frequency range tend to improve the classification outcome. Statistics of prosody contours and Mel-frequency cepstral coefficients (MFCC) have been used for feature set construction and support vector machine (SVM) has been used as a classifier, yielding high recognition rates.

Keywords: speech emotion recognition, pitch extraction, prosody contours.

1 Introduction

Speech emotion recognition is an important part of the human-computer interaction, along with speech recognition, speech synthesis and various applications of image analysis. Successful emotion recognition connected with expert systems and decision support systems can lead to decreasing the semantic gap in speech signal content understanding and interpretation.

Speech emotion recognition may be seen as a typical example of a pattern recognition task. It consists of feature extraction and selection, followed by the classification stage. In this work, the feature extraction process is analysed, particularly for one of the most important feature sources – the fundamental frequency of the speech signal.

The fundamental frequency (F_0) detection is a classical problem for which a number of methods, based on both time-domain and spectral representations, have been proposed [1][2]. As a great number of successful speech emotion recognition approaches is partially or sometimes even completely based on the F_0 contour analysis (cf. [3]), it is interesting to know how the parameters and the effectiveness of the underlying F_0 detection algorithm influence the classification results.

A. Dziech and A. Czyżewski (Eds.): MCSS 2012, CCIS 287, pp. 292–303, 2012.
© Springer-Verlag Berlin Heidelberg 2012

The pitch estimation accuracy, for example, seems to be of minor significance here, in contrast to e.g. music information retrieval applications (MIR) in which the precise transcription to MIDI note numbers is often required. On the other hand, the octave errors may have significant impact on statistical descriptors of the F_0 contour, such as mean, variance, etc. Another problem, specific to speech processing, is the reliability of voiced/unvoiced segments detection. Assigning pitch value for some unvoiced consonants, e.g. for voiceless fricatives, may introduce extremely high values into the obtained F_0 sequences, similarly as in the case of octave errors. This may in turn lead to errors in estimation of the global F_0 declination or trend, which is an important indicator of the speaker's emotional state [3].

Approaches in which the speech segmentation is based on voiced/unvoiced segments detection are especially vulnerable to the quality of partitioning obtained from the F_0 detection algorithm. It should be also noted that the number and lengths of the voiced segments may be used to construct some useful features, e.g. the speech rate.

In this work we demonstrate some problems connected to the reliability of one of the popular pitch determination algorithms used by the speech-processing community [4]. Several parameters of the algorithm are described and their influence on the overall result of speech emotion recognition is investigated on the example of a well-known EMO-DB [6] dataset.

The rest of the paper is organized as follows: the next section presents some basic facts about the speech emotion recognition problem, the approaches found in literature and dedicated speech databases available. The principles of the chosen F_0 detection algorithm are then presented along with examples of speech analysis done with various configurations of the values of its parameters. The last two sections present the results of the emotion recognition tests, their analysis and some recommendations about adjusting the pitch determination procedure adequately.

2   Review of Previous Work

Following over twenty years of the research, emotional speech recognition is a mature field of pattern recognition with many specialised feature sets and reliable analytical methods developed. There exist also numerous speech databases reported in literature (for extensive list see [7]), some of them publicly available for testing different methodologies and comparing the results. It should be noted, however, that while many of the proposed approaches, tools and algorithms are coincident with those applied by the speech-processing community in other research areas, most notably in speech recognition, the data collections used for emotional speech classification have to be designed and built specially for this specific task.

2.1 Datasets

Several speech datasets are available and can be used in speech analysis in research [7]. There are many aspects, which can make some datasets particularly suitable for specific tasks:

- Language of dataset – various languages can have specific accents or even be tonal – this fact is crucial in emotion recognition.
- Kinds of speech – natural, simulated and elicited. Speech samples created by professional actors simulating emotions in given sentences are believed to be the most reliable [7].
- Represented emotions – number of emotions presented in speech samples. This number can vary from 2 emotions (binary classification tasks – e.g. recognizing whether customers of call centers are angry or not) up to 7 and more.

2.2 Feature Extraction

Emotional states of the speaker are typically manifested in prosodic features of the utterances, such as intonation or stress. Pitch and energy contours are therefore among the most commonly analysed characteristics in this context. In addition, contours of vocal tract features, such as formant frequencies and their bandwidths, as well as sequences of some other short-time spectral characteristics, e.g. Mel-frequency cepstral coefficients (MFCC) [8][9] or log-frequency power coefficients (LFPCs), are often considered [7].

The prosody contours, once obtained, may be used in several ways for emotion recognition. Further analysis, apart from hidden Markov models, artificial neural networks and other more specialised tools [10][11], is often based on some elementary statistics, such as min/max value, mean, standard deviation and higher-order moments, median and other quantiles [12][13].

The problem which has been given much attention recently is how to segment the utterance and choose its most valuable parts w.r.t. the emotional state recognition. The preliminary stage of the analysis may include e.g. dividing the speech signal subsequences into phoneme groups [7] or consonant – vowels (CV) units [14]. In this latter case the reliable detection of vowel onset points (VOP) is required [15]. Separate regions of the CV units (consonant, vowel and transitional) may be then considered in the emotion classification context.

A different approach introduced by Koolagudi and Rao [14] is based on pitch synchronous analysis. For this purpose the glottal closure instants (GCI) must be found, which – in the absence of accompanying electroglottogram record – must be estimated directly from the speech signal [16][17].

A simpler approach involves analysis of the voiced segments of speech only, as reported by the pitch extraction algorithm [18]. It may be shown [12] that it yields significantly better results than a blind frame-based segmentation.

3 F_0 Recognition Algorithm

In this work we investigate an autocorrelation-based algorithm proposed by Paul Boersma [4], which has been implemented in Praat [5] – a popular software tool for speech analysis and synthesis. Two operations introduced in this algorithm enable to significantly improve the F_0 estimation results: a) reliably estimating the autocorrelation of the input signal on the basis of its short-term analysis by special application of the autocorrelation of the windowing (Hanning) function, and b) using the sampling theorem in the lag domain for precise reconstruction of the autocorrelation function.

The maxima of the autocorrelation function, once found, represent the candidate values of F_0 and higher-order harmonics. One additional "unvoiced candidate" is always considered in case no salient maxima have been found. The final decision is taken by analysing the candidates in past frames and choosing the lowest-cost path in the corresponding state space by means of dynamic programming. Several parameters are defined for controlling this procedure:

Silence Threshold. This parameter determines whether the given sound frame is considered silent or not. The lower the value is, the more silent sound will be classified as loud.

Voicing Threshold. This parameter describes the strength of the unvoiced candidate. The higher the value is, the less frames will be classified as voiced.

Pitch Ceiling. This parameter introduces an upper bound of analysed spectrum – frequencies higher than this parameter's value will be ignored.

Octave Cost. This parameter determines a degree of choosing high-frequency candidates. Increasing this value will favor higher candidate frequencies.

Octave-Jump Cost. This parameter affects a possibility of accepting large frequency jumps. Increasing this value decreases a degree of acceptance of large frequency jumps.

Voiced/Unvoiced Cost. This parameter affects a possibility of accepting voiced – unvoiced transition. Increasing this value decreases a degree of acceptance of voiced – unvoiced transitions.

Max. Number of Candidates. This parameters determines how many frequency candidates are considered for each frame.

In order to illustrate the influence of some of these parameters on the obtained pitch contours, two different utterances has been chosen from the EMO-DB dataset. The problem with the first one (Fig. 1) lies in low phonation – the sentence is almost whispered – and in assigning high pitch values to voiceless sibilants and plosives (e.g. the last two segments marked with dashed lines correspond to a word fragment "stück"). Computing statistics from such contours

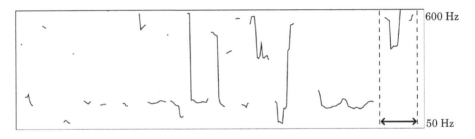

Fig. 1. Pitch contour obtained from the "13a05Tc.wav" file (emotion: sadness; duration: 4.45s) with default parameter values (voicing threshold = 0.45; silence threshold = 0.03)

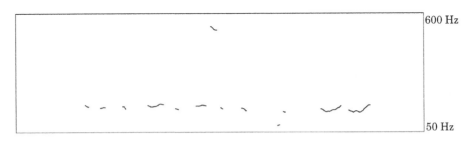

Fig. 2. Pitch contour obtained from the same file with modified parameter values (voicing threshold = 0.7; silence threshold = 0.1)

Fig. 3. Pitch contour obtained from the "10b03La.wav" file (emotion: boredom; duration: 3.83s) with three different values of the voicing threshold. The middle value (0.45) is the default. Downward trend, typical for boredom, may be clearly observed.

seems to yield highly unreliable results. Increasing the voicing threshold and also the silence threshold helps significantly, as may be observed in Fig. 2.

The influence of the voicing threshold alone is presented in Fig. 3, where an opposite problem may be seen – almost whole utterance is reported as voiced. Manipulating the voicing threshold enables to obtain segmentation, although it would be probably difficult to automatically find the optimal values.

4   Experiment Model

Berlin emotion database EMO-DB has been chosen as a dataset in our experiments, as it is widely used in speech-related research. This dataset contains 535 samples covering seven basic emotions (anger, boredom, disgust, fear, happiness, neutral and sadness), ten speakers (five males, five females) and ten sentences.

The feature extraction process, performed on each speech sample, begins with extraction of pitch and intensity contours and the contours of the first four formant frequencies and 12 MFCC coefficients. The Praat tool is used for this purpose [5] with time step between consecutive sound frames set to 10 ms. Next we compute the mean and standard deviation for every contour, considering the voiced segments only, as reported by the pitch detection algorithm. Additional features are the maximum difference and trend of the pitch contour, the mean duration of unvoiced segments and the speech rate estimated from the mean length of the voiced segments [7].

The four formant frequencies, obtained on the basis of LPC algorithm preceded by preemphasis stage, are searched for within the range 50 Hz – 5450 Hz [5].

The filterbank used for MFCC coefficients computation comprises triangular filter functions spaced equidistantly on the mel scale [5][9]. Each filter function $H_k(f)$ is defined as:

$$H_k(f) = \begin{cases} \frac{f-l_k}{c_k-l_k} \; ; \; \text{for } f \in [l_k, c_k] \; , \\[2mm] \frac{r_k-f}{r_k-c_k} \; ; \; \text{for } f \in [c_k, r_k] \; , \\[2mm] 0 \; ; \; \text{otherwise} \; , \end{cases} \tag{1}$$

where f is the frequency and the parameters l_k, c_k, r_k are given as:

$$\begin{aligned} c_k &= \mu(kd) \; , \\ l_k &= \mu((k-1)d) \; , \\ r_k &= \mu((k+1)d) \; , \end{aligned} \tag{2}$$

where k is a positive integer (filter function number) and $\mu(m)$ is a mel-to-Hertz conversion function:

$$\mu(m) = 700 \left(10^{\frac{m}{2595}} - 1 \right) \; . \tag{3}$$

The parameter d in formulas (2), which defines the distance between consecutive filters and their width in the mel scale, has been set to 100. The output sequence of the filterbank (1), for any given sound frame, is transformed with the discrete cosine transform and the first twelve DCT output values define our MFCC coefficients.

As for the trend of the F_0 contour, several approaches have been considered, from the regression line coefficient computation [3] to some simple estimations based on e.g. comparing the mean pitch value of the first and the last one third

Table 1. Best parameter values

Parameter	Best	Default
Silence threshold	0.02	0.03
Voicing threshold	0.5	0.45
Pitch ceiling	400 Hz	600 Hz
Octave cost	0.02	0.01
Octave-jump cost	0.7	0.35
Voiced/Unvoiced cost	0.14	0.14
Max. candidates	n/a	15

of the utterance. Linear regression method proved the best as a single feature, although in conjunction with all other prosody features the simpler methods were almost equally efficient, which probably indicates high correlation within the feature set.

All the parameters of the F_0 extraction algorithm discussed in section 3 have been tested independently. For every parameter we performed exhaustive search through a wide range of possible values. As the criterion we took the overall emotion recognition result obtained with support vector machine (SVM) on the whole EMO-DB dataset with ten-fold cross-validation scheme. We used libSVM package [19] available with Weka software [20]. Several parameter settings have been tested. Radial basis function (RBF) kernel type:

$$K(\mathbf{u}, \mathbf{v}) = e^{-\gamma ||u-v||^2} , \qquad (4)$$

has proved the most effective with ν-support vector classification. The values of the relevant parameters used were: $\gamma = 0.58$ and $\nu = 0.3$. Some tests performed with the classical C-support vector classification type indicated that applying the exhaustive grid search method [19] on the (C, γ) parameter space would possibly lead to further improvement in the classification result.

5   Results and Discussion

The best values of the pitch extraction algorithm we found are presented in Table 1. They are usually not far from the defaults, and the difference in classification outcome is often quite small.

The most interesting fact is that the suggested value of the pitch ceiling parameter, much lower than the default one, enables to obtain significantly better results. The default value of 600 Hz generally reflects what we know about typical ranges of male and female voices [21]. Nevertheless, in this case, setting the lower value seems to have beneficial effect on the emotion recognition results. Figures 4 and 5 present the dependence of the recognition errors on the maximal pitch value.

Several tests have been done, with various subsets of the feature set and some additional operations such as median filtering of the pitch contour. The results

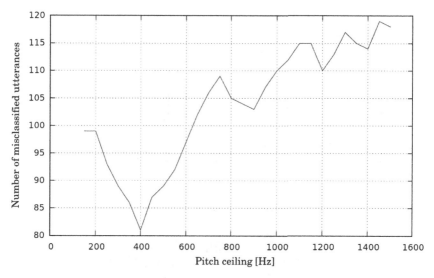

Fig. 4. Misclassified utterances for different pitch ceiling values and default values of all the remaining parameters

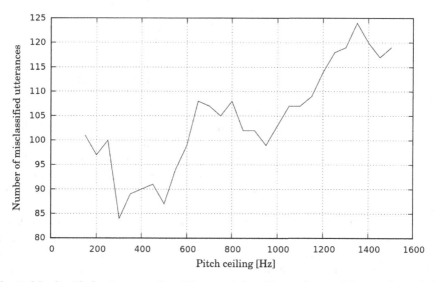

Fig. 5. Misclassified utterances for different pitch ceiling values and best values of all the remaining parameters

were similar to what may be seen in figures 4 and 5 leading to a conclusion that the optimal range of maximum analysed pitch values for the emotion recognition task varies between 300 Hz and 500 Hz.

Table 2. Recognition results per class

Correct recognitions (true positive rate)	Emotion
82.3%	Neutral
91.3%	Anger
84.0%	Boredom
76.1%	Disgust
85.5%	Fear
74.6%	Happiness
93.5%	Sadness
84.9%	**Weighted average**

Table 3. Confusion matrix

Assigned class							Real class
(N)	(A)	(B)	(D)	(F)	(H)	(S)	
65	0	9	1	2	0	2	(N)eutral
0	116	0	1	0	10	0	(A)nger
7	0	68	0	2	0	4	(B)oredom
0	3	1	35	3	3	1	(D)isgust
2	0	1	2	59	3	2	(F)ear
0	13	0	1	4	53	0	(H)appiness
0	0	4	0	0	0	58	(S)adness

The recognition results obtained for the pitch ceiling of 400 Hz and all other parameters set to default values are presented in Table 2 and the corresponding confusion matrix is shown in Table 3.

The obtained TPR values outperform many of the results reported in literature for the same database and for similar approach based on statistics of prosody contours [12][13]. Koolagudi and Rao reported better results for the EMO-DB dataset [14] (96% for the pitch-synchronous analysis and linear prediction cepstral coefficients) but they rejected two of the seven emotions: boredom and disgust, which makes the results incomparable.

As for the pitch ceiling parameter, it is not completely clear why the lower value yields better results. For some emotions, e.g. boredom, most of the pitch contours lie below 300 Hz, also for female voices (Fig. 6), so even decreasing the pitch ceiling by half does not affect the result. However, in other emotions such as anger or happiness the voiced vowels with pitch values over 350 – 400 Hz appear quite often (Fig. 7) and the situations similar to that in Fig. 1 are in fact infrequent, hence the analysis up to 600 Hz seems justified. The possible solution would be that the values of other features are not much reliable in high-pitched speech fragments, so excluding those fragments from computations is beneficial. This hypothesis seems to deserve closer attention in future research.

f [Hz]

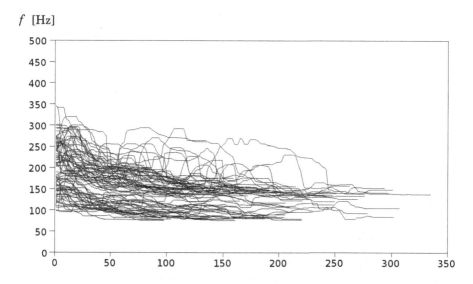

Fig. 6. Pitch contours (unvoiced fragments rejected) for all utterances of emotion: boredom. Two gender-dependent clusters may be easily observed.

f [Hz]

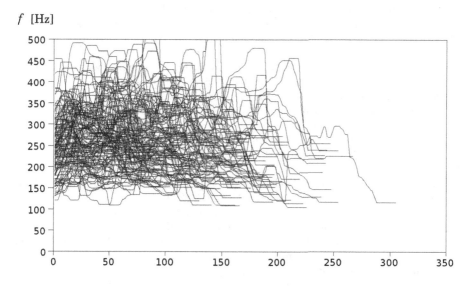

Fig. 7. Pitch contours for all utterances of emotion: anger

6   Conclusion and Future Work

In this paper the pitch extraction problem has been considered in the context of speech emotion recognition. The obtained results are good in terms of recognition rates per each class and comparable or better than those known from literature.

The parameters of an autocorrelation-based algorithm for F_0 detection have been analysed and the best values, w.r.t. the emotion recognition results have been reported. The most interesting observation was that the upper limit of pitch frequencies considered should be set lower than expected in order to yield best classification outcome. This fact will be further investigated on different emotional speech databases and with different pitch extraction and classification algorithms in future work.

References

1. Dziubiński, M., Kostek, B.: High accuracy and octave error immune pitch detection algorithms. Archives of Acoustics 29(1), 1–21 (2004)
2. Gerhard, D.: Pitch Extraction and Fundamental Frequency: History and Current Techniques. Technical Report TR-CS 2003-06, Dept. of Computer Science, University of Regina (2003)
3. Paeschke, A.: Global Trend of Fundamental Frequency in Emotional Speech. In: Proceedings of Speech Prosody, Nara, Japan (2004)
4. Boersma, P.: Accurate short-term analysis of the fundamental frequency and the harmonics-to-noise ratio of a sampled sound. In: IFA Proceedings 17 (1993)
5. Boersma, P.: Praat, a system for doing phonetics by computer. Glot International 5(9/10), 341–345 (2001)
6. Burkhardt, F., Paeschke, A., Rolfes, M., Sendlmeier, W., Weiss, B.: A Database of German Emotional Speech. In: Proceedings Interspeech, Portugal (2005)
7. Ververidis, D., Kotropoulos, C.: Emotional speech recognition: Resources, features, and methods. Speech Communication 48(9) (2006)
8. Neiberg, D., Elenius, K., Karlsson, I., Laskowski, K.: Emotion Recognition in Spontaneous Speech. Working Papers 52, University of Lund (2006)
9. Niewiadomy, D., Pelikant, A.: Digital Speech Signal Parametrization by Mel Frequency Cepstral Coefficients and Word Boundaries. Journal of Applied Computer Science 15(2), 71–81 (2007)
10. Mao, X., Chen, L., Zhang, B.: Mandarin speech emotion recognition based on a hybrid HMM/ANN. International Journal of Computers 1(4) (2007)
11. Nogueiras, A., Moreno, A., Bonafonte, A., Mariño, J.B.: Speech Emotion Recognition Using Hidden Markov Models. In: 7th European Conference on Speech Communication and Technology, Aalborg, Denmark (2001)
12. Mansoorizadeh, M., Charkari, N.M.: Speech emotion recognition: comparison of speech segmentation approaches. In: IKT 2007 (2007)
13. Datcu, D., Rothkrantz, L.J.M.: The recognition of emotions from speech using GentleBoost classifier. A comparison approach. In: International Conference on Computer Systems and Technologies (2006)
14. Koolagudi, S.G., Rao, K.S.: Real life emotion classification using VOP and pitch based spectral features. In: India Conference (INDICON) Annual IEEE (2010)
15. Prasanna, S.R.M., Reddy, B.V.S., Krishnamoorthy, P.: Vowel onset point detection using source, spectral peaks, and modulation spectrum energies. IEEE Trans. Audio, Speech, and Language Processing 17, 556–565 (2009)
16. Murty, K.S.R., Yegnanarayana, B.: Epoch extraction from speech signals. IEEE Trans. Audio, Speech, Language Processing 16(8), 1602–1615 (2008)
17. Hahn, M., Kang, D.G.: Precise glottal closure instant detector for voiced speech. IEE Electronics Letters 32(23) (1996)

18. Shami, M.T., Kamel, M.S.: Segment-based approach to the recognition of emotions in speech. In: ICME (2005)
19. Chang, C.C., Lin, C.J.: LIBSVM: a library for support vector machines. ACM Transactions on Intelligent Systems and Technology 2, 27:1–27:27 (2011)
20. Hall, M., Frank, E., Holmes, G., Pfahringer, B., Reutemann, P., Witten, I.H.: The WEKA Data Mining Software: An Update. SIGKDD Explorations 11(1) (2009)
21. Xuedong, H., Acero, A., Hon, H.W.: Spoken Language Processing. Prentice Hall PTR (2001)

Security Infrastructures: Towards the INDECT System Security

Nikolai Stoianov[1], Manuel Urueña[2], Marcin Niemiec[3],
Petr Machník[4], and Gema Maestro[5]

[1] Technical University of Sofia, INDECT Project Team, 8, Kliment Ohridski St.,
1000 Sofia, Bulgaria
nkl_stnv@tu-sofia.bg
[2] Universidad Carlos III de Madrid, Department of Telematic Engineering,
Avda. de la Universidad, 30 E-28911 Leganés (Madrid), Spain
muruenya@it.uc3m.es
[3] AGH University of Science and Technology,
Department of Telecommunications, Mickiewicza 30 Ave., 30-059 Krakow, Poland
niemiec@kt.agh.edu.pl
[4] VSB-Technical University of Ostrava, Department of Telecommunications,
17. Listopadu 15, 708 33, Ostrava, Czech Republic
petr.machnik@vsb.cz
[5] APIF Moviquity SA Madrid, Madrid, Spain
gmm@moviquity.com

Abstract. This paper provides an overview of the security infrastructures being deployed inside the INDECT project. These security infrastructures can be organized in five main areas: Public Key Infrastructure, Communication security, Cryptography security, Application security and Access control, based on certificates and smartcards. This paper presents the new ideas and deployed testbeds for these five areas. In particular, it explains the hierarchical architecture of the INDECT PKI, the different technologies employed in the VPN testbed, the INDECT Block Cipher (IBC) – a new cryptography algorithm that is being integrated in OpenSSL/OpenVPN libraries, and how TLS/SSL and X.509 certificates stored in smart-cards are employed to protect INDECT applications and to implement the access control of the INDECT Portal. All the proposed mechanisms have been designed to work together as the security foundation of all systems being developed by the INDECT project.

Keywords: Public Key Infrastructure (PKI), Virtual Private Network (VPN), symmetric block ciphers, application security, smartcard, certificates, access control, Transport Layer Security (TLS).

1   Introduction

Nowadays the requirements of any ICT system regarding data protection and information security are constantly increasing. INDECT (*Intelligent information system supporting observation, searching and detection for security of citizens in*

A. Dziech and A. Czyżewski (Eds.): MCSS 2012, CCIS 287, pp. 304–315, 2012.

urban environment) [4] is a Collaborative Research Project funded by the 7th EU Framework Program whose objective is to develop advanced tools for Police forces. In particular the Work Package 8 (WP8) of the INDECT project is focused on increasing the security of the information stored, exchanged or accessed by INDECT systems and users.

2   Architecture of INDECT Public Key Infrastructure

One of the main characteristics of the INDECT project is that is composed by multiple heterogeneous systems that exchange sensitive information among them. Therefore it is necessary to fulfill all requirements for information security: Access Control, Authentication, Non-Reputation, Data Confidentiality, Communication Security, Data Integrity, Availability and Privacy [1]. The main element of the security infrastructures being deployed to provide these security properties is the INDECT Public Key Infrastructure (PKI). This PKI is the base for creating a heterogeneous and secure environment, based on X.509 certificates, public keys and asymmetric cryptographic. The INDECT PKI architecture has a hierarchical, two-level structure:

- Level I – only the Root Certification Authority (Root CA) operates at this level. This CA is offline to prevent attacks to the PKI.
- Level II – there are two CAs at this level: one for issuing certificates for users (Users CA), and other CA for issuing certificates for devices (Devices CA). Between these two CAs a trusted connection is established.

The Root CA only issues certificates for two main CAs, the Users CA and the Devices CA.

The Users CA manages (create, issue, revoke etc.) all the certificates related to the users of INDECT systems. Users use these certificates to log into the individual systems or the INDECT web portals, sign documents or encrypt connections and e-mails. These X.509 certificates can be installed in web browsers or securely stored in a smart-card.

The Devices CA manages all aspects of certificates issued for devices (PCs, PDAs, CCTVs, etc.). Each certificate is assigned to a specific device, thus each device can be uniquely identified and managed based on its certificate. Devices certificates' are used for creating secure communication channels, for signing streams and documents, and for identification.

Table 1 shows the appropriate key sizes for the proposed CA's and the certificates they issue.

Table 1. Suggested size of the RSA private key

CA role	Key size
Root CA	8192 bits
User CA	4096 bits
Device CA	4096 bits
User certificate	2048 bits
Device certificate	1024 bits

Moreover, Figure 1 shows the different levels of the INDECT PKI and the relation between its different CAs.

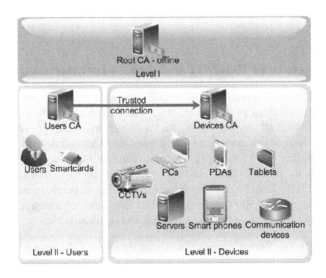

Fig. 1. Sample presentation of PKI infrastructure

In order to test the feasibility of the proposed INDECT PKI, a PKI testbed based on OpenCA [10] and EJBCA [9] has been deployed. A sample user certificate issued by this PKI is shown below:

```
Certificate:
    Data:
        Version: 3 (0x2)
        Serial Number: 0 (0x0)
        Signature Algorithm: sha1WithRSAEncryption
        Issuer: emailAddress=nkl_stnv@tu-sofia.bg,CN=Nikolai
         Stoianov,OU=Technical University of Sofia,O=INDECT TestCA,C=BG
        Validity
            Not Before: Apr  1 03:57:23 2011 GMT
            Not After : Apr  4 03:57:23 2016 GMT
        Subject: emailAddress=nkl_stnv@tu-sofia.bg,CN=Nikolai
         Stoianov,OU=Technical University of Sofia,O=INDECT TestCA,C=BG
        Subject Public Key Info:
            Public Key Algorithm: rsaEncryption
            RSA Public Key: (4096 bit)
                Modulus (4096 bit):
                    00:b2:d1:41:d2:b6:48:22:59:...
```

```
          Exponent: 65537 (0x10001)
    X509v3 extensions:
        X509v3 Basic Constraints: critical
            CA:TRUE
        X509v3 Subject Key Identifier:
            C8:A9...
        X509v3 Authority Key Identifier:
            keyid:C8:A9:...
            DirName:/C=BG/O=INDECT TestCA/OU=Technical University of
            Sofia/CN=Nikolai Stoianov/emailAddress=
            nkl_stnv@tu-sofia.bg
            serial:00
        X509v3 Key Usage:
            Digital Signature, Non Repudiation, Certificate Sign,
            CRL Sign
        X509v3 Subject Alternative Name:
            email:nkl_stnv@tu-sofia.bg
        X509v3 Issuer Alternative Name:
            email:nkl_stnv@tu-sofia.bg
        Netscape Cert Type:
            SSL CA, S/MIME CA, Object Signing CA
        Netscape Comment:
            OpenCA LiveCD Demo CA Certification Authority
            Certificate
        X509v3 CRL Distribution Points:
            URI:http://10.0.2.15/pub/crl/cacrl.crl
        Netscape CA Revocation Url:
            http://10.0.2.15/pub/crl/cacrl.crl
        Netscape Revocation Url:
            http://10.0.2.15/pub/crl/cacrl.crl
    Signature Algorithm: sha1WithRSAEncryption
        8f:9f:a6...
```

Currently the work is focused on creating fully functional infrastructure based on EJBCA, implementing specific extensions in X.509 certificates, and finally to employ the certificates issued by the PKI in other INDECT systems and in the remaining security infrastructures.

3   Communication Security

One of the main components of the secure communication infrastructure within the INDECT system is a Virtual Private Network (VPN) framework that will enable the secure communication among multiple remote nodes and servers interconnected over public networks. Nowadays virtual private networks are usually based on two different technologies – IPsec and SSL. IPsec VPN creates encrypted tunnels for all IP traffic, independently of the applications being used. On the other hand SSL VPNs ensure the secure transmission of messages for particular applications, mainly HTTP. While IPsec VPN is more suitable for site-to-site connections where distant networks are mutually interconnected, SSL VPN is usually the best solution for remote access connections where individual users communicate securely with a VPN gateway.

For the implementation of VPNs in the INDECT system, only open-source solutions will be used. The StrongSwan software package seems to be a convenient open-source IPsec VPN solution. StrongSwan is intended primarily for devices using Linux. It is fully compatible with other standard IPsec VPN implementations, and thus can be used in networks with mixed equipment.

As an open-source SSL VPN solution, the best option appears to be the OpenVPN software package. OpenVPN can be installed in computers with either Linux or Windows operating systems. OpenVPN is a very flexible and scalable VPN. For example, it works well with NAT in contrast to IPsec VPN. On the other hand, it has problems with compatibility with other VPN solutions.

Both VPN types, StrongSwan and OpenVPN, support PKI and authentication based on X.509 certificates. In such a case, each VPN client obtains a certificate from a certification authority which is subsequently used to authenticate the client when a secure tunnel has to be created between the client and the VPN gateway (see Figure 2). To support Authentication, Authorization and Auditing (AAA) services, an additional LDAP/RADIUS server can be employed, which must be located inside the private network.

Within the INDECT system, users will employ mainly OpenVPN to securely communicate between their terminals (desktop, laptop, PDA, smartphone, etc.) and servers located in the police headquarters. The INDECT Devices CA will authenticate the individual terminals.

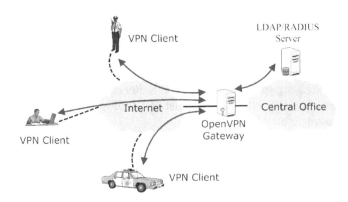

Fig. 2. Example of an INDECT Virtual Private Network based on OpenVPN

4   Cryptographic Security

One of the main activities performed by WP8 of the INDECT project refers to modern cryptographic techniques. The research on this field allows developing new algorithms and protocols which ensure high-level of data confidentiality. Usually, the new algorithms are evaluated by means of proper simulators and tested to check the resistance on several attacks.

The INDECT Block Cipher (IBC), invented by AGH, is an algorithm that is able to fast encrypt data, and thus meet the stringent confidentiality requirements of Police forces. The cipher transforms a message in order to make it unreadable to anyone except some proper entities (i.e., the sender and recipient). It is a symmetric block cipher, so both encryption and decryption processes transform a message by means of the same key (secret key).

Deliverable D8.3 "Specification of new constructed block cipher and evaluation of its vulnerability to errors" [2] describes the IBC algorithm in detail. Below, only the major features are presented:

- Substitution-permutation structure
- Architecture depends on key
- Huge number of non-linear S-boxes (about $5,35*10^{18}$)
- Block size: 256 bit
- Key lengths: 128, 192, 320, 576 bit
- Number of rounds: 8, 10, 12, 14

The IBC algorithm was initially tested by means of a new simulator. The simulator checks the main security features of the cipher, which decide the strength of cryptographic algorithm. All simulations confirmed that IBC ensures the high-level of data confidentiality. The following features were tested:

- Balancing,
- Non-linearity,
- Strict Avalanche Criterion (SAC),
- Completeness,
- Diffusion order, and
- Structure of XOR table.

The functionality of the new cipher was also verified by means of a IBC implementation. The IBC application was presented in deliverable D9.13 "New block ciphers" [3]. The C++ programming language was chosen for the implementation of the algorithm itself, whereas the Graphical User Interface (GUI) was built by means of C++/CLI language under .NET platform. The visual interface of IBC application with example of encryption process is presented in Figure 3.

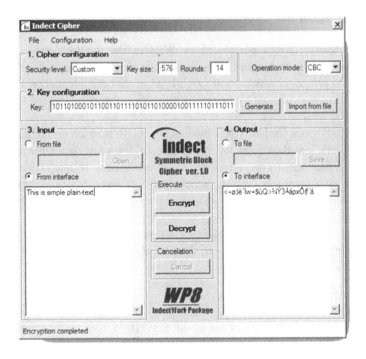

Fig. 3. The interface of IBC application with an example of the encryption process

Now, the current work of WP8 connected with cryptography is focused on the integration of the new ciphers with popular security environments such as OpenSSL and OpenVPN. This integration will allow to use the developed algorithms in practice. By means of such applications as OpenSSL, the IBC cipher can be used by other work packages to secure different INDECT subsystem. In particular, some simple substitution-permutation ciphers have been integrated within the OpenSSL environment by modifying crucial files of the OpenSSL library (i.e.: apps/dsa.c, apps/speed.c, apps/progs.pl and different files from crypto/objects directory).

5   Application Security

One of the main challenges of WP8 is how to protect in a secure manner the diverse set of applications being developed by the INDECT project, without designing a specific security mechanism for each application. The main design insight is that most networked[1] INDECT applications have got either a web interface or are based on web

---

[1] The INDECT project is also developing standalone, non-networked applications. We won't consider it security here, since its usage is confined to particular systems and the information that can be disclosed by a security breach is limited to the local data of the application.

services. Therefore we should start studying the standard security mechanisms for the Hypertext Transfer Protocol (HTTP). The secure version of HTTP is commonly known as "https" since this is the protocol name that appears at the beginning of the Uniform Resource Locator (URL) of secure web sites. However "https" is not a new protocol itself, but specifies that the HTTP protocol runs on top of a secure protocol. This secure protocol is usually called TLS/SSL, and it will be the foundation of the common security mechanism of web-based INDECT applications.

The Transport Layer Security (TLS) [7] and its predecessor, and more known, the Secure Sockets Layer (SSL) [5], are client-server protocols that provide communications security on top of the Transmission Control Protocol (TCP). Although SSL was originally designed for the web, they are application-agnostic, meaning that any application protocol running on top of TCP may run on top of TLS/SSL. It is an advanced security protocol featuring symmetric-cryptography encryption, asymmetric-cryptography key exchange, end-point authentication based on X.509 certificates, and message integrity protection by means of message authentication codes. Moreover, TLS/SSL is an extensible protocol since peers are able to negotiate which version of the protocol and what cipher suite (e.g. TLS_RSA_WITH_AES_256_CBC_SHA) will be used during the communication session. However it is worth noting that TLS/SSL does not provide digital signature or non-repudiation services, thus these security mechanisms must be implemented by the applications that require it.

Usually in a web TLS/SSL session only the server is authenticated, that is, the web server sends its X.509 certificate to the client. After validating the certificate (i.e. checking server's name, the expiration date, the whole certificate chain, the revocation list, etc.) the client encrypts the session key exchange message with the public key of the certificate, thus the communication can only progress if the server has the associated private key. This way, web browsers can check that they are actually communicating with the intended web server (e.g. the bank website), thwarting any kind of man-in-the-middle attack. However INDECT applications will also authenticate the client of the TLS/SSL session by means of the X.509 certificate stored in the Smart Card or the browser of the user. Figure 4 details the setup of a TLS/SSL session with mutual authentication. The extra steps for authenticating the user are shown in bold.

The advantage of using TLS/SSL to enhance the security of INDECT applications is that it is already implemented by all major web browser and web servers, and it is enforced even before the application is called. Therefore, when TLS/SSL is properly configured in the server, the application is certain that the client has been authenticated by means of the X.509 certificate stored in a smart-card or at the user's browser. Thus, it provides from the start an authentication service based on a "something you have" and/or "something you know" credentials.

From a cryptographic point of view, and after several revisions, TLS version 1.2 is considered a secure protocol, although many attacks have been proposed against practical details of its implementation. Recently the security breaches of some trusted Certificate Authorities, including the issue of fraudulent server certificates, have called into question the security of PKI, and thus the TLS/SSL authentication security. However, we argue that these attacks do not pose a threat to the usage of TLS/SSL by INDECT systems, since secure terminals do not trust other Certification Authorities (CAs) than INDECT's one. Moreover INDECT servers do only request and accept client certificates issued by the INDECT Users' CA. Therefore, even if the certificate of an INDECT device (e.g. a node station) is compromised, it cannot be employed neither to supplant an application server, nor a user.

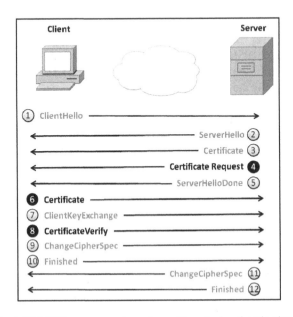

Fig. 4. TLS/SSL secure session setup with mutual authentication [6]

6   Access Control Based on Certificates and Smartcards

The Federated Identity Concept, where the Service Providers (SP) delegate the identity management to an entity referred as Identity Provider (IdP) within a controlled environment called Circle of Trust (CoT), has been the starting point for the identity management solution implemented within INDECT project. The system formed by (at least) one IDP and (at least) one SP is called a federation, and it is characterized by having a relationship of trust among its members, simplifying data communication and validation of the user in a secure way. However, Identity Federated solutions are focused on managing and protecting the end user's information employed by the different services. INDECT, and therefore, its systems, must provide a controlled and secure access to the different services. To maximize the compliance of these requirements so as to meet this scenario, the Federated Identity Paradigm has been extrapolated to a model that also relays on a Public Key Infrastructure.

Within INDECT solution, the role of the IdP is played by the INDECT PKI. The different CAs of the PKI are in charge of issuing the certificates that guarantee the identities of all the INDECT elements: users, devices and service providers, pointing out their unique identifier and attributes as well as managing their validity. The CA is a trustworthy element in which all the other INDECT elements have to trust. Based on certificates, any exchange of information is carried out between authenticated and trustworthy elements through secure channels.

The certificate generation, and therefore the PKI, is linked to the user registration process. When a user is registered in the system, the User CA creates the corresponding certificate, including part of the information provided in the registration process. The information stored in the certificate is relevant for the user management process but it does not have to be updated frequently (i.e. the user identifier).

This certificate enables checking the identity of the user. When a user accesses any INDECT systems through the INDECT Portal, its certificate is required by means of TLS/SSL. The authentication solution is able to check the validity of the certification with the corresponding CA and process its data, informing the Portal (the system) about the users' identity and attributes (i.e. role).

The access control solution is composed by four main elements:

- INDECT Security Local Tools, which allow to generate, a key pair (private and public) as the previous step to the generation of a certificate, and on the other hand, the certificate container (pkcs12), necessary to load the certificate in a smart card or in the browser.
- INDECT PKI Certification Authority, which issues the X509 users, service providers and access device certificates and manages their validity. The certificates, through their extensions, contain the information that has been considered relevant for the access control process such as the security level of the user.
- Certification Holder, where the certificate is store and from which it can be loaded, such as a smart card.
- Authentication Filter Libraries, that, once integrated in the INDECT application servers (service providers), will allow INDECT applications to manage the request information where the certificate is sent, validating the certificate and recovering the certificate information, identity or any other extension.

This design also supports the possibility of users without a certificate. In this case, a login and password will be required. This fact allows providing two levels of security by default, which can be also aggregated depending on the levels associated to the users' attributes.

The users' identifier and attributes are synchronized with the user manager, which manages a repository of users based on a LDAP directory. This repository is controlled by the application's administrator that manages the (local) system users and their privileges.

Based on the user's attributes (contained in the certificate and in the LDAP repository) the users' rights are built. According to this information, the user will access the Portal dashboard through which her authorized INDECT services will be reachable. Depending on the security access level of the services or information, not only the user's certificate but also the access device certificate will be required. Both of them will be processed by the authentication service so that this information can be checked at Portal level and also at service level if required, by the user manager that manages authorized clients.

Fig. 5. The interface of INDECT Security Local Tools for the certificate container generation

The interoperability among the different INDECT systems is also based on certificates. When an INDECT user accesses other INDECT system (i.e. Spanish user accessing Polish system) its certificate is required and validated with the CA that issued the certificate. This way, the user is authenticated in the target system and its access is registered. Based on the users' identity and attributes pointed out in its certificate, the user has a 'basic' access (i.e. according to its role). These rights can be updated by the 'local' administrator following the proper policies (e.g. after some written authorization).

Therefore, the PKI architecture is a key element in INDECT, not only for the provision of a secure and trustworthy environment (nodes and communications) but also for managing the control access to the system, authenticating access devices and users as well as for facilitating subsequent authorization.

7   Conclusion

An enterprise security architecture based on PKI functionality is one of the most preferred and reliable solutions for data protection [8]. By building a PKI infrastructure, the INDECT project sets way of working with state-of-the-art security technologies. Moreover, the separation of INDECT PKI in two hierarchical levels allows both, securing the main PKI element – the ROOT CA - and to manage and operate systems and users in different way. For instance defining two CAs for users and devices enables the possibility of identifying each device in the systems and to manage each user individually. This type of organization of PKI gives us the possibility to secure the data in different ways on different points of the creating, transmitting, editing and storing process.

Communication security is basic a tool for creating a secure environment. Therefore VPNs based on X.509 certificates have been selected for INDECT. In this case key management and key negotiation doesn't required additional secure channel. This way of creating a secure communication environment simplifies the cryptography key infrastructure and minimizes the number of keys.

Nowadays cryptography is the only way to guarantee the confidentiality of data. The development of new protocols and algorithms based on symmetric ciphers like the INDECT Block Cipher (IBC) gives users the possibility to use, exchange and store data in secure way. Newly developed tools, protocols and applications in INDECT project may implement the new IBC algorithm to protect sensitive information when communicating with other entities.

The TLS/SSL protocol is the foundation of the security of networked INDECT applications. It enables communications' security, including encryption, message integrity and mutual authentication between clients and servers. However TLS/SSL just provides a base security layer, INDECT applications may implement further security mechanisms such as password-based authentication, or digital signatures for non-repudiation services.

Access control based on certificates and smart-cards is one mechanism that ensures non-repudiation. Also, by using a smart-card access control, an additional level of authentication can be enabled. INDECT application should implement this kind of access control in order to have the possibility to use issued and stored certificates together with additional extensions built into these certificates.

Acknowledgments. This work has been funded by the EU Project INDECT (*Intelligent information system supporting observation, searching and detection for security of citizens in urban environment*) — grant agreement number: 218086.

References

1. INDECT Consortium. D8.1: Specification of Requirements for Security and Confidentiality of the System, http://www.indect-project.eu/files/deliverables/public/INDECT_Deliverable_D8.1_v20091223.pdf/view
2. INDECT Consortium. D8.3: Specification of new constructed block cipher and evaluation of its vulnerability to errors (December 2010), http://www.indect-project.eu/files/deliverables/public/deliverable-8.3
3. INDECT Consortium. D9.13: New block ciphers (December 2010), http://www.indect-project.eu/files/deliverables/public/deliverable-9.13
4. INDECT project web site, http://www.indect-project.eu
5. Hickman, K.: The SSL Protocol. Netscape Communications Corp. (February 1995)
6. Thomas, S.A.: SSL and TLS Essentials: Securing the Web. Wiley Computer Publishing (2000)
7. Dierks, T., Rescorla, E.: The Transport Layer Security (TLS) Protocol Version 1.2, RFC 5246 (August 2008)
8. Zhelyazkov D., Stoianov, N.: PKI Infrastructure in the BA – Prerequisite for Minimization of the Risk and Enhancement of the Information Security, CIO, Special issue Communication & Information Technologies for the Defense, pp. 19-20 (September 2009) ISSN 13112-5605
9. http://www.ejbca.org/
10. http://www.openca.org/

Application of Virtual Gate for Counting People Participating in Large Public Events

Krzysztof Kopaczewski[1], Maciej Szczodrak[1],
Andrzej Czyżewski[1], and Henryk Krawczyk[2]

[1] Multimedia Systems Department
[2] Computer Architecture Department
Gdansk University of Technology, ETI Faculty, Gdansk, Poland
{kkop,szczodry,andcz}@sound.eti.pg.gda.pl

Abstract. The concept and practical application of the developed algorithm for people counting in crowded scene is presented. The aim of the work is to estimate the number of people passing towards entrances of a large sport hall. The details of implemented the Virtual Gate algorithm are presented. The video signal from the camera installed in the building constituted the input for the algorithm. The most challenging problem was the unpredicted behavior of people while entering the building. A series of experiments during real sport events and concerts was made. The case of improved organization of people passing is described and the influence on the counting results is shown. The results of the studies are shown and achieved outcomes are discussed.

Keywords: crowd behavior, image processing, crowd counting.

1   Introduction

People attending large public events could incur various risks evoked by the excessive number of persons gathered in a specific place, particularly during a sport game or a concert. Exceeding the value regarded as the safe limit may cause that in some emergency situations people would suffer injuries or death [1]. The organizers should know the number of people that are gathered in the building or in the enclosed outdoor space. Similarly to many objects of this type, the building considered in this work is not equipped with the people counting systems such as mechanical gates. Besides, people often feel concerned crossing such an installation, as to what would happen in case of necessity of rapid leaving the building. Moreover, the behavior of the crowd while entering an object through wide door makes it impossible to use other optical or mechanical means such as radiation beam systems. The infrared barriers are ineffective because they often count as a single person the group of people being close each to other [2]. The problem is also that such solutions usually are not able to distinguish the pedestrian movement direction. Especially, when the large group of people appears, it gets very hard to determine their number by the majority of image processing methods which commonly use object detection and tracking. Background extraction-based approaches, such as the ones developed at the Multimedia Systems Department of the Gdansk University of technology [3] cannot separate objects properly when people walk at very small distances separating them or their hands are

A. Dziech and A. Czyżewski (Eds.): MCSS 2012, CCIS 287, pp. 316–327, 2012.

connected. Other methods use multiple cameras to deal with the segmentation problem [4], or apply models of human figures obtained during observing a foreground of image [5]. Moreover, installing numerous cameras would be impractical in the considered building.

The commercial systems vary in the technologies applied and the actual target to solve. An example of a laser counter offered is SICK LD-PeCo [6] and visible light systems are very common [7] [8] [9]. These kinds of products often achieve best performance in some specific conditions, only. Meanwhile, the situations found while gathering the experimental data proven to be difficult to interpret algorithmically. An example showing the character of people motion is presented in Fig. 1.

Fig. 1. Recorded crowd behavior near building entrance

The proposed algorithm is expected to work in a system with centralized architecture, where the video signals gathered from multiple cameras is being processed by an efficient computer cluster. The aim of the system is to show the estimated number of people while they are incoming to the hall through several gates. The KASKADA supercomputing platform is the algorithm working environment [10].

In this paper we present the design and practical utilization of the developed algorithm for counting the people entering a large object. The experiments were made while the spectators were entering and exiting the object through a set of wide doors. The next section presents a description of the virtual gate algorithm, then the setup and results of experiments are shown, whereas Section 5 concludes the paper.

2   The Virtual Gate Algorithm

The Virtual Gate algorithm we developed is based on the modified Optical Flow method. The motivation for employing the Optical Flow method is its immunity to image background changes in difficult light condition such as illumination visible through open doors. The method developed for counting people does not involve classifying modules because the aim of the algorithm is to detect size and direction of motion of objects in video sequences having dimensions similar to the size of an average human body. Moreover, the Virtual Gate is used in places where the human motion is expected, especially at entrances, passes, etc. The algorithm consists of two separate modules: the main module performing low-level image processing tasks and the calibration module. The algorithm is implemented in C++ programming language with application of OpenCV libraries.

2.1 Main Module

Virtual Gate is devoted to counting people in crowd passing through the scene observed with the camera. The illustration of the sample setup of the virtual gate is presented in Fig. 2a. The Virtual Gate distinguishes two directions of people motion, namely "in" and "out" (see Fig. 2a).

The detailed structure of the Virtual Gate is depicted in Fig. 2b. It is composed of a set of rectangle regions (R_i) situated next to each other and overlapping. Rectangles have the same shape and their size is corresponding to the size of an average human body contour, with respect to its height and width (for a particular camera view).

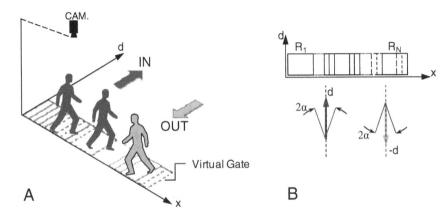

Fig. 2. Virtual Gate: setup for counting people (a) and some details illustrating its principle of working (b)

The motion of objects is estimated in each region R_i using the Dense Optical Flow method [11] [12]. The set of vectors representing the direction and the velocity of the motion detected is obtained in the result of above operation. Two directions of people motion through the virtual gate are considered, namely forward and backward ("in" and "out", $+d$ and $-d$, as in Fig. 2b). Moreover, a small divergence of the direction (α) is allowed, because usually people do not maintain bearing while walking. The tolerance should not be too large, because of the need to discard those walking along the gate.

The block diagram of the algorithm is presented in Fig. 3. For each input video frame, vectors representing motion speed and direction are calculated. Vectors pointing to the direction outside the range of $d\pm\alpha$ and $-d\pm\alpha$ are rejected. Then the number (V) of origins of vectors lying in each region R_i is obtained. In the next step, this number is compared to a threshold value. The threshold is proportional to the average area of human silhouette at given camera view and is obtained experimentally. If V is greater than the threshold value, the "in" or "out" people counter is increased and rectangle R_i enters an inactive ("hold") state for the period of C frames. The inactive state means that R_i is not performing calculations. This operation is done in order to avoid counting errors and to allow leaving the area of R_i by the moving object while not being counted more than once. The period of the "hold" state is obtained experimentally during the calibration and for the setup discussed in this paper is about 0.6 s.

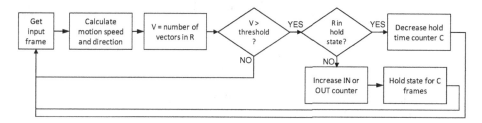

Fig. 3. Block diagram of the virtual gate main algorithm

2.2   Calibration

The calibration module is a part of the Virtual Gate algorithm. The aim of this module is to find the optimal counting threshold (T) in order to improve the effectiveness of the people counting algorithm. Searching is based on the bisection method which approaches the optimal result in sequential iterations. The input data for the calibration process are:

a) video sequence presenting passage of individuals and groups of people,
b) the Virtual Gate geometry and parameters (i.e. dimensions of regions R_i, number and distance between regions),
c) real number of people passed through the gate in selected time moments.

The error of counting is calculated in successive iterations as presented in eq. 1, 2, 3:

$$S_0 = w_0 \cdot (y_0 - x_0) \tag{1}$$

$$S_i = w_i \cdot [(y_i - y_{i-1}) - (x_i - x_{i-1})] \tag{2}$$

$$S = \sum_{i=1}^{n} S_i \tag{3}$$

where the input data are:
n- number of selected time moments,
$f_i = f_0, f_1, ..., f_n$- frame number in selected time moment,
$x_i = x_0, x_1, ..., x_n$- real number of people passed through the gate in selected time moment i (obtained using the Ground Truth method),
$w_i = w_0, w_1, ..., w_n$- weight coefficient.
The variables are defined as follows:
$y_i = y_0, y_1, ..., y_n$- number of motion objects counted in i-th selected time moment by the Virtual Gate algorithm,
$S_i = S_0, S_1, ..., S_n$- counting error in selected time moment,
S- total counting error.

The weight coefficient is added in order to favor either passage of individuals or groups of people. In each step, partial errors and the total error are minimized and then the counting threshold is increased or decreased respectively according to equations 4 and 5. The calibration process stops when $T_{corr} \leq 1$.

$$T = \begin{cases} T + T_{corr} \text{ if } S > 0 \\ T - T_{corr} \text{ if } S \leq 0 \end{cases} \tag{4}$$

$$T_{corr} \leftarrow \frac{T_{corr}}{2} \tag{5}$$

where:
   T - counting threshold (value within range 0, 1...100),
   T_{corr} - counting threshold correction.

If the obtained counting threshold value is not typical (for the considered setup it means that the value is outside the range of 55 to 85), two reasons are possible. The parameters of the Virtual Gate may be wrong e.g. the size of rectangle R_i does not correspond to the size of human silhouette in image. The second reason could be invalid motion detections obtained by the Optical Flow algorithm.

3   The Test Environment

Experiments location. The experiments were made in sports and entertainment hall which maximum capacity is 15.000 of people ("Ergo Arena" located in Gdansk). The image acquisition hardware was deployed at the main entrance which consists of 6 symmetrical doors, 3 meters wide each. The cameras were set to observe only 3 doors because the others were not being used during the events, frequently. Fig. 4 presents the detailed camera setup. The cameras were installed at the height of 6.5 meters above the floor, whereas the width of the door is 2.9 m.

   The experiments have been conducted between 8th December 2011 and 13th January 2012. In this period, several recordings were made. Table 1 presents some details of the test recordings discussed in this paper. Each contains a scene of crowd movement in specific direction (in or out), which is either entering or leaving the building. The number of people in each test recording was counted manually and treated as the Ground Truth reference value, which was later compared to the Virtual Gate algorithm output. Moreover, the optimal settings for cameras and necessary lighting conditions, presented in Table 1, were obtained experimentally.

Hardware and software settings. The image acquisition hardware was IP-cameras Axis P1346. White balance was set in such way that the camera was calibrated automatically to the light conditions and then held the settings. The exposure control was set to "flicker-free 50Hz" mode with the motion priority. To achieve the maximum image brightness the Iris was set to the minimal value. All scenes excluding the excerpt No. 5 were recorded with the backlight compensation mode. The additional lighting of the scene was applied in various configurations.

Fig. 4. Main entrance view – symbols A1, A2, A3 enumerate doors, C.1, C.2, C.3 represent symbols of cameras

Table 1. Test recordings and camera setup. Recordings No. 6, 7, 8 contain the same scene, but different test settings and parameters

Test set, recording	Direction	Door	Duration [hh:mm]	Shutter [s]		Gain [dB]		Lighting [W]
				Min	Max	Min	Max	
1, 1	out	A2	01:24	1/17000	1/100	0	41	600
2, 2	out	A2	00:22	1/17000	1/100	0	41	600
3, 3	out	A3	00:22	1/17000	1/100	0	41	none
4, 4	in	A2	01:17	1/17000	1/100	0	41	600
5, 5	in	A2	01:05	1/50	1/50	0	0	600
6, 6	in	A1	01:16	Auto	Auto	Auto	Auto	2x600+40
7, 6	in	A1	01:16	1/50	1/50	6	6	2x600+40
8, 6	in	A1	01:16	Auto	Auto	Auto	Auto	2x600+40

4   Results of Experiments

For each recording, in the preliminary step, several experiments were made in order to achieve the optimal parameters of the Virtual Gate algorithm with the desired camera settings. The final results of people counting are presented in Table 2 and details concerning experiments are described in the following text. The efficiency is calculated as the proportion of the real people counting result to the number produced by the algorithm.

Table 2. Summary results of Virtual Gate efficiency for all test recordings

Test set, recording	No. of people, Ground Truth	Counting threshold	Efficiency [%]
1, 1	388	62	91.2
2, 2	278	62	99.3
3, 3	297	67	98.0
4, 4	641	62	84.0
5, 5	481	84	72.0
6, 6	699	75	69.7
7, 6	699	75	26.7
8, 6	699	68	97.8

The algorithm calibration was made for the recording No. 2 (representing the case when people were leaving the building), where the observed motion of the crowd was stable and orderly. The calibration was made for the frames in the range between 4046 and 6050.

Then the computed counting threshold was applied to the Virtual Gate and then recordings 1 and 2 were processed. The efficiency of the Virtual Gate for both recordings results from data gathered in Table 2. The resultant efficiency for the recording No. 2 is higher than that one obtained for the recording No. 1, because the motion of crowd was more chaotic in the recording No. 1. Moreover, the video quality of the latter was lower (more frame droppings appeared than for the recording No. 2). The detailed results of the comparison of the number of people observed (Ground Truth) and counted by the Virtual Gate algorithm for the recordings No. 1 and No. 2 are presented in Fig. 5 and in Fig. 6, respectively.

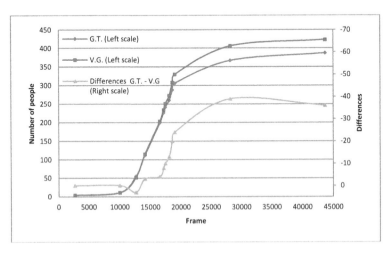

Fig. 5. Number of people in recording No. 1: G.T. – Ground Truth, V.G. – Virtual Gate algorithm result

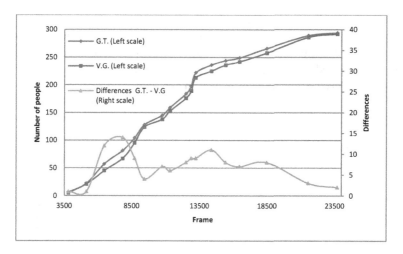

Fig. 6. Number of people in recording No. 2: G.T. – Ground Truth, V.G. – Virtual Gate algorithm result

The recording No. 3 contains the scene of people leaving the object. This recording was different from the other ones, because of the non-conventional camera view direction rotated by 45 degrees (see Fig. 3). Such a setup was imposed by the difficulty of placing the camera above the door A3, caused by the building construction. Nevertheless, a high efficiency of counting was achieved, as show the results are presented in Table 2 and in Fig 7.

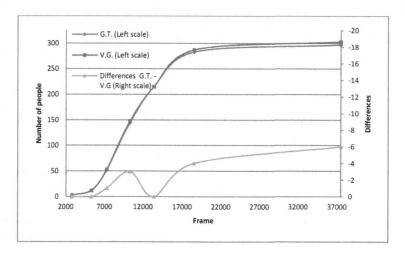

Fig. 7. Number of people in recording No. 3: G.T. – Ground Truth, V.G. – Virtual Gate algorithm result

In the next experiment, the result of calibration done for the recording No. 2 was applied to calculation related to the recording No. 4. In this case, the efficiency was not very high, because the calibration has been made for the case of people exiting the building, and applied to the recording of people entering the building. The different crowd behavior has the prevailing influence on outcomes.

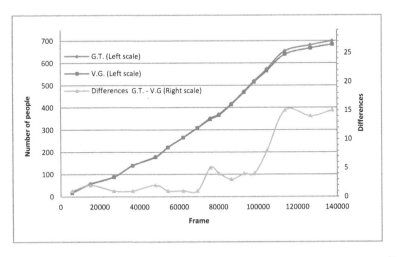

Fig. 8. Number of people in recording No. 6 at test set No. 8: G.T. – Ground Truth, V.G. – Virtual Gate algorithm result

The main problem was to gain a high quality image and favorable conditions of camera settings due to unrecognized camera flaw. Consequently, the frame acquisition rate was decreased to 15 from 30 in order to eliminate frame drops.

In the last experiment, two recordings were done simultaneously by two cameras for the same place and scene, but for different camera settings. In the first camera automatic settings (see Table 1, test set No. 6) and the frame rate of 30 fps were applied. The second camera parameters were set manually (see Table 1, test set No. 7) and the frame rate was decreased to 15 fps, in order to eliminate drops. Moreover, a barrier was installed in order to force the crowd direction, as is shown in Fig. 10. A large difference of efficiency between these two recordings was noticed. Test set No. 8 configuration parameters was the same as in test set No. 6 and counting threshold was different. The results obtained in test set 7 are visibly worse, because decreased frame rate caused that displacement of moving people observed in consecutive frames was high. In consequence, utilized optical flow algorithm was not working correctly.

Furthermore, the example images from the video recordings together with the results of the Virtual Gate processing are presented in Figs: 9, 10, 11, 12, and 13. The number in the upper left corner of the Virtual Gate represents the total number of people counted and the numbers printed just above the rectangles (with the smaller font size) mean detailed person count results.

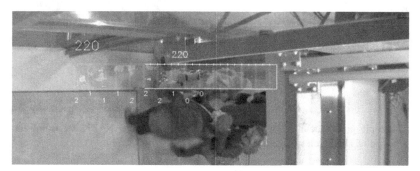

Fig. 9. Result of people counting for recording No. 3 made with camera 3 (tilt angle 45 degrees). Door A3 is under the footbridge and unconventional camera setup was applied

Fig. 10. Installation of barriers (marked with green arrows) for regulating the crowd motion. Barriers were perpendicular to the Virtual Gate and parallel to main crowd direction. Recording No. 6, test set No. 8

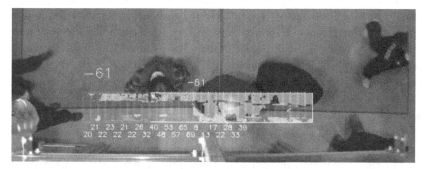

Fig. 11. People groups counted by Virtual Gate. Processed frame No. 12994 from recording No. 1

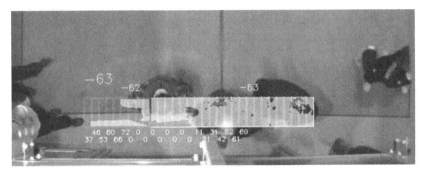

Fig. 12. People groups counted by Virtual Gate. Processed frame No. 12997 from recording No. 1. Individuals being close together were counted properly (see Fig. 11)

Fig. 13. Real people behavior with lots of collisions and chaotic direction of movement. Processed frame No. 13505 from recording No. 1

5   Conclusions

The concept and practical application results of the algorithm application for people counting in a crowd passing the virtual gate were presented in this paper. The achieved results are satisfactory considering the real people behavior which was far from organized movement. A further work will focus on extending the system to operate at other entrances to the hall of the "Ergo-Arena" object being used less frequently. The practical application of the system can increase the security in the discussed object.

Acknowledgements. Research funded within the project No. POIG.02.03.03-00-008/08, entitled "MAYDAY EURO 2012 - the supercomputer platform of context-depended analysis of multimedia data streams for identifying specified objects or safety threads". The project is subsidized by the European regional development fund and by the Polish State budget.

References

1. Mollen, M.: A Failure of Responsibility - Report to Mayor David N. Dinkins on the December 28, 1991 Tragedy at City College of New York (January 1992)
2. Mathews, E., Poigne, A.: Evaluation of a "Smart" Pedestrian Counting System Based on Echo State Networks. EURASIP Journal on Embedded Systems, Article ID 352172, 1–9 (2009)
3. Czyzewski, A., Dalka, P.: Moving Object Detection and Tracking for the Purpose of Multimodal Surveillance System in Urban Areas. In: Tsihrintzis, G.A., et al. (eds.) New Directions in Intelligent Interactive Multimedia. SCI, vol. 142, pp. 75–84. Springer, Heidelberg (2008)
4. Yang, D.B., Gonzalez-Banos, H.H., Guibas, L.J.: Counting People in Crowds with a Real-Time Network of Simple Image Sensors. In: Proc. of the Ninth IEEE International Conference on Computer Vision (2003)
5. Ge, W., Collins, R.T.: Marked point processes for crowd counting. In: Proc. CVPR, pp. 2913–2920 (2009)
6. Laser Measurement Technology LD-PeCo-5.5, https://www.mysick.com/PDF/Create.aspx?ProductID=33787&Culture=en-US (access February 2, 2012)
7. Honeywell, People Counter, http://www.honeywellvideo.com/products/ias/va/160978.html (access February 2, 2012)
8. Peco modules – VisualTools, http://www.visual-tools.com/en/products/video-analysis-people-counting-and-pos-control/peco-units/documentation (access February 2, 2012)
9. Video Turnstile, People counting systems, http://videoturnstile.com/counting.html (access February 2, 2012)
10. Krawczyk, H., Proficz, J.: The task graph assignment for KASKADA platform. In: Proc. 5th International Conference on Software and Data Technologies, July 22-24 (2010)
11. Atcheson, B., Heidrich, W., Ihrke, I.: An evaluation of optical flow algorithms for background oriented schlieren imaging. Experiments in Fluids 46, 467–476 (2009)
12. Burt, P.J., Adelson, E.H.: The Laplacian pyramid as a compact image code. IEEE Trans. Commun. 31, 532–540 (1983)

A New Method for Automatic Generation of Animated Motion

Piotr Szczuko

Multimedia Systems Department, Electronics, Telecommunications and Informatics Faculty,
Gdańsk University of Technology, Poland
szczuko@sound.eti.pg.gda.pl

Abstract. A new method for generation of animation with a quality comparable
to a natural motion is presented. Proposed algorithm is based on fuzzy description of motion parameters and subjective features. It is assumed that such
processing increases naturalness and quality of motion, which is verified by
subjective evaluation tests. First, reference motion data are gathered utilizing a
motion capture system, then these data are reduced and only main poses of the
action are left. The resulting motion is simplified and its quality is considerably
decreased. Then, utilizing the automatic motion enhancement system,
ANIMATOR, a new version of the action is generated, based on input poses
and subjective descriptors given by the user. Finally, a comparison between the
original and recreated motion is performed. Presented method is useful for automatic motion generation and can be paired with motion data reduction procedure for regaining naturalness. Moreover the reduced version can easily be
edited in the ANIMATOR system, and in this way a new action can be created.

Keywords: computer animation, motion capture, fuzzy logic.

1 Introduction

Animated computer characters are commonly used in virtual reality (VR)
applications, computer games (as so-called avatars), movies, and educational
software. Current trends in VR aim at providing full interaction and personalization of
the avatar's look, outfit, gender, or age. Therefore adding a new aspect of personality
for adjustment – *movement style* – seems to be a logical step in avatar development.
The most advanced method for acquiring animated movement is motion capture,
though it has high technical requirements [5][8]. An important drawback is that
capturing motion of a real actor does not allow for achieving an exaggerated
movement, typical for animated movies and cartoons, and changing the animation
style is practically impossible. While developing a VR application variants of actions
for different types of avatars could be captured, but the work would be very tedious,
and such an approach seems impractical. Conversely, a *keyframe* animation method
can be used for creating an avatar animation by hand [13][14]. It doesn't require
expensive hardware, the result motion can be either natural or exaggerated in style,

A. Dziech and A. Czyżewski (Eds.): MCSS 2012, CCIS 287, pp. 328–339, 2012.

but this technique is time-consuming and quality greatly depends on the animator's skills. However, in this method the representation of animation data is intuitive, clear, and easy to edit. Therefore there is a persisting need for developing new methods of motion style adjustments based either on captured performance of a real actor or on a *keyframe* animation, and as a result introducing new quality and range of styles defined by the user.

Methods of artificial intelligence were already employed for realistic motion generation, namely genetic algorithms and neural networks, for simulation of human motoric system. A prominent example in this field is a commercial system *Endorphin* [16], with neural networks trained accordingly to perform realistic actions, such as remaining in an upright position, balancing, reacting to external force, jumping, grasping, and falling. This technology is used for special effects in movies. Nevertheless, it lacks complex performance, e.g. gesticulation or facial expressions.

In the research presented below, a new method is proposed for generation of animated movement which combines motion capture computer animation and traditional cartoon creation techniques. It is a continuation of the work by the author, et al. [4][9][10][11]. In this approach, motion capture is employed to collect interesting and meaningful aspects of a character's action. The data is then simplified to reduce motion representation and facilitate animation editing by hand. Finally, the data is processed so that new, high quality motion can be automatically generated.

For the processing a fuzzy logic approach is utilized [15]. Fuzzy description of motion subjective features combined with reference subjective test results provide fuzzy rules implemented in the ANIMATOR system, employed for calculation of output motion parameters. The result is a method for a high quality animation generation, capable of producing a range of styles from natural to exaggerated, depending on the user input.

2   Animation Parameterization

In this research it is assumed that input animation contains only the most important poses of the action. Such animation can be a result of **simplification of real motion recordings** obtained from motion capture systems. For the presented experiments three sequences of natural motion were registered utilizing motion capture system [6], and then main poses of those actions were extracted to achieve animations only with key poses and linear transitions between them. A proposed animation processing method deals with this type of motion representation (Fig. 1). In classical *keyframe* animation it is called *pose-to-pose* approach [1][12]. For further processing any *pose-to-pose* animation can be segmented into parts containing first a still pose, then one or more transitions and finally the second still pose. Each animation segment is parameterized and processed separately utilizing the described methodology.

Each single segment is described by timing, and location of animated objects (parameterized later by vector **A**). These animated objects are limbs of the character, and parameters describe a sequence of the character's poses, defining the avatar's action.

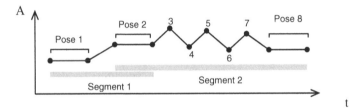

Fig. 1. Pose-to-pose animation segmentation process. Changes in time of one animation parameter (location of a hand) are shown. Segment 1 contains one transition, segment 2 contains six
transitions. Boundaries of segments are determined by poses held in time (pose 1, 2, and 8).

In the case of a handmade animation, a sequence of poses prepared by the animator
is also accompanied with additional motion phases. These phases do not change the
meaning of the action, but add subtle variations to transitions between defined poses,
influencing character *personality*, and motion *fluidity*, *quality*, and *style*. Additional
phases used by the animator are: 1) *anticipation* (a motion preceding main transition
between two poses adjacent in time), 2) *overshoot* (a motion after transition, when
motion stops not abruptly but slowly), 3) *moving hold* (is a slight and subtle change in
pose over a number of frames). Anticipation displays preparation for the action, e.g.
squat before jump, overshoot portrays inertia, and moving hold is responsible for
showing balancing and maintaining aliveness. Correct utilization of these additional
phases influences naturalness and fluency of motion, related to a high subjective
quality of animation. Animation rules involving the utilization of phases were
formulated by animators of the Walt Disney Studios [12].

To simplify the problem, the following assumptions have to be made:

1. *anticipation* and *overshoot* are alike, i.e. their times (t) and limbs rotations amplitudes (A) are assumed to be equal, therefore for parameterization only two values
   are used: $dA=A_a=A_o$ and $dt=t_a=t_o$;
2. *moving holds* are calculated as random movements, with times and amplitudes tak-
   en from continuous uniform distribution limited to ranges of $1/10 \cdot dA$ and $1/10 \cdot dt$.

Motion parameters for a single segment of animation are presented in Fig. 2. As mentioned before, parameters related to poses, are stored in vector **A**. Parameters connected to subjective features and additional motion phases are contained in **B**:

$$\mathbf{A}=[V, A, t] \qquad \mathbf{B}=[dA, dt] \qquad (1)$$

where: $A=a_3-a_2$, $t=t_3-t_2$, $V=A/t$.

Decisions on amplitudes and lengths of phases usually come from experience, and
it is the animator's task to portray the character's personality and the animation style
by these phases. To help the animator calculate these parameters an ANIMATOR
system was proposed. The task is to decide which values of B should be used for
achieving the best result. Here that task is accomplished by fuzzy inference.

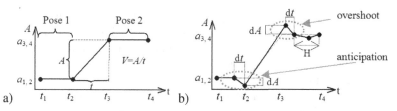

Fig. 2. Motion phases parameterization: a) a transition between two poses is described with $\mathbf{A}=[V, A, t]$, b) additional phases of anticipation, overshoot and moving hold characterized by $\mathbf{B}=[dA, dt]$. "H" depicts range of subtle random variations of the target pose

3   Motion Data Processing

Traditional animation rules [1][12] describe a way of additional phases utilization for achieving particular subjective results. *Fast motion with large amplitude should be preceded by a large anticipation and finished with a large overshoot; a long motion should be preceded by a long anticipation and finished with a long overshoot.* Taking this into consideration it was assumed that proportionality occurs between these parameters, which can be described as (2):

$$dA = alpha \cdot V \cdot A \qquad dt = beta \cdot V \cdot t \qquad (2)$$

where *alpha* and *beta* are the new proposed proportionality coefficients. In case dA or dt exceeds the assumed maximal value[1], the following saturation functions are used (3):

$$dA = g(alpha \cdot V \cdot A), \text{ where } g(x) = 0,265\pi \cdot tgh(x / 0,22\pi)$$
$$dt = h(beta \cdot V \cdot t), \text{ where } h(x) = 10 \cdot tgh(x / 0,125) \qquad (3)$$

Reference Subjective Tests

While considering Eqs. (2) and (3), output variables dA and dt depend on coefficients *alpha* and *beta*. If one agrees that changes of dA and dt influence subjective meaning of the result motion then a relation should exist between subjective features and *alpha* and *beta*. These relations were examined during data mining of the preliminary subjective tests results. In these tests reference animations where used, containing two poses and a transition, with *anticipation*, *overshoot* and *moving hold* phases as shown in Fig. 2. This results in *style* and *fluidity* changes (Fig. 3). Discrete values of coefficients were used: $alpha=\{0.3, 0.4, ..., 1.2, 1.3\}$, and $beta=\{1,3,5,7\}$[2]. The participants task[3] was to assess features of motion with discrete scales: *style*={*natural, middle, exaggerated*}; *fluidity*={*fluid, middle, abrupt*}; *quality*={1, 2, 3, 4, 5}. Results of *fluidity* and *style* evaluation are contained in a vector $\mathbf{Q}=[style, fluidity]$, and *quality*

---

[1] During tests maximum values of dA and dt were subjectively verified by the viewers.

[2] It was verified which ranges of values return subjectively acceptable motion, and then how large discretization steps should be taken to return significant changes of animation features. Results are <0.3; 1.3> with step 0.1 for *alpha* and <1; 7> with step 2 for *beta*.

[3] Participants were first familiarized with the background of the research by watching sample animations with possible variations of the movement, with comments from the authors.

scores (**QS**) are processed individually as an additional criterion. Evaluation of visual stimuli was performed with respect to recommendations [2][3] (Sec. 4.1).

The results were strongly correlated which indicates that certain connections between selected subjective features and the proposed coefficients exist, therefore rules describing such relations can be created. For example: *style* value increases with the *alpha* value, and *fluidity* level increases with the *beta* value. Moreover, changing *style* does not cause *quality* changes, and strong positive correlation exists between *fluidity* and *quality*. These observations are reflected by values of particular correlations (Tab. 1). Overall subjective quality **QS** rated in a test is averaged giving a mean opinion score value (MOS) for every variant of animation, characterized with a particular pair of (*alpha*, *beta*) values (Tab. 2).

Table 1. Correlation between subjective and objective parameters of assessed animations

	beta-style	beta-fluidity	beta-quality	alpha-style	alpha-fluidity	alpha-quality	style-fluidity	style-quality	fluidity-quality
R	-0.14	0.86	0.81	0.82	0.16	0.09	-0.21	-0.27	0.94

Table 2. Preferences of test participants, MOS for animations described with pairs of coefficients (*alpha*, *beta*): gray – cases with MOS higher or equal 3.2, black – higher or equal 3.8

		alpha									
		0.3	0.4	0.5	0.6	0.7	0.8	0.9	1.0	1.1	1.2	1.3
	1	3.21	3.16	2.68	2.91	2.92	3.3	2.8	3.11	3.08	3.06	3.07
beta	3	3.22	3.19	3.2	3.14	2.92	3.41	2.41	3.42	3.23	3.04	3.28
	5	3.7	4.03	3.8	3.5	3.69	3.8	3.82	3.35	3.41	3.47	3.53
	7	4.17	4.04	3.92	3.7	3.8	4.1	3.9	3.5	3.5	3.85	3.69

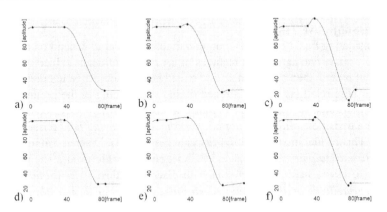

Fig. 3. A transition between two values processed utilizing various *alpha* and *beta*: (a) original, (b) motion with anticipation and overshoot added, *natural style* and *fluid motion*, (c) large *alpha* value results in *exaggerated* motion, (d) small *alpha* results in motion close to the unprocessed one, (e) large *beta* results in elongation of phases, (f) small *beta* results in *abrupt* motion

For utilization of the above observations the mapping between subjective requirements (*fluidity*, *style*) and animation parameters (*alpha*, *beta*) should be determined.

Once the mapping is discovered, then the ANIMATOR system can process any animation described with (V, A, t), and based on (fluidity, style) requirements calculate (alpha, beta) coefficients. This finally results in (dA, dt) parameters of motion phases that must be added to the animation according to Eqs. (2)(3). It is assumed that the relation between motion attributes (fluidity, style) and coefficients (alpha, beta) can be defined based on these results. The methodology of processing the test results is described in Sec. 3.1.

3.1   Data Mining of Subjective Test Results

The relations discussed above are used for creating rules that connect objective and subjective parameters of animation. During subjective evaluation tests participants rated animations with subjective features $Q=\{fluidity, style\}$ and *quality score* QS, and this results in ambiguous information about considered relations. A particular animation characterized by $A=[V, A, t]$ and $B=[dA, dt]$ is not evaluated identically by every test participant. It is assumed that instead of a pair (A, B), the animation will be described with (alpha, beta), because Eqs. (2)(3) precisely define relations between given A and result B. Relation: $f:(alpha, beta)\rightarrow(fluidity, style)$ is not objective. This ambiguous function f is called the evaluation function and reflects viewers' answers. Inverse function is being sought, f^{1}: $Q\rightarrow(alpha, beta)$, which for given required values of features Q returns correct (alpha, beta). That function is also ambiguous. Therefore, for each Q a result is first generated as a set of objects – animations that were subjectively evaluated as having values matching given $Q=[style, fluidity]$, but are differentiated by (alpha, beta). From this set one object is finally selected based on additional *mean quality score* QS. Therefore, for any Q it is possible to generate unambiguous rules referencing Q with (alpha, beta) of the best rated animation (Tab. 3). Obtained mappings are modeled employing fuzzy logic processing, due to its useful features discussed below. For the described purpose a Mamdani fuzzy inference model was chosen [7] for its clarity (it can be comprehended and modified easily by the user being an animator not a scientist), and flexibility (it can be effortlessly extended with rules or attributes other than current ones). It provides fuzzy output which after defuzzification results in continuous values applicable as motion parameters.

Table 3. Result mapping between required *fluidity* and *style* of animations and *alpha* and *beta* proportionality coefficients

alpha		fluidity			beta		fluidity		
		abrupt	medium	fluid			abrupt	medium	fluid
	natural	0.7	0.5	0.3		natural	3	5	7
style	medium	0.9	0.7	0.5	style	medium	1	5	5
	exaggerated	1.3	1.1	0.9		exaggerated	3	5	7

The extrapolation and interpolation capabilities are also of essence. The engineered fuzzy system can accept input values from out of the initially considered ranges, and the processed results will still remain meaningful. The center of gravity (COG) defuzzification method is used, and output values are characterized by the first order continuity: small changes in input values always generate a small change of the output value, which is highly expected and facilitates finding the most suitable settings by the trial and error method.

Interpolation surfaces (Fig. 4) visualize relations between two of selected input attributes values and one of the output values. While constructing a fuzzy inference system a common approach is to emphasize values obtained by measurements of input-output values of some real processes. Then, membership functions and rules are formed in such a manner that each measurement becomes a support point for the interpolation surface. Finally, for the particular input a desired output is achieved, consistent with the real process output. In the considered application, the real process (subjective assessment of animation) was "measured" by means of subjective tests. It was assumed that the fuzzy model should implement values obtained in that process, therefore interpolation surfaces are supported by values from Tab. 3. For describing fuzzy membership triangular-shaped functions were chosen (Fig. 5).

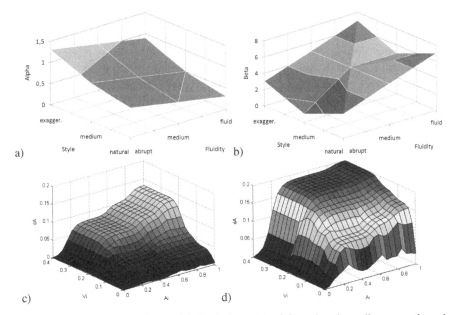

Fig. 4. Interpolation surfaces for modeled relations: (a) *alpha* value depending on *style* and *fluidity*, (b) *beta* value depending on *style* and *fluidity*, (c) dA value depending on *V* and *A* for *natural fluid motion*, (d) dA value depending on *V* and *A* for *exaggerated fluid* motion

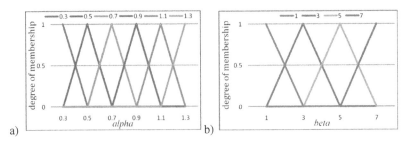

Fig. 5. Membership functions for: (a) *alpha*, (b) *beta*

3.2   Test animation Creation Process

For evaluation of the ANIMATOR system performance two scenarios are considered, each based on processing of *pose-to-pose* animation comprising: poses created from scratch, or extracted from motion capture recordings of natural motion.

Scenario 1: Five prototype animated character actions were prepared by the animator employing *pose-to-pose* approach, containing actions of: reaching, waving, squatting, stretching, and neck rubbing.

Scenario 2: Reference natural motion data were recorded utilizing a motion capture system. Three sequences were recorded portraying actions of: lifting an object from the floor, waving to the camera, and throwing the object (Fig. 6). Motion data were next reduced by hand: from these sequences only main poses of the actions were extracted and arranged in time accordingly to not change the meaning of the actions. The resultant motions resemble simple animations with *pose-to-pose* approach, their qualities were considerably decreased (Fig. 7).

Fig. 6. Frames form motion capture sequence and result poses of the animated character

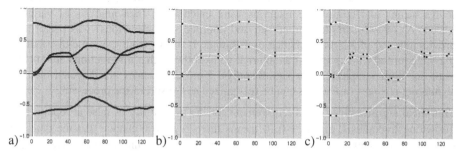

Fig. 7. Data of motion sequences: a) recorded motion capture sequence, b) reduced version complying *pose-to-pose* approach, c) final version recreated utilizing the ANIMATOR system. Presented motion curves show rotation quaternions of character's right arm

Employing the automatic motion enhancement system, new versions of actions are generated, based on input poses and subjective descriptors given by the user. The goal of the processing was to recreate a natural motion, as close as possible to the reference one, therefore only values *fluidity=high* and *style=natural* were chosen.

4   Verification Experiments

Animation subjective features were rated with respect to ACR tests (*Absolute Category Rating*) [3]. It advises to use 5-grade scale for quality ratings, with text labels or numerical values: *Excellent* = 5; *Good* = 4; *Fair* = 3; *Poor* = 2; *Bad* = 1. Statistical processing of test results is performed on a numerical scale, therefore it is possible to calculate e.g. arithmetic mean value, MOS (*Mean Opinion Score*), and perform a test for equal means, variance analysis or verify correlation.

Test Signals: For elimination of unwanted impact of environment on results, tests were performed in normalized conditions: LCD screens with brightness set to 40 and contrast to 80, neutral color temperature, screen resolution 1024x768, refresh rate 60Hz, viewer distance with respect to recommendation [2] set to eight times of the image height, i.e. 1,15m[4]. Video test sequences were replayed from digital video files, therefore providing 100% repeatability of stimuli. Signals were played in random order for each of the participants, therefore the sequence did not influence overall ratings. Motions were performed by a simplified stick figure character, therefore a naturalness of appearance and/or likeability did not influence quality scores. Test signals portrayed scenarios discussed in Sec. 3.2.

Test Participants: A test group of 10 people was carefully selected, assuring no impact of personal differences, e.g. experience in computer animation on final ratings: the group was not connected to video quality domain, did not perform similar tests in last 6 months, and had not seen the presented animations before.

Statistical Analysis of Results: It is advisable to verify if unprocessed sequences do not differ significantly in quality, therefore not disturbing final mean scores. For that purpose the *t-student* statistical test for equal means was performed. Next for gathered scores means and confidence intervals were calculated. Then analysis of variance (ANOVA) of two samples (processed and unprocessed animations) was performed, providing a test of a hypothesis that each sample is drawn from the same distribution (having similar scores) against an alternative hypothesis that distributions are not the same for both samples (differ significantly). If no differences are detected then no significant changes of features can be recognized as a result of the processing.

Statistical significance. α is a requirement for the probability of false acceptance of the wrong hypothesis, e.g. $\alpha = 0,10$ means there is a 10% chance that the detected relation is a coincidence. Usually $\alpha = 0,05$ is taken into account for subjective evaluation tests. Calculated statistics (e.g. ANOVA) provide value of false probability for processed samples, denoted as p, that is expected to be lower than α. For mean values a 95% confidence interval is used [2]. Confidence interval is defined as:

---

[4] Video of 480 pixels in height, c.a. 14cm, should be watched from a distance of 8·0,14=1.15m

$$[\bar{u} - \delta, \bar{u} + \delta] \qquad (4)$$

where:
$$\delta = 1{,}96\frac{S}{\sqrt{N}} \qquad (5)$$

and standard deviation S is calculated as:

$$S = \sqrt{\sum_{i=1}^{N}\frac{(\bar{u} - u_i)^2}{(N-1)}} \qquad (6)$$

Finally, with probability of 95% the real value of the mean is within the calculated confidence interval (4), with the center value $\bar{u}$.

4.1   *Pose-to-Pose* Animation Enhancement Results

It is assumed the processing increases naturalness and quality of motion, and this assumption is verified by the final subjective evaluation tests. *Pose-to-pose* animations prepared by animator were enhanced using the ANIMATOR system with all combinations of input descriptors for *style* and *fluidity*. Mean Opinion Score values for all result animations are presented in Tab. 4. Processed animations obtained statistically valid higher scores than non-processed. Moreover fluid motion was always rated higher than the abrupt one. Finally, variation of animation style does not influence quality scores (Fig. 8), therefore the developed method can be applied for generation of many versions of a prototype action, matching required style and fluidity.

Table 4. MOS for animations depending on values of descriptors used for style enhancement

Mean Opinion Score		style		
		natural	medium	exaggerated
fluidity	abrupt	2,08	2,06	2,08
	medium	3,28	3,22	3,1
	fluid	4,24	3,92	4,02
non-processed		1,5		

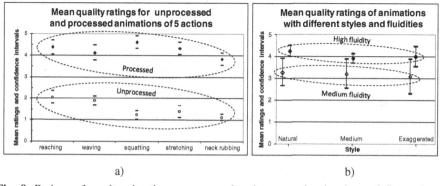

a)   b)

Fig. 8. Ratings of result animation: unprocessed and processed animations of five actions: (a) processed animations have significantly higher quality; (b) style does not influence quality

4.2   Comparison of Natural and Generated Motion

The next performed test employed blind rating of three samples per action: 1) original not reduced motion, recorded with motion capture hardware, 2) its reduced version, with only main poses of motion, 3) a recreated motion, based on reduced version processed by the ANIMATOR system. Obtained scores (Fig. 9) suggest that a substantial increase of quality between reduced and recreated versions is obtained. It can be concluded that quality of the recreated version matches the original one.

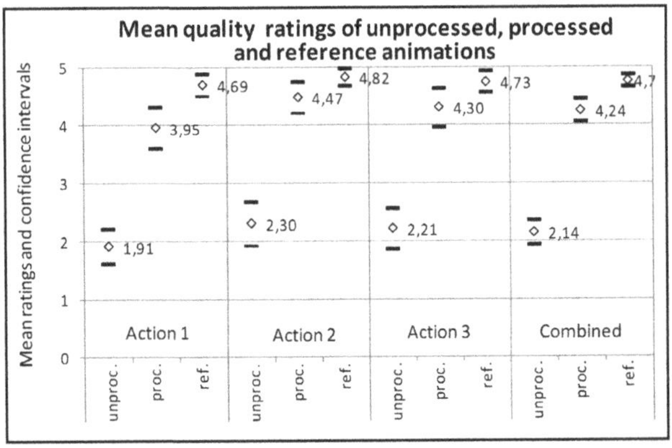

Fig. 9. Ratings of unprocessed, processed, and reference motion capture animations. Actions are: 1) lifting an object from the floor, 2) throwing an object, 3) waving to the camera

5   Conclusions

The ANIMATOR system for animation enhancement was developed utilizing methodology and rules of traditional animation, combined with fuzzy processing. In the system fuzzy rules are used for calculation of parameters of additional motion phases that should be inserted for alteration of subjective features such as stylization and fluidity of motion. New proportionality coefficients *alpha* and *beta* were defined that are strongly correlated with subjective features of animation. This approach was verified in two scenarios, one based on the automatic enhancement of *pose-to-pose* animations prepared by the user, and the second one enhancing reduced *motion capture* recordings. It was confirmed that in both cases a significant increase in motion quality is obtained, with recreated motion quality very close to the original one.

The approach described can effectively be utilized for an automatic animation enhancement and rapid development of virtual character's actions.

Acknowledgements. Research is subsidized by the European Commission within FP7 project "INDECT" (Grant Agreement No. 218086).

References

1. Blair, P.: Cartoon Animation. Walter Foster Publishing, Laguna Hills (1995)
2. ITU-R BT.500-13 Recommendation: Methodology for the Subjective Assessment of the Quality of Television Pictures (2012)

3. ITU-T P.800 Recommendation: Methods for Subjective Determination of Transmission Quality. Geneva, Switzerland (1996)
4. Kostek, B., Szczuko, P.: Rough Set-Based Application to Recognition of Emotionally-Charged Animated Character's Gestures. In: Peters, J.F., Skowron, A. (eds.) Transactions on Rough Sets V. LNCS, vol. 4100, pp. 146–166. Springer, Heidelberg (2006)
5. Menache, A.: Understanding Motion Capture for Computer Animation and Video Games. Morgan Kaufmann, San Francisco (1999)
6. OPTITRACK: Optical Motion Capture and Tracking,
http://naturalpoint.com/optitrack/ and,
http://www.motioncapture.com (accessed March 31, 2010)
7. Pedrycz, W., Gomide, F.: Fuzzy Systems Engineering: Toward Human-Centric Computing. Wiley-IEEE Press, New Jersey (2007)
8. Remondino, F., Schrotter, G., Roditakis, A., D'Apuzzo, N.: Markerless Motion Capture from Single or Multi-Camera Video Sequence. In: Proceedings of International Workshop on Modeling and Motion Capture Techniques for Virtual Environments (CAPTECH), Zermatt, Switzerland (2004)
9. Kostek, B., Szczuko, P.: Analysis and Generation of Emotionally-Charged Animated Gesticulation. In: Ślęzak, D., Yao, J., Peters, J.F., Ziarko, W.P., Hu, X. (eds.) RSFDGrC 2005, Part II. LNCS (LNAI), vol. 3642, pp. 333–341. Springer, Heidelberg (2005)
10. Szczuko, P.: Application of Fuzzy Rules in Computer Animation. PhD Thesis. Gdańsk University of Technology, Poland (2008),
http://sound.eti.pg.gda.pl/animacje (accessed March 31, 2010)
11. Szczuko, P., Kostek, B., Czyżewski, A.: New Method for Personalization of Avatar Animation. In: Cyran, K.A., Kozielski, S., Peters, J.F., Stańczyk, U., Wakulicz-Deja, A. (eds.) Man-Machine Interactions. AISC, vol. 59, pp. 435–443. Springer, Heidelberg (2009)
12. Thomas, F., Johnston, O.: Disney Animation – The Illusion of Life. Abbeville Press, New York (1981)
13. Whitaker, H., Halas, J.: Timing for animation. Focal Press, Oxford (2002)
14. Williams, R.: The Animator's Survival Kit: A Manual of Methods, Principles, and Formulas for Classical, Computer, Games, Stop Motion, and Internet Animators. Faber & Faber, London (2002)
15. Zadeh, L.A.: Fuzzy Logic = Computing with Words. IEEE Transactions on Fuzzy Systems 4, 103–111 (1996)
16. Endorphin Software Natural Motion Ltd.,
http://www.naturalmotion.com (accessed February 19, 2012)

Resolving Conflicts in Object Tracking in Video Stream Employing Key Point Matching

Grzegorz Szwoch

Gdansk University of Technology, Multimedia Systems Department
Narutowicza 11/12, 80-233 Gdansk, Poland
greg@sound.eti.pg.gda.pl

Abstract. A novel approach to resolving ambiguous situations in object tracking in video streams is presented. The proposed method combines standard tracking technique employing Kalman filters with global feature matching method. Object detection is performed using a background subtraction algorithm, then Kalman filters are used for object tracking. At the same time, SURF key points are detected only in image sections identified as moving objects and stored in trackers. Descriptors of these key points are used for object matching in case of tracking conflicts, for identification of the current position of each tracked object. Results of experiments indicate that the proposed method is useful in resolving conflict situations in object tracking, such as overlapping or splitting objects.

Keywords: video analysis, object tracking, Kalman filters, image matching.

1   Introduction

Object tracking is a common procedure in most of the modern automatic video surveillance systems. The main task of such system is automatic detection of important security threats in video streams from surveillance cameras. Detection of such events requires performing several steps of video stream analysis [1]. One of them is a procedure which tracks movement of each object in the consecutive camera images. Using the data obtained from the object tracking module, together with data from further processing steps such as object classification, movement analysis, etc., it is possible to detect defined events based on rules interpreting behavior of the tracked moving objects. However, in order to perform efficient event detection, each object has to be tracked continuously as long as it is present in the area covered by the camera, even if it is obstructed from the camera view (either partially or completely).

Numerous methods for object tracking were proposed in the literature. The first group of methods tracks changes between the complete consecutive frames in the video stream. This group includes a mean shift algorithm [2] and methods based on the optical flow approach, both the dense (e.g. Horn-Schunck [3]) and the sparse ones (e.g. Lucas-Kanade [4]). The second group of methods analyzes moving objects detected in the image, not the whole camera frame. These algorithms require performing object detection before the tracking is performed, using algorithms such as Gaussian

A. Dziech and A. Czyżewski (Eds.): MCSS 2012, CCIS 287, pp. 340–349, 2012.
© Springer-Verlag Berlin Heidelberg 2012

Mixtures Models (GMM) [5], or the Codebook [6] and post-processing stages that include morphological cleaning and shadow removal [1]. The detected objects are then tracked on a frame-by-frame basis. Kalman filters that predict positions of tracked objects in the current frame and are updated with current positions of detected moving objects, are commonly used for object tracking [7].

None of the abovementioned approaches, in their basic form, are optimal for efficient and accurate object tracking in the real-time system for event detection in video streams in varying conditions. Kalman filters perform well in case of small number of moving objects, but they fail in complex situations, such as tracking a person in the crowded hall. Algorithms based on optical flow provide more accurate tracking in crowded scenes, but their real-time implementation and identification of individual objects are problematic. In our previous experiments we used Kalman filters to track moving objects [1]. However, this approach requires solving tracking conflicts caused by ambiguous relations between trackers and objects, e.g. a group of persons. Efficient handling of such cases is not a trivial task. Previously, we experimented with using the predicted states of Kalman filters to estimate the tracker position within a blob, and using color and texture descriptors to match trackers with blobs [8]. However, Kalman state estimation works only in case of short-term and infrequent tracking conflicts. Therefore, a combined approach: an extension of object tracking procedure based on Kalman filters with local descriptors matching, is proposed in this paper. The details of the algorithm are provided in the following sections.

2   The Object Tracking Algorithm

The algorithm for tracking of moving objects is based on Kalman filters that track movement of each object detected with a background subtraction algorithm, on a frame-by-frame basis. Handling ambiguous relations between detected objects and trackers requires a conflict resolving procedure. In order to improve accuracy of conflict resolving, we propose a method based on matching distinctive key points in the image. The details of the proposed procedure are given in the following subsections.

2.1   Overview of the Algorithm

The tracking procedure shown in Fig. 1 processes each image frame from the camera successively. The processing begins with detection of moving objects, using either GMM or the Codebook algorithm. As a result, a binary mask is obtained, with non-zero values marking image pixels that do not belong to the background. Additionally, shadows of objects are removed and morphological processing removes noise and small holes. Connected components (blobs) are extracted and those having area below the threshold are discarded. The remaining blobs represent detected moving objects.

The movement of objects in the consecutive camera images is tracked using Kalman filters [7]. Relationship between predicted trackers positions and detected moving objects is established [1]. Next, trackers have to be updated with data on the moving objects (position, size) obtained from the current camera image. However, in order to ensure that the procedure is able to track all the objects in the camera view, ambiguous relations between trackers and detected objects (tracking conflicts) have to be resolved. Otherwise, tracking errors (trackers updated with incorrect data) may propagate from one image frame to another, resulting in a loss of tracked object. The proposed procedure for conflict resolving works as follows. Image features (key points) are found in both the image of the detected object and in image of the tracked object (stored in the tracker). For each key point, a vector of descriptors is calculated. Next, matching pairs of key points (one from the detected object and one from the tracker) are found by calculation of distance between the vectors of key point descriptors. Matched key points are analyzed and a region of the current image containing the tracked object (or its part) is extracted and used for updating the tracker. The resolving procedure is applied to all conflicting trackers and detected objects, it is also used for verification of unambiguous relations (one detected object related to a single tracker). The algorithm is repeated for the consecutive camera images.

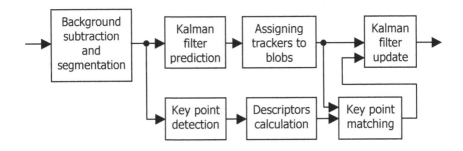

Fig. 1. Structure of the proposed object tracking algorithm

2.2 Tracking with Kalman Filters

A tracker is a structure based on the Kalman filter [7], assigned to each newly detected object and intended to track its movement through the camera view. The state of a tracker is described using a vector of parameters representing position and size of the tracked and a difference from the previous tracker state:

$$\mathbf{t} = [x_t, y_t, w_t, h_t, dx_t, dy_t, dw_t, dh_t], \tag{1}$$

where x_t and y_t define the center point of the tracked object in the image, w_t and h_t are the width and the height of the object's bounding box and the remaining parameters describe an inter-frame change of the position and size of the object [1]. For each analyzed camera image, two steps are performed: prediction (estimation of the current state) and updating of the filter state with a current measurement (ideally, the actual position and size of the object).

The trackers of all objects are maintained throughout the processed camera video frames. If a tracked object is lost, its tracker is deleted. For each analyzed image, the predicted state of each tracker is calculated internally by its Kalman filter. In the next step, relationship between the detected objects and the active trackers are established.

Let the matrix $\mathbf{I}$ be the currently analyzed video image and $\mathbf{M}$ represent a single channel moving objects mask, having non-zero values only for pixels belonging to moving objects. Using a connected component extraction procedure, a set of blobs is found. A single blob $\mathbf{B}$ is described using its bounding rectangle b_{box}, with position of a center point (x_b, y_b), width w_b and height h_b:

$$\mathbf{B}(x_b, y_b, w_b, h_b) = \left\{ \mathbf{I}(x, y) \middle| x_b - \frac{w_b}{2} \leq x < x_b + \frac{w_b}{2} \wedge y_b - \frac{h_b}{2} \leq y < y_b + \frac{h_b}{2} \wedge \mathbf{M}(x, y) > 0 \right\} \quad (2)$$

where $\mathbf{I}(x, y)$ and $\mathbf{M}(x, y)$ denote the value of input image and the object detection mask at the pixel position (x, y), respectively. The predicted position of each tracker in the current frame may also be described using a bounding box t_{box}:

$$t_{box} = \left\{ (x, y) \middle| x_t' - \frac{w_t'}{2} \leq x < x_t' + \frac{w_t'}{2} \wedge y_t' - \frac{h_t'}{2} \leq y < y_t' + \frac{h_t'}{2} \right\} \quad (3)$$

using notation from Eq. 1, where $\mathbf{t'}$ represents a prediction of the Kalman filter state.

Assuming that N_B blobs were found in the current frame and N_T trackers are active, the relationship between trackers and blobs may be expressed using a binary relationship matrix $\mathbf{R}$, constructed by examining whether bounding rectangles of trackers and blobs intersect with each other:

$$R(i, j) = \begin{cases} 1, & b_{box}(i) \cap t_{box}(j) \neq 0 \\ 0, & \text{otherwise} \end{cases}, i = 1, \ldots, N_B, \ j = 1, \ldots, NT \quad (4)$$

Next, conflict situations, which are not a 'one blob to one tracker' cases, need to be resolved. We cannot use a simple measure distance, such as Mahalanobis metric or the Munkres algorithm, because in general there are no 'single blob to single tracker' relations in all cases. Object blobs may be fragmented or merged. Therefore, we propose the following procedure.

1. Merge all the blobs related with j-th tracker into a new blob:

$$\mathbf{B}(j) = \sum_i \left(\mathbf{B}_i \middle| \mathbf{R}(i, j) \neq 0 \right), \ i = 1, \ldots, N_B, \ j = 1, \ldots, N_T \quad (5)$$

2. Determine position of the tracked object within the merged blob.
3. Use the found object position for tracker updating.

Proper realization of the step 2 requires analysis of the merged blob content, comparison with the images of objects stored in the related trackers and estimation of the current object position in the image. In the proposed approach, key points and their descriptors are used for this task.

2.3   Key Points Extraction, Description and Matching

The main task of the conflict resolving procedure is matching the blob representing a moving object (or its part) to the tracked object. The image of the detected object is compared with the object image stored in its tracker during the last non-conflict update. Alternatively, the tracker may collect images of the tracked object from the successive images and use them all for matching with the detected objects. However, comparing images on a pixel-by pixel basis is inefficient. The proposed approach is based on finding distinctive pixels (key points) that are stable, invariant to scaling and rotation and allowing for easy calculation of local image descriptors. Various feature extraction methods have been proposed in the literature and the SURF (Speeded Up Robust Features) algorithm [9] is probably the most popular and efficient one. Different key point extraction algorithms were compared and the results are presented further in the paper. After the initial testing, the SURF algorithm was chosen for key point extraction in the image masked out by the object detection result and for computation of local descriptors of each key point.

Given the image of multiple, overlapping moving objects (in tracking conflict) and stored images of each single tracked object that takes part in this conflict, the matching procedure finds sections of the current image in which an object is visible. Matching the current image to all tracked objects is done by calculating a distance between each pair of key point descriptors. In order to reduce the risk of errors and the computation time, key points belonging to the analyzed blob are compared only with key points of related trackers. A similarity of two key points is computed as a distance between descriptor vectors of these key points. For i-th key point of the blob and j-th key point of the tracker image, described with SURF vectors $\mathbf{v}_i$ and $\mathbf{x}_j$, respectively, a distance is calculated using the Euclidean norm:

$$d(i,j) = \sqrt{\sum_{k=1}^{k=64}\left(\mathbf{v}_{i,k} - \mathbf{x}_{j,k}\right)^2} \tag{6}$$

Distance values have to be calculated for each pair of the blob-tracker key points. A tracker key point that matches i-th key point in the related blob is given by:

$$j_{match}(i) = j \mid d(i,j) = \min\{d(i,1),...,d(i,J)\} \wedge d(i,j) \le d_{max} \tag{7}$$

where J is number of key points in the examined tracker image. The threshold d_{max} is introduced in order to discard weak matches between key points. If none of the tracker key points fulfill this condition, the key point of the detected object has no matching tracker key point. If a single blob has N associated trackers in a conflict situation, it is matched against each individual tracker. Therefore, for a key point in the detected object, a matching point may be detected in each tracker image. The key point that is finally selected as a match is the one with the smallest d value.

2.4   Updating Kalman Trackers Using the Obtained Key Point Matches

As a result of the previous step, matches between key points of trackers and blobs are established. An estimated position of each tracked object, found with both prediction of the Kalman filter state and the analysis of matching key points, is used for Kalman filter update. The algorithm works as follows. Suppose that as a result of analysis of relations between trackers and detected objects, a group of N_T trackers related with N_B

blobs ($N_T > 1$ and/or $N_B > 1$) was found. First, key points and their descriptors are found in each detected object. For each tracker in the group, key points of stored image are matched against key points of all the related blobs. For each blob in the group, found matches between key points of this blob and all the related trackers are collected. If a blob key point was matched with more than one tracker key points, only the strongest match is retained. The remaining matches are analyzed for the unprocessed blob-tracker pairs. The matched key points remaining after post-processing are used to establish a region of the detected object covering the object represented by the tracker. If the object is partially obscured, estimation of the position of the whole object may be done by analysis of the matched key points position. The established image region covering the tracked image is used as a new measurement for updating the tracker. The described procedure is repeated until all the groups of related trackers and detected objects in the current image are processed.

3   Experiments and Examples of Use

Experiments were performed on a PC (quad-core CPU at 2.80 GHz, 6 GB RAM), using recordings from PETS 2006 (Performance Evaluation of Tracking and Surveillance) dataset [10] and own recordings. The first test was done in order to assess the performance of state-of-art key point extraction algorithms and to select an optimal algorithm for the proposed object tracking problem. Table 1 shows results of key point extraction using different algorithms, performed on one of PETS recordings (720 x 576 pixels, 15 fps). Average processing time per frame and a number of found key points are given. The FAST (Features from Accelerated Segment Test) method was fastest, but it found too many clustered key points that were not distinctive enough. MSER (Maximally Stable Extremal Region) and Star detected too few points. SIFT (Scale Invariant Feature Transform) was the most accurate one, but the computation time was too long. The Shi-Tomasi and SURF methods extracted similar number of key points, but in case of SURF, these points were more distributed across the moving objects, while the Shi-Tomasi method found almost all key points only on edges of the objects. Computation time using the object detection phase in case of SURF was marginally longer than when full frame processing was used, but key points belonging to the background were eliminated, making further key point matching significantly easier. Therefore, SURF was selected for extraction of key points in moving objects for the purpose of the presented algorithm.

The remaining experiments tested validity of the proposed method of matching trackers with detected moving objects, using key point matching. Some examples of difficult tracking situations are presented below.

During the conflict situation in object tracking, no measurement for Kalman filter updating is available. After the conflict is finished, trackers of respective objects need to be updated using the correct blobs (objects must not swap their trackers). The proposed approach based on key points matching is useful for this task. If the objects are tracked without a conflict, their images are stored in the tracker (Fig. 2a) and used for key point detection and SURF descriptor calculation during the conflict, when images in trackers are not updated (Fig. 2b). After the conflict is finished, key points in the current image are matched with key points of the object images remembered from the state before the conflict, using the proposed procedure. Using the result of key point matching, it is possible to identify the tracker that should be updated with the detected object (Fig. 2c). This approach is especially helpful in case of objects that rapidly change their direction of movement during the conflict.

Table 1. Average processing time of one video frame in ms and a number of detected key points using various algorithms, full frame analysis vs detection only in moving objects regions

Algorithm	Full frame		Only moving objects		
	Time	No. of KP	Time w/o object det.	Time with object det.	No. of KP
FAST	7.74	2539	7.46	22.10	568
Shi-Tomasi	29.77	1000	29.29	43.93	446
MSER	246.69	272	60.10	74.74	57
Star	48.04	247	49.28	63.92	104
SIFT	489.29	1506	207.19	221.83	349
SURF	59.29	1549	49.34	63.98	283

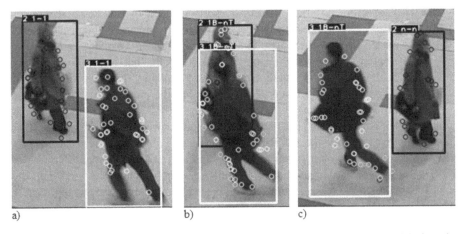

a)   b)   c)

Fig. 2. Verification of trackers after the tracking conflict: a) key points detected before the conflict, b) during conflict, c) after the conflict (only matched key points shown)

In case of object splitting, e.g. when a person leaves a luggage and leaves, a tracker should be divided into two separate trackers. A simple approach is to test whether the detected objects covered by the same tracker are separated by a distance larger than a defined threshold [8]. The proposed algorithm improves handling of such cases by using key point matching. Previously detected key points, stored in the tracker, are compared with key points from each of the related objects after the splitting. The best matching object is used to update the tracker and the remaining objects receive new trackers. In the example shown in Fig. 3, more matching key points were found in the bag image than in the person image, therefore the tracker remained with the bag.

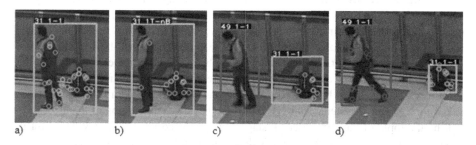

Fig. 3. Detection of splitting objects: a) a person with a bag, b) person leaving a bag, but still tracked with the same tracker, c) using the key point matching procedure, the object split is detected, the tracker remains with the left bag and a new tracker is created for the person, c) both objects have their own trackers

Fig. 4 presents an attempt of applying the proposed method for tracker updating in conflict situation. It can be seen that matching the current image of conflicting object with images of all objects obtained a number of image frames earlier, is problematic. Only a small number of key point pairs with strong match was found and in some cases, incorrect key points were matched. This is caused mainly by the method of SURF descriptors calculation and by the fact that the most distinctive key points are usually found on the object edges. Different neighborhood of the key point found in the image of an object before and during the tracking conflict results in difference of descriptors calculated from these key points, which causes the distance between descriptor vectors to be large.

Figs. 2–4 illustrate some tracking conflict situations in which key point matching may be helpful. Quantitative comparison between the original and the proposed method requires an accurate ground truth data and such tests will be performed in the future, after the tracking algorithm is improved.

4   Conclusions

In the previous work on object tracking in video, researchers focused mainly on either global image features matching or on using trackers updated with object detection results. In this paper, an approach augmenting the tracker algorithm with key point matching for resolving difficult situations in object tracking, was proposed. The algorithm improves tracking accuracy by using SURF key points for finding the parts of the image that match tracked objects. By using results of object detection and discarding the static background from analysis, the risk of incorrect key point matching was reduced. Using key points of images of moving objects stored in trackers during their non-conflict updates, it is possible to find positions of these key points in the image of

348   G. Szwoch

conflicting objects. Therefore, Kalman filters may be updated with estimated positions of the tracked objects. Several improvements of the presented procedure are planned. The main problem is that the number of matched key points is often too small because of large difference between the current and the past image of an object. This may be addressed by using more than one image frame for key point matching. Moreover, a more robust procedure for discarding incorrect key point matches is needed and a procedure for estimation of object position has to be extended so that the occluded object part is taken into account. Another practical problem is that commercial usage of the SURF algorithm is protected by a patent. Therefore, alternative and patent-free key point extraction methods have to be evaluated and the optimal algorithm has to be selected for a practical system. These issues, as well as extensive testing of the algorithm, will be the topic of the next research stage. The proposed algorithm may be useful in video surveillance systems for automatic detection of important security threats.

Fig. 4. Results of the test in which key points of all three objects being in the tracking conflict (b – e) are matched with key points of tracked objects obtained before the conflict (a). Circles mark key points matched with the tracker (only matches with distance < 0.2)

Acknowledgements. Research is subsidized by the European Commission within FP7 project INDECT, Grant Agreement No. 218086.

References

1. Czyżewski, A., Szwoch, G., Dalka, P., et al.: Multi-Stage Video Analysis Framework. In: Lin, W. (ed.) Video Surveillance, pp. 147–172. InTech, Rijeka (2011)
2. Fukunaga, K., Hostetler, L.D.: The Estimation of the Gradient of a Density Function, with Applications in Pattern Recognition. IEEE Trans. on Information Theory 21, 32–40 (1975)
3. Horn, B.K.P., Schunck, B.G.: Determining Optical Flow. Artificial Intelligence 17, 185–203 (1981)
4. Lucas, B.D., Kanade, T.: An Iterative Image Registration Technique with an Application to Stereo Vision. In: Proc. of Imaging Understanding Workshop, pp. 121–130 (1981)
5. Stauffer, C., Grimson, W.E.L.: Adaptive Background Mixture Models for Real-time Tracking. In: Proc. of IEEE Conference on Computer Vision and Pattern Recognition (CVPR), pp. 246–252 (1999)
6. Kim, K., Chalidabhongse, T.H., Harwood, D., Davis, L.: Real-time Foreground-Background Segmentation using Codebook Model. Real-time Imaging 11, 172–185 (2005)
7. Welch, G., Bishop, G.: An Introduction to the Kalman Filter. Technical report TR 95-041. UNC-Chapel Hill (2006), http://www.cs.unc.edu/~welch/kalman/
8. Szwoch, G., Dalka, P., Czyżewski, A.: Resolving Conflicts in Object Tracking for Automatic Detection of Events in Video. Elektronika 52, 52–55 (2011)
9. Bay, H., Ess, A., Tuytelaars, T., Van Gool, L.: SURF: Speeded Up Robust Features. Computer Vision and Image Understanding 110, 346–359 (2008)
10. PETS, Benchmark Data (2006), http://www.cvg.rdg.ac.uk/PETS2006/data.html

Performance of Basic Spectral Descriptors and MRMR Algorithm to the Detection of Acoustic Events

Eva Vozarikova, Martin Lojka, Jozef Juhar, and Anton Cizmar

Technical University of Kosice
Dept. of Electronics and Multimedia Communications, FEI TU Kosice
Park Komenskeho 13, 041 20 Kosice, Slovak Republic
{eva.vozarikova,martin.lojka,jozef.juhar,anton.cizmar}@tuke.sk
http://kemt.fei.tuke.sk

Abstract. This paper is focused on the detection of abnormal situations via sound information. As a main feature extraction algorithm, basic spectral low - level descriptors defined in MPEG-7 standard were used. Various settings for spectral descriptors such as Audio Spectrum Envelope, Audio Spectrum Flatness, Audio Spectrum Centroid and Audio Spectrum Spread were used and many experiments were done for finding the limits of using them for the purpose of acoustic event detection in urban environment. For improving the recognition rate we also applied the feature selection algorithm called Minimum Redundancy Maximum Relevance. The proposed framework of recognizing potentially dangerous acoustic events such as breaking glass and gun shots, based on the extraction of basic spectral descriptors through well known Hidden Markov Models based classification is presented here.

Keywords: MPEG -7, feature selection, MRMR, acoustic events.

1 Introduction

In last few years a huge interest was concentrated to the detection of selected sound, where one category of sounds can include potential dangerous ones, for example gun shots, sounds of car crash accidents, calling for help, etc. The detection of the mentioned sound category is relevant for surveillance or security systems. They are created these times by using the knowledge of different scientific areas, e.g. signal processing, pattern recognition, machine learning, etc. As was indicated, surveillance or security systems can help to protect lives and properties. In some cases these systems can be very helpful for police because they operate according to the given instructions and they are not affected by any irrational behavior in comparison with people, e.g. the bystander effect [1], [2].

This effect is sometimes present in today's society. Experiments focused on it brought interesting results - the presence of other people declines the will to help. The reasons can be different, e.g. the threat of danger, troubles from the

A. Dziech and A. Czyżewski (Eds.): MCSS 2012, CCIS 287, pp. 350–359, 2012.

investigation or the threat that somebody did not understand the particular situation well.

In one experiment, people waited in a waiting room for an interview. Some of them waited alone and others in small groups. During the waiting, smoke started to get into the room from the ventilation system. The results show that 75% of people who waited alone announced the smoke up to two minutes. But only 10 % of people who waited in small groups announced the smoke up to the given time. In other similar experiment, researchers investigated how many people are willing to respond to calling for help. About 70% of people who waited alone reacted opposite to 40% of people waiting in pairs and only 7% of those which waited with passive confederates [3]. But the most famous story is about the murdered Kitty Genovese [1], which would be alive if somebody had helped her. There are similar experiments and sad stories that indicate the need of surveillance systems focused on the detection of abnormal or dangerous situations.

In an ideal case, surveillance system [4] processes and evaluates input audio-visual information stream and if some sound or visual object is recognized as dangerous, a surveillance system should generate an alert. Some dangerous situations are easier to detect via audio information then video information, for example calling for help, breaking glass, shouting, crying, gun shots, etc. When light conditions are bad or critic situation is out of camera coverage, video based system will probably fail. On the other hand, extreme weather conditions can limit the audio based surveillance system.

Each intelligent surveillance system can be divided into the feature extraction block and the classification block. In the feature extraction block, the input audio signal is transformed into the feature vectors according to the chosen feature extraction algorithm in time, frequency, cepstral or any other domain. In this paper basic spectral descriptors defined in MPEG-7 standard were used [5]. Some interesting results were achieved by the Mel-Frequency Cepstral Coefficients - MFCC too [6], [7], which were primarily developed for speech or speaker recognition tasks. This method was created according to the human perception system.

The effective feature extraction should highlight the relevant information and reduce the number of input data by removing irrelevant information using various methods [8]. Discrete Cosine Transform - DCT is used in MFCC extraction for decorelating coefficients. Other kinds of methods such as Principal Component Analysis - PCA, Linear Discriminant Analysis - LDA, etc. transform original feature space to another space with lower dimension. A feature selection algorithm such as Minimum Redundancy Maximum Relevance - MRMR, Forward Selection - FS, Backward Selection - BS, RefiefF, etc. can be used to select relevant subset of features from full set of features. Some of them (PCA, LDA) change the values of coefficients, other (FS, BS, RefiefF) select the relevant ones or sort them to the new sequence according to their priority (MRMR).

In the classifier block the input patterns are classified into predefined classes according to the knowledge obtained in the training process of classifier. There are several methods that can perform it e.g. Hidden Markov Models (HMM), Gaussian Mixture Models (GMM), Bayesian Networks (BN), Support Vector Machines (SVM) and Neural Networks (NN), etc. In this work one, two and three states HMM up to 256 PDFs were used.

The research described in this paper is focused on the feature extraction based on MPEG-7 and post processing of extracted features by the MRMR selection algorithm. This approach is applied to the acoustic event detection framework as a partial task in the complex intelligent surveillance system.

The rest of the paper has the following structure: section 2. describes the background of work, section 3. presents feature extraction methodology. Section 4. gives information about used feature selection algorithm - MRMR. Section 5. includes the description of experiments and results, finally the conclusion and future work proposal follows in Section 6.

2   Background

MPEG-7 is an ISO/IEC standard developed by the Moving Picture Experts Group (MPEG) [5], [9], [10]. One part of the standard deals with audio information, it is MPEG-7 Audio.There 17 low - level descriptors are defined.

We focused on the basic spectral descriptors i.e. Audio Spectrum Envelope (ASE), Audio Spectrum Spread (ASS), Audio Spectrum Centroid (ASC) and Audio Spectrum Flatness (ASF). They are computed from input audio file with the frame length equal to hopSize. ASE and ASF generate "m" - dimensional feature vector per signal frame and not all of extracted coefficients have the same importance like other ones. We have supposed that included information can be reduced by feature selection algorithms. For sorting the coefficients, we chose MRMR (Maximum Relevance Minimum Redundancy) algorithm [11], which is mostly used in genetic. We applied the same approach,which was used for selection of relevant genes, for selection the relevant subset form full set of extracted features. This strategy is explained and performed in this work.

3   Feature Extraction

The feature extraction method is the essential part of any detection system. The efficient feature extraction is a very important phase of overall process, because the recognition performance directly depends on the quality of extracted feature vectors.

In our experiments basic spectral descriptors namely Audio Spectrum Envelope (ASE), Audio Spectrum Centroid (ASC), Audio Spectrum Spread (ASS) and Audio Spectrum Flatness (ASF) were used to extract relevant features.

3.1   Audio Spectrum Envelope - ASE

The audio spectrum envelope (ASE) [5] is obtained by summing the energy of the original power spectrum within a series of frequency bands which are logarithmically distributed between lower edge and higher frequency edges. The spectral resolution r of the frequency bands can be choose between within $1/16$ of an octave to 8 octaves. The sum of power coefficients in band $b[loF_b, hiF_b]$ gives the ASE coefficient for this frequency range. The coefficient for the band b is computed by the following way:

$$ASE(b) = \sum_{k=loK_b}^{hiK_b} P(k) \qquad (1 \leq b \leq B_{in}), \tag{1}$$

where $P(k)$ are the power spectrum coefficients, loK_b (resp. hiK_b) correspond to the lower edge of the band loF_b (the higher edge of the band hiF_b).

The ASE provides a compact representation of the spectrogram of the input acoustic signal.

3.2   Audio Spectrum Centroid - ASC

The audio spectrum centroid (ASC) [5] corresponds to the centre of gravity of a log-frequency power spectrum. For a given frame of signal, ASC descriptor is computed from the modified power coefficients and their frequencies. In the Eq. (2) $P'(k')$ represents the power spectrum and $f'(k')$ represent corresponding frequencies.

$$ASC = \frac{\sum_{k'=0}^{(N_{FT}/2)-K_{low}} log_2\left(\frac{f'(k')}{1000}\right) P'(k')}{\sum_{k'=0}^{(N_{FT}/2)-K_{low}} P'(k')}. \tag{2}$$

The ASC informs on the shape of the power spectrum. It indicates whether in a power spectrum are dominated lower or higher frequencies and can be regarded as an approximation of the perceptual sharpness of the signal.

3.3   Audio Spectrum Spread - ASS

In MPEG-7, audio spectrum spread (ASS) [5] is defined as the second central moment of the log-frequency spectrum. It is also called instantaneous bandwidth. For a given signal frame ASS is computed following way:

$$ASS = \frac{\sum_{k'=0}^{(N_{FT}/2)-K_{low}} \left[log_2\left(\frac{f'(k')}{1000}\right) - ASC\right]^2 P'(k')}{\sum_{k'=0}^{(N_{FT}/2)-K_{low}} P'(k')}. \tag{3}$$

ASS descriptor is extracted by taking the root-mean-square (RMS) deviation of the spectrum from its centroid ASC. The ASS gives indications about how the spectrum is distributed around its centroid. A low ASS value means that the spectrum may be concentrated around the centroid, whereas a high value reflects a distribution of power across a wider range of frequencies. ASS is a measure of the spectral shape.

3.4 Audio Spectrum Flatness - ASF

The audio spectrum flatness (ASF) [5] reflects the flatness properties of the power spectrum. ASF consists of a series of values where each one represents the deviation of the signals power spectrum from a flat shape inside a predefined frequency band. The power coefficients are computed from non-overlapping frames where the spectrum is divided into $1/8$ octave resolution logarithmically spaced overlapping frequency bands. For each band b, a spectral flatness is estimated as the ratio between the geometric mean and the arithmetic mean of the spectral power coefficients within this band:

$$ASF(b) = \frac{\sqrt[hiK'_b - loK'_b + 1]{\Pi_{k'=loK'_b}^{hiK'_b} P_g(k')}}{\dfrac{1}{hiK'_b - loK'_b + 1} \displaystyle\sum_{k'=loK'_b}^{hiK'_b} P_g(k')}, \qquad (1 \le b \le B). \qquad (4)$$

For all bands under the lower edge (e.g. 1 kHz), the power coefficients are averaged in the normal way. For all bands above this edge, power coefficients are grouped $P_g(k')$. The terms $hiK'b$ and $loK'b$ represent the high and low limit for band b. High values of ASF coefficients reflect noisiness, on the other hand, low values indicate a harmonic structure of the spectrum.

4 Feature Selection

A feature extraction can consist of two different steps, more precisely extraction of characteristic features according to the extraction algorithm, then the selection of relevant subset of features. A two phase feature selection was applied in work [11], where first algorithm selects candidate feature set and the next one selects final subset of features. Some of algorithms need to cooperate with particular classifier, others work for overall classifier. Minimum Redundancy Maximum Relevance - MRMR belongs in to the last group.

MRMR algorithm [11],[12] selects a compact set of superior features at very low cost. Algorithm includes two independent criterions:

- *maximal relevance*,
- *minimal redundancy*.

First criteria *maximal relevance*: features are selected according to the highest relevance (dependency) to the target class c. Relevance can by interpreted such as correlation or mutual information, which defined dependencies between variables.

For discrete variables, the *mutual information* I of two variables x and y is based on their joint probabilistic distribution $p(x, y)$ and their probabilities $p(x)$ and $p(y)$:

$$I(x, y) = \sum_{i,j} p(x_i, y_j) log \frac{p(x_i, y_j)}{p(x_i)p(y_j)}. \tag{5}$$

Searching according *maximal relevance* chooses the feature with high relevance to the target class c, so that these features have strong dependency to this class, therefore *maximal relevance* is associated with *maximal dependency*. *Maximal relevance* (maximal dependency - $maxD$) of feature set S with features x_i can be described with formula:

$$maxD(S, c), \qquad D = \frac{1}{|S|^2} \sum_{x_i \in S} I(x_i, c), \tag{6}$$

where $maxD(S, c)$ is computed with the mean values of all mutual information values between individual feature x_i and corresponding c class.

Features selected by *maximal relevance* usually have rich redundancy (we can say that there is a strong dependency between features). When two features x_i, x_j are heavily dependent on each other, their class-discriminant power would not change much if one of them will be removed. For this reasons *minimal redundancy* ($minR$) criteria is applied by formula:

$$minR(S), \qquad R = \frac{1}{|S|^2} \sum_{x_i x_j \in S} I(x_i, x_j). \tag{7}$$

MRMR combines these two criterions *maximum relevance* and *minimum redundancy* by the operator $\phi(D, R)$, which will optimize D and R simultaneously:

$$max\phi(D, R), \qquad \phi = D - R. \tag{8}$$

MRMR feature selection framework were successfully applied to genes selection, where selected genes led to the improvement of class prediction on five gene expression data sets such as Lymphoma, Lung, Leukemia, Colon, NCI [11].

It should be noticed that the combination a very effective features with another very effective features does not necessarily lead to the better feature set.

In general for feature selection algorithms, there are these advantages:

- a dimension reduction for reducing the computational cost,
- a reduction of irrelevant features (noise features) for improving the classification accuracy.

5   Description of Experiments and Results

Acoustic events used in experiments in a quiet environment were recorded [13]. Recordings (48 kHz, 16 bits/sample) were cut and manually labeled using Transcriber. Then the training and the testing set of data was created. For training process 153 shots, 141 breaking glass and 53 min. of background were used, 11 shots and 13 breaking glass were mixed into the 50 seconds of background and then used in offline tests.

Feature extraction algorithm was inspired by basic spectral descriptors. Our attention was focused on ASE, ASC, ASS and ASF descriptors. Each of them were extracted with different settings from the input acoustic signal. ASE and ASF descriptors generated "m"-dimensional feature vectors, therefore MRMR algorithm was applied for sorting them under their importance. ASS and ASC generated one parameter per signal frame.

5.1 ASE Experiments

Many experiments were done by ASE descriptor, which describe the log-frequency power spectrum of acoustic signal. We investigated the performance of ASE for the detection of acoustic events.

The influence of frame length was investigated by the hopSize parameter, which was set to 10, 20, 30, 50, 100 ms. The best experimental value for hopSize parameter was 20 ms. Then several tests for finding the suitable frequency limits for lower edge (loEdge) and higher frequency edge (hiEdge) were done. According to MPEG-7 recommendation loEdge and hiEdge should be set to 62.5 Hz and 16000 Hz. The extraction of ASE was also done by the 62.5, 1000, 2000 for loEdge and 16000, 22000, 23000 Hz for hiEdge frequency limit.

The most suitable loEdge was 1000 Hz and hiEdge was 23000 Hz. ASE with 1/8 frequency resolution was extracted. The best results with two states HMM (4 PDFs) were reached. Originally ASE had 38 coefficients per frame. This dimension was reduced according to MRMR algorithm to the 32 and 34 parameters per frame.

5.2 ASF Experiments

Many experiments with ASF descriptor for the purpose of appropriate spectral flatness description were performed [14]. Each change of the input acoustic signal causes the change in ASF coefficients. Generally, ASF descriptor generates $m-$dimension feature vector per signal frame depending on the frequency band, which width is appointed according the $loEdge$ and $hiEdge$ parameters. The $loEdge$ parameter was set to 250, 1000, 2000, 2500, 4000, 10000 and $hiEdge$ was set to 22000. Also we investigated the appropriate size of $hopSize$ parameter. We extracted ASF with 3, 4, 5, 10, 20, 30 ms of hopSize.

The best results of ASF were reached by two states HMM (128 and 256 PDFs), where ASF features from 1000 Hz to 22000 Hz frequency range and with 5 ms of hopSize were extracted. The feature vector had size of 17 coefficients. This vector size was reduced by MRMR to size 7.

5.3 ASC and ASS Experiments

ASC describes the centre of gravity of a log-frequency power spectrum and ASS gives information about how the spectrum is distributed around its centroid.

They were computed separately from input signal with the same hopSize. It was set to 2, 3, 4, 5, 10, 20, 30, 50, 100 ms. These descriptors generated one feature per frame.

The best results of ASC were obtained by the hopSize set to 3 ms with one state HMM (4 PDFs - 256 PDFs) and for ASS with hopSize equal to 2 ms, with one states HMM (2, 4, 16 - 256 PDFs). For the fusion of all extracted descriptors, ASC and ASS with hopSize equal to 5 ms was chosen.

5.4   Selection and Fusion of Features

The creation of final set of features is depicted in the Fig. 1.

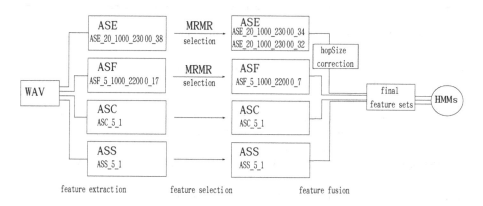

Fig. 1. Principial block scheme of feature processing

In the first step descriptors with the best settings were extracted according to the previous lines. Then the MRMR selection algorithm was applied to the searching and sorting them from the best to the worst one. ASE feature vector had 38 parameters per frame. After MRMR selection the order of ASE did not change. Therefore last 4 and 6 coefficients from each feature vector were removed. This way 34 and 32 dimension feature vectors were created. MRMR was also applied to the ASF, which full feature vector consisted of 17 parameters. The normal order (from 1 to 17-th) was changed by MRMR to the different sequence of parameters. You can see proposed orders in the Fig. 2.

ASE_20_1000_23000_38 (full set) | 1 | 2 | 3 | 4 | 5 | 6 | 7 | 8 | 9 | 10 | 11 | 12 | 13 | 14 | 15 | 16 | 17 | ... | 30 | 31 | 32 | 33 | 34 | 35 | 36 | 37 | 38 |
ASE_20_1000_23000_34 (selected set) | 1 | 2 | 3 | 4 | 5 | 6 | 7 | 8 | 9 | 10 | 11 | 12 | 13 | 14 | 15 | 16 | 17 | ... | 30 | 31 | 32 | 33 | 34 | | | | |
ASE_20_1000_23000_32 (selected set) | 1 | 2 | 3 | 4 | 5 | 6 | 7 | 8 | 9 | 10 | 11 | 12 | 13 | 14 | 15 | 16 | 17 | ... | 30 | 31 | 32 | | | | | | |

ASF_5_1000_22000_17 (full set) | 16 | 15 | 12 | 17 | 14 | 11 | 13 | 10 | 8 | 7 | 9 | 6 | 5 | 4 | 3 | 1 | 2 |
ASF_5_1000_22000_7 (selected set) | 16 | 15 | 12 | 17 | 14 | 11 | 13 | | | | | | | | | | |

Fig. 2. Feature selection by MRMR

The creation of final feature vectors followed after MRMR selection. Features were extracted with different hopSize (20, 5, 5, 5) therefore their feature matrices

Model type	PDFs								
	1	2	4	8	16	32	64	128	256
ASE_ASF_ASC_ASS_full_set (2 state)	72,58	61,29	64,52	61,29	**75,81**	69,35	**75,81**	—	—
ASE_38_ASF_17 (2 state)	59,68	**79,03**	**85,48**	**83,87**	**75,81**	**87,1**	**75,81**	**75,81**	—
ASE_select_34_ASF_17 (2 state)	62,9	**75,81**	70,97	**79,03**	74,19	**83,87**	**90,32**	—	—
ASE_select_32_ASF_17 (2 state)	66,13	**75,81**	66,13	70,97	**75,81**	**75,81**	**83,87**	**87,1**	**91,94**
ASE_select_32_ASF_select_7_ASC_ASS (2 state)	11,29	11,29	43,55	50	58,06	53,23	56,45	69,35	72,58
ASE_select_32_ASC_ASS (2 state)	74,19	64,52	54,84	48,39	43,55	54,84	56,45	56,45	—
ASF_select_7_ASC_ASS (2 state)	**77,42**	**79,03**	69,35	67,74	56,45	66,13	69,35	69,35	69,35
ASE_ASF_ASC_ASS_full_set (3 state)	59,68	66,13	61,29	58,06	61,29	58,06	45,16	37,1	30,65
ASE_38_ASF_select_7 (3 state)	62,9	74,19	**79,03**	**87,1**	**82,26**	**90,32**	**82,26**	70,97	67,74
ASE_select_34_ASF_select_7 (3 state)	69,35	70,97	69,35	74,19	**79,03**	**88,71**	**88,71**	**79,03**	—

Fig. 3. Recogntion results [ACC%] of shots and breaking glass by different feature sets

had different size. This problem was solved by duplications of each vector of X recording m - times, $m = hopSizeX_i$ / $hopSizeX_j$, $(i > j)$. The same numbers of vectors were achieved this way.

After mentioned processing of features, various fusion sets of descriptors were created, but only results of selected combinations of features are presented (results higher than 75% are bold). The rest of the combination did not bring considerable improvement: ASC-ASS, ASE-select-32-ASF-select-7, ASE-select-32-ASF-ASC-ASS but in some cases the worst results were occurred in comparison of the baseline models ASE-ASF-ASC-ASS trained with full sets of 57 parameters. HMM based classification for one, two and three states acoustic models was performed, but in some cases the training process failed for HMMs with higher number of PDFs. In general two states models reached better results than one and three states models.

6   Conclusion and Future Work

This work presented the feature extraction based on basic spectral descriptors from MPEG-7 and application the MRMR selection for the creation of final feature sets. For each descriptor, the best settings were found, then MRMR selection for choosing the most relevant features was applied to the ASE and ASF. Various final sets of features were created and evaluated. The use of ASS and ASC with combination of ASE and ASF did not bring better recognition results, but the interesting results were achieved by various combinations of ASE and ASF. The best result 91,94% occurred with ASE-select-32-ASF-17 with two states HMM (256 PDFs).

In several cases the mentioned approach brought better performance in comparison with baseline models.

In the future we would like to continue with the combination of various feature extraction algorithms and also we would like to perform two stage feature selection for choosing the relevant feature subset with the purpose to create a reliable detection system.

Acknowledgments. This work has been performed partially in the framework of the EU ICT Project INDECT (FP7 - 218086) and by the Ministry of Education of Slovak Republic under research VEGA 1/0386/12.

References

1. Latané, B., Darley, J.M.: Bystander apathy. American Scientist. 57(2) (1969)
2. Latané, B., Darley, J.M.: Group inhibition of bystander intervention in emergencies. Journal of Personality and Social Psychology 10(3), 215–221 (1968)
3. Bystander "Apathy", `http://faculty.babson.edu/krollag/org_site/soc_psych/latane_bystand.html`
4. Ntalampiras, S., Potamitis, I., Fakotakis, N.: On acoustic surveillance of hazardous situations. In: ICASSP, Taiwan, pp. 165–168 (2009) ISBN: 978-1-4244-2353-8
5. Kim, H.G., Moreau, N., Sikora, T.: MPEG-7 audio and beyond: Audio content indexing and retrieval, p. 304. Wiley (2005) ISBN: 978-0-470-09334-4
6. Mesaros, A., Heittola, T., Eronen, A., Virtanen, T.: Acoustic event detection in real life recordings. In: EUSIPCO 2010, Denmark, pp. 1267–1271 (2010) ISSN: 2067-1465
7. Ghulam, M., Yousef, A.A., Mansour, A., Mohammad, N.H.: Environment recognition using selected MPEG-7 audio features and Mel-Frequency Cepstral Coefficients. In: International Conference on Digital Telecommunications, pp. 11–16 (2010)
8. Vozarikova, E., Pleva, M., Juhar, J., Cizmar, A.: Surveillance system based on the acoustic events detection. Journal of Electrical and Electronics Engineering 4(1), 255–258 (2011)
9. Casey, M.: General sound classification and similarity in MPEG-7. Organised Sound 6(2), 153–164 (2001)
10. Muhammad, G., Alghathbar, K.: Environment recognition from audio using MPEG-7 features. In: International Conference on Embedded and Multimedia Computing, pp. 1–6 (2009) ISBN: 978-1-4244-4995-8
11. Peng, H., Long, F., Ding, C.: Feature selection based on mutual information: criteria of max-dependency, max-relevance, and min-redundancy. IEEE Transactions on Pattern Analysis and Machine Intelligence 27(8), 1226–1238 (2005)
12. Ding, C., Peng, H.: Minimum redundancy feature selection from microarray gene expression data. In: Bioinformatics Conference, CSB 2003, pp. 523–528 (2003) ISBN: 0-7695-2000-6
13. Pleva, M., Vozarikova, E., Dobos, L., Cizmar, A.: The joint database of audio events and backgrounds for monitoring of urban areas. Journal of Electrical and Electronics Engineering 4(1), 185–188 (2011)
14. Vozarikova, E., Juhar, J., Cizmar, A.: Study of audio spectrum flatness for acoustic events recognition. In: Digital Technologies 2011, Žilina, pp. 295–298 (2011) ISBN: 978-80-554-0437-0

Video Watermarking Based on DPCM
and Transformcoding Using Permutation
Spectrum and QIM[*]

Jakob Wassermann and Peter Wais

Dep. of Electronic Engineering
University of Applied Sciences Technikum Wien
Hoechstaedplatz 5, 1200 Wien
jakob.wassermann@technikum.at
wais_peter@hotmail.com

Abstract. Robust Video watermarking techniques based on DPCM Encoder and permutated DCT spectrum combined with QIM is introduced. For this purpose the pixels of incoming frames undergo permutation, followed by block wise DCT transform. The DPCM encoder generates differential spectral images. The significant coefficients of these differential spectral images are selected for embedding procedure by quantization index modulation (QIM) method. By this technique the watermark is more robust against attacks and compression. To further increase robustness and to avoid visible degradation additionally the frame order and the watermark pixels were rearranged. To extract the watermark from the video the same operations have to be done on the decoder side. The incoming video frames are permutated followed by block wise DCT transform. After the decoder builds the differential spectral image the embedded coefficients can be selected and watermark extracted by inverse QIM.

Keywords: Videowatermarking, DPCM, Transformcoding, DCT, QIM.

1 Introduction

Watermarking of videos is a new and very quickly developing area. It can be used to guarantee authenticity, to identify the source, creator and owner or authorized consumer. Also it could be used to transport hidden information and can be used as a tool for data protection. There are a lot of water-marking technologies that specially fit to the demand of video requirements. Most of them are frame based, it means they use the I-Frame only of the GOP's(Group of Picture's) [1] and the DCT (Discrete Cosine Transformation) approach of still images [2]

In this paper a new method for robust video watermarking based on classical DPCM (Differential Pulse Code Modulation) and permutated DCT Transform using quantized index modulation (QIM) as an embedding technique is presented. Instead of

---

[*] This research was supported by European Commission FP7Grant INDECT Project. No.FP7-218086.

A. Dziech and A. Czyżewski (Eds.): MCSS 2012, CCIS 287, pp. 360–368, 2012.

ingest the watermarks into the spectrum of every frame (frame based approach) the hidden information is embedded into the differences between the spectra of two sequential frames. To enhance robustness the pixel of the incoming frames are rearranged before they undergo the DCT transform. The selected coefficients of the spectral differences are used for the embedding process, which is realized by using QIM method.

2   Classical DPCM

The classical DPCM is a data compression method that removes the existing redundancy between the frames [3]. In Figure 1 closed loop DPCM encoder is depicted.

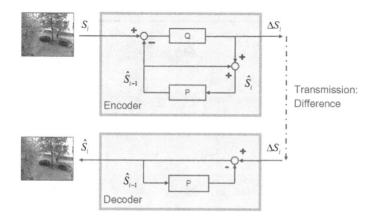

Fig. 1. Classical DPCM

From the incoming video frame S_i the previous frame $\hat{S}_{i-1}$, which is stored in the predictor memory P, is subtracted. We obtain the difference between these two frames that undergoes the quantization process by quantizer Q (Eq. (1)).

$$\Delta S_i = S_i - \hat{S}_{i-1} + \delta_i \tag{1}$$

δ_i stands for the added quantization noise. Simultaneously the quantized difference image is used to reconstruct the original frame on the encoder side, by adding the previous reconstructed frame to it.

$$\hat{S}_i = \Delta S_i + \hat{S}_{i-1} \tag{2}$$

From the transmitted difference image the DPCM decoder is able to reconstruct the frames. The reconstruction procedure is shown in Eq. (2). The closed loop is used to avoid the accumulation of the quantization errors.

2.1   W-DPCM in Time Domain

To use the DPCM techniques for watermarking the encoder and decoder were modified and well adapted for the needs of watermarking technology. Instead of hiding watermarks in every frame separately, the differences between the frames are watermarked. To realize this, a modification of closed loop DPCM was done. In the following Figure 2 new developed DPCM based watermarking algorithm in time domain is depicted.

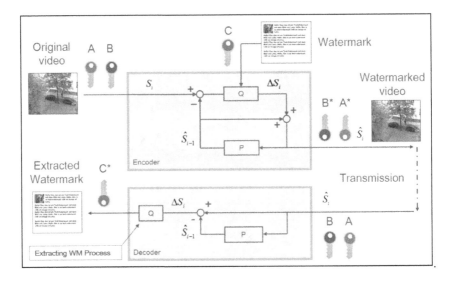

Fig. 2. DPCM based Watermarking Algorithm

From the actual frame the predicted frame from P is subtracted and the so called differential image is undergoing the quantization process Q (Eq. (1)). Through steering the quantizer functionality by watermark content, it is possible to embed watermarks into these differential images. It is done by modifying the LSBs (Least Significant Bits) of the differential values in dependency of the watermark content.

Very important is to underline the fact, that in this scheme the predictor on the encoder side works identically as the predictor on the decoder side (property of closed loop DPCM). The obtained differential image on the decoder side undergoes the extraction procedure by analysing their LSB values according to the codebook of the watermark. The extraction procedure is very simple and is reverse to the embedding process. If $|\Delta S_i|$ is even, the watermark pixel is black otherwise it is white.

The permutation can improve the performance dramatically. Three permutation keys are introduced: Key A is a permutation of the order of the frames, key B is the permutation of the pixel order inside of the frames and key C is the permutation of the watermark itself. The results for W-DPCM Watermarking in time domain were already introduced in [4].

The results show that W-DPCM in time domain enables the high capacity watermarking, however with low robustness.

3   Proposed Watermarking Scheme

The big disadvantage of watermarking in the time domain is the lack of robustness. The robustness can be increased by introducing the watermarking procedure in spectral domain. In [4] and [5] such algorithms were introduced. They work in DCT domain. The watermarks are inserted into the so called significant coefficients of DCT spectrum. Unfortunately the number of such significant coefficients is not very high.

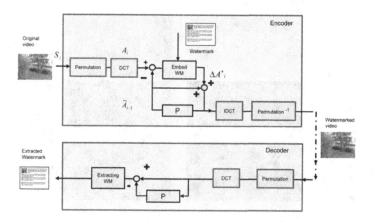

Fig. 3. Watermarking Scheme of W-DPCM in DCT Domain with Rearrangement of Frame Pixels

To overcome this limitation a combination of permutation and DCT Transform is introduced. Zhao, Chen and Liu [6] were the firsts who presented the idea of spectrum permutation in their paper "robust wavelet-domain watermarking algorithm based on chaotic map". They have shown that the robustness due to rearrangement of pixel position followed by Wavelet Transform could be increased significantly.

In this publication the idea of combination of permutation followed by the DCT transform was integrated into W-DPCM. The embedding procedure was done by QIM techniques.

In Figure 3 this new watermarking algorithm is depicted. Because the whole procedure works with Luminance Channel Y at the beginning of encoding procedure the original RGB video sequence is transformed into YCrCb colorspace. The incoming frames undergo at first the permutation and then the block wise DCT transform. In figure 4 the influence of the permutation on DCT spectrum is visualized. Without permutation the spectrum of the original frame has a visible structure. It's even possible to discover the outlines of the original image. In contrast the DCT spectrum of permutated frame looks like a noise. The value coefficients are distributed uniformly. This "noisy" image lies as input for W-DPCM encoder which generates the differential spectral image, whose coefficients are selected for embedding procedure by quantization index modulation (QIM) techniques.   A good overview of QIM is given in [7]

Original Frame   Permutated Frame

DCT Spectrum of Orginal Frame   DCT Spectrum of permutated Frame

Fig. 4. Block wise DCT Spectrum of Original and Permutated Frame

After the watermark was embedded into a differential spectral image, it is used (differential spectral image) to reconstruct the spectrum of the original frame. The inverse DCT transform and then the inverse permutation finally create an embedded frame.

Selection of the Embedding Coefficients

To realize the watermarking embedding procedure some coefficients from the DCT spectrum of the differential image should be selected. Because the spectral significant coefficients are distributed uniformly, (see Figure 4) actually they can be chosen freely. The experimental results showed, that the best performance could be achieved by coefficients from the gray area (Coefficients Nr.:2,9,10, 17, 18, see Figure 5)

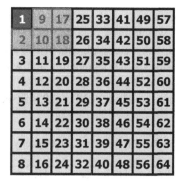

Fig. 5. The Ordering Structure of DCT Coefficients

The number of embedded coefficients determines the embedding data rate but impaires the quality of the image. The tradeoff between capacity and quality lies by one coefficient per block of 8x8 pixels. (See chapter Results.)

Quantization Procedure
To make the embedding procedure more robust against distortions and attacks the following quantization algorithm is proposed. Is the embedded value of the watermark "one" then the last 7 bits of the selected coefficients are substituted by value 32. In case the watermark value is "zero" the last 7 bits of selected coefficients are substituted by value 96. The corresponding decoding process is very simple. By selecting the embedded coefficients from differential spectral image, the last 7 bits are investigated. If the calculated value is inside the interval [0-64], so "one" is detected. Otherwise (interval [65-128]) the detected value is "zero". The sign and the MSB of the selected coefficients were preserved.

4   Results

The investigation was done with raw video format with the resolution of 640*360 and 25fps. The watermarking processing was performed only for luminance channel (after converting RGB into YCrCb colorspace), because it is more robust against distortions, than any other channels. It was investigated how many bits of the watermark could be embedded into the video without significant impairments. The degradation of the watermarked output video was measured with SSIM (Structural Similarity) index. SSIM is based on the human eye perception and so the expressiveness about distortion is better than in the traditional methods like PSNR (Peak Signal to Noise Ratio) or MSE (Mean Square Error) [8].

In the second investigation the robustness of this watermarking algorithm was tested. For this purpose several video codecs were applied and the watermark extracted.

In Table 1 the results of capacity and robustness measurements are presented. The video sequence "drift.avi", which has a data rate of 5Mbit/s in moving JPEG format (MJPEG). It was embedded with watermark data of 180 kbit/s. The watermarks can be completely extracted with any visible degradation and with SSIM index of 0.9956. It means that original and embedded are similar to 99,56%. The embedded video shows hardly any visible degradations and has SSIM of 0,9245 (see Figure 7).

Table 1. Results of Robustness and Capacity investigation

Format	Video Data Rate	Number of Blocks per Frame	Number of embedded Coefficients	Embedded Data Rate	SSIM Video	SSIM Watermarking
MJPEG	5Mbit/s	3600	2	180kbit/s	0,9245	0,9956
Cinepack	0,5Mbit/s	3600	1	90Kbit/s	0,8843	0,7345
XVID	0,7Mbit/s	3600	1	90 Kbit/s	0,9012	0,7796

Fig. 6. Original Frame and Detailed view

Cinepack and XVID have much higher compression ratio that MJPEG. The embedded data rate is reduced to 90 kbit/s and the output video has a SSIM of 0,8843 (Cinepack) and 0,9012 (XVID). It is slightly below the quality of MJPEG. (See Figures 8 and 9)

The extracted watermarks are impaired by the compression procedure but still are good visible. See Figure 10.

Fig. 7. Embedded and MJPEG coded

Fig. 8. Embedded and Cinepack coded

Fig. 9. Embedded and XVID coded

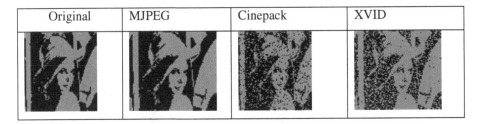

Original	MJPEG	Cinepack	XVID

Fig. 10. Extracted Watermarks after Compression

5   Conclusion

A new method for embedding watermarking into video sequences based on DPCM and permutated DCT spectrum with quantization index modulation (QIM) was introduced. The robustness of this system was tested by different compression codecs like MJPEG, XVID, Cinepack. The reasonable embedded data rate is 90 kBit/s, which is about 12% of the total data rate. The output video has a good quality with SSIM value of 0,9012(XVID). The extracted watermarks are impaired by the compression procedure but still good visible with SSIM value of   0,7796.

References

1. Benham, D., Memon, N., Yeo, B.-L., Yeung, M.M.: Fast Watermarking of DCT-based compressed images. In: Proceeding of International Conference and Imaging Science, Systems and Applications, pp. 243–252 (1997)
2. Hartung, F., Girod, B.: Digital watermarking of raw or compressed video. In: Proceedings of European EOS/SPIE Symposium on Advanced Imaging and Network Technologies, Digital Compression Technologies and Systems for Video Communication, pp. 205–213 (1996)
3. O'Neal Jr., J.B.: Differential pulse code modulation with entropy coding. IEEE Trans. Inform. Theory IT-21, 169–174 (1976)
4. Wassermann, J., Moser, G.: New Approach in High Capacity Video Watermarking based on DPCM Coding and MPEG Structure. In: Proceeding of Multimedia Communications, Services and Security, MCSS 2010, Krakow, pp. 229–233 (May 2010)
5. Wassermann, J., Dziech, A.: New Approach in Video Watermarking based on DPCM and Transformcoding. In: Proceeding of Eleventh International Conference on Pattern Recognition and Information Processing, PRIP 2011, Minsk, May 18-20, pp. 165–171 (2011)
6. Zhao, D., Chen, G., Liu, W.: A Chaos-Based Robust Wavelet-Domain Watermarking Algorithm. Chaos, Solutions and Fractals 22, 47 (2004)
7. Chen, B., Wornell, W.G.: Quantization Index Modulation for Digital Watermarking and Information Embedding of Multimedia. Journal of VLSI Signal Processing 27, 7–33 (2001)
8. Wang, Z., Bovik, A.C., Sheikh, H.R., Simoncelli, E.P.: Image Quality Assessment: From Error Visibility to Structural Similarity. IEE Transaction on Image processing 13(4) (April 2004)

Internet Identity Analysis
and Similarities Detection

Krzysztof Wilaszek, Tomasz Wójcik, Andrzej Opaliński,
and Wojciech Turek

AGH University of Science and Technology, Krakow, Poland
{wikrzysz,wotomasz}@student.agh.edu.pl,
{andrzej.opalinski,wojciech.turek}@agh.edu.pl

Abstract. Growing popularity of Web 2.0 systems created huge set of publicly available data, which is continuously expanded by users of the Internet. The anonymity of publications in Web systems encourages some users to publish false or illegal statements. Tools for identifying portal users, who publish such posts could result in higher quality of information and could be useful for law enforcement services. In this paper a method for finding similar Internet identities is introduced. Detected similarities can be used for finding several accounts of the same person. The method is based on calculating various measures characterizing forums users. It uses Web crawling system to collect data from forums. A prototype system for finding similar users is described and tests results are presented.

Keywords: Web crawling, identity analysis, text processing.

1 Introduction

The World Wide Web is the greatest public base of electronic data. It contains large amounts of information on almost every subject. Last decade brought fast development of "Web 2.0" solutions, which allowed users to become authors of Web pages content. This results in many side effects, including partial anonymousness of the source of information present on the Web. Verification of information correctness becomes very hard, which can lead to lower quality of information. Moreover, the sense of anonymity may encourage some users to publish illegal data. As a result law enforcement services are becoming more and more interested in the content of Web pages.

Probably the most widespread type of Web 2.0 system is an Internet forum. It provides a convenient environment for discussions which can involve many people. Typically a forum is dedicated to a particular topic, organized in threads and composed of messages, which are stored in a database and can be browsed or searched. This form of knowledge exchange system turned out to be suitable for many purposes. Forums are used for gathering groups of people with similar interests, building systems for reporting bugs in software products or collecting opinions concerning products.

A. Dziech and A. Czyżewski (Eds.): MCSS 2012, CCIS 287, pp. 369–379, 2012.
© Springer-Verlag Berlin Heidelberg 2012

Each message posted to a forum is always associated with a forum user, who wrote the message. Typically a person is required to create a user account before posting any messages. This process requires providing some personal data, therefore theoretically each post can be associated with a particular person in the real world. In practice it is relatively easy to create accounts anonymously, by providing false personal data and temporary email address. Such account can be used for sending messages with false information, insults or illegal content. Finding real-world identity of such user is a very hard task.

People, who send messages to different Internet forums often have several Internet identities, several accounts with different identifiers (nicknames). Some of these identities may be used for normal activities and can be identified easily. Other identities of the same person can be used for performing malicious or criminal acts. Associating two or more such accounts with a single person could help identifying identities or finding people distributing false or illegal information.

In this paper a concept of a method for finding similarities between different Internet identities is presented. The basic idea is to define several measures characterizing particular users of forums. The measures have to be calculated during Web pages crawling and stored in a database using formats suitable for fast searching. Finding similarities between identities is based on comparing values of different measures.

A prototype system for finding similarities between forum users has been successfully implemented and tested. It is integrated with a crawling system, which has been developed at the Department of Computer Science, AGH UST [1].

The paper starts with a short introduction to the problems concerning Internet identities. In the next sections the general idea and the architecture of the Internet Identities Analysis System is presented. The following section presents implemented algorithms for similarities detection. Then selected aspects of implementation are discussed together with some results of performed tests. Final section contains analysis of further development of the system.

2   Internet Identities Anonymity

Visible effect of development of the Internet Network WEB 2.0 solutions, where everyone can be a content author or an editor, is rapid increase of number of people active on Web pages. There is no need to know HTML syntax nor possess own server or domain. One can easily publish some information using blogs, forums, or social network portals. These factors caused vast number of people who exist on the Web. Although social activity on the Internet is significantly different than in real life. First difference is that on the Web one can create his virtual identity arbitrarily, without any relation to reality. Furthermore, one can create multiple identities witch significantly different characteristics and aims. All those facts result with colossal phenomenon of virtual anonymity, and increasing danger of concealment of real identity.

However, this is a fundamental feature of the Internet and it is usually treated as it's huge advantage. However, there are also some circumstances, when it becomes a drawback. This problem appears usually when we consider public security aspects, such as child pornography and molestation, drugs distributing, racist hate spreading or terrorists activities which is common known as CyberCrime [3] and developed with Internet evolution. There are also no strictly illegal negative aspects of virtual identity anonymity. As an example: multiple fake opinions, generated by many virtual identities of one person, published on experience exchange and recommendation portals. It could create false reality state, which can mislead unaware users. It's known as "deceptive opinion spam". [4][5].

The issue of detection and identification of multiple virtual identities is currently under many researches, but the problem is so complex, that it remains still unsolved. Some researches focus on criminal activities and bases on data mining model and solutions.[2] Considering different aspects of identities features, a few main models of identification can be presented. Some based on users' text comparison, focusing on text semantic, style or emotional aspect. [6], [7], [8],[9]. Another approach is to use behavior or activity features of virtual identity, as a base to comparison. [10]. Important approach used to deal with that issue is to apply some characteristic features from Social Network domains. [11]. After analysis and features extraction some attributes could also be helpful in identity comparison.

Despite many researches in this domain, virtual identity anonymity issue still remains unsolved. Moreover, with an evolution of WEB 2.0 and Internet Social Networks, this problem also evolve becoming more complex and interesting to study.

3   Internet Identities Similarities Detection

The main concept of the Internet identities similarities detection method is based on finding features of users, which cannot be easily hidden. People, who want to hide real identity would not use the same login name or email address on several forums, however they typically cannot get rid of particular habits or skills. For example, a person does not know correct spelling of particular words, the spelling errors will occur in all posts of the person.

The features that are used for finding similarities have to be represented in comparable form. The comparable form of a feature is called a measure. A measure express for example how often certain word occurred in user's posts or what time the posts were published. A measure can be calculated for given user and returned as vector of numbers. In order to find similar users, subtraction of measure vectors are calculated for each pair of users. Norm of that subtraction indicate how similar the users are in terms of particular measure.

In order to compare two users, several measures (of several different features) should be used. Integration of similarity factor from several measures requires providing weights for all used measures. The values of weights should be determined experimentally.

There are many different measures that can be calculated basing on information about users and their posts, which are retrieved from forum Web pages. The kind of data used during calculation of measure values can be used as criterion of measures classification. Three main classes can be identified:

1. based on analysis of dates of posts publication,
2. based on the posts content,
3. based on information about users.

The following measures have been defined, implemented and tested:

Measures Based on Analysis of Dates of Posts Publication. This group of measures was prepared because of assumption that people can be identified by periods of time they are usually active. It was supposed that people are often browsing Web pages routinely in similar periods of the time:

- a measure calculating for the user a seven-dimensional vector describing posts count in particular days of the week,
- a measure describing the user by a vector representing posts count in following hours of the day,
- a measure evaluating for the user a vector describing posts count in particular months,
- a measure describing user activity since 1. January 1995. It is a date near to establishment of the first Polish Web portal. This date was considered to be start of popularization of the virtual conversations groups.

Measures Based on the Post's Contents. During the development of these measures it was assumed that similarities according to used words, mistakes in posts or other characteristic elements contained in the posts can indicate similarities between users:

- a measure calculating percentage, average participation of specific tokens in all posts of the user,
- a measure evaluating average occurrence of specific "emoticons" in posts of the user,
- a measure describing average usage of untypical words by the user. The words are categorized as untypical when they are not in the basic dictionary of the SWAT system,
- a measures enabling to monitor average participation of the post fragments matching one of the specified regular expressions,
- a measure calculating average count of references to specific Web pages,
- a measure calculating average occurrence of the references to the specified file formats among all the links in user's posts,
- a measure calculating a vector indicating how often the user publishes posts with the references to pages with addresses containing the specific domain,
- a measure describing average count of the references to Web pages in posts of the user,
- a measure calculating average quotations count per one user's post,
- a measure evaluating average quoted text length in posts of the user,

- a measure calculating average sentence length in all posts of the user,
- a measure enabling to calculate frequency of the usage of numbers in posts of the user.

Measures Based on the Information About User. These measures were developed to observe similarities between users derived from the information about a person which was delivered by Web forum application (e.g. posts count or date of joining the forum group) or was provided by users about themselves (e.g. name, user description, location):

- a measure calculating relation between user's activity (defined as posts count) and average activity among all users of the forum,
- a measure describing average length of user's posts.

Obviously, many different measures can be calculated and used for comparing profiles. Research into usefulness of different measures is a very interesting direction of further work.

4 The System for Identities Analysis

In order to verify the idea, the prototype system has been implemented and tested. The system performed analysis of Web pages content during Internet crawling. Information extracted from forums Web pages was stored in a database. The system provides a Web GUI for testing implemented algorithms for calculating similarities between profiles.

The Web crawler used by the system has been created in the Department of Computer Science, AGH UST [1]. It is designed to handle hundreds of millions of Web pages. It runs on a cluster of several computers and handles up to a hundred pages on a single node per second. Performance of the system can be improved easily, by adding more nodes to the cluster.

The system consist of two main modules:

Data Acquisition Module – it is based on crawling mechanism and forum detection parser. Whenever crawler's parser detects forum content it indexes all valuable information (virtual identity details, textual content, etc.) in a database.

Search Module – it provides an access to the data stored in the database. It compute similarities and allows to select the features that should be taken into account during virtual identities matching.

Main functionalities offered by the system are:

- Crawl specified forum and store it's content in database structures.
- Find most similar users from selected forums, based on selected characteristics. User may define set of characteristics used during comparison process and it's priorities.
- Find most similar users to selected user, based on selected characteristics.

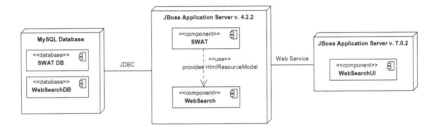

Fig. 1. Component diagram of the prototype system

4.1   Architecture

The created search module system consists of three main components, which can be deployed to independent nodes, as figure 1 shows.

WebSearch - main component, implementing crucial functionalities (gathering data from web site, measures calculation, comparison between users) which are available through web service

WebSearchUI - the component that constitutes graphical user interface, making use of interface provided by WebSearch's web service

MySQL Database - stores data about forums, users, their posts and calculated measures

SWAT, SWAT DB - web crawler and its database, it provides WebSearch component with parsed model of a website; SWAT uses WebSearch module as a plugin, directly invoking appropriate methods

4.2   WebSearch Component Overview

The successive steps taken in a web page analysis are:

1. downloading and parsing of a web page,
2. classification,
3. data extraction,
4. data serialization.

After downloading and parsing a Web page, the system tries to detect, if the page contains forum posts. If this process is successful, the Web page model, which can be processed by further forum page analysis strategies, is created.

The internal structure of the HTML documents of forum pages can be significantly various. It was the root cause of difficulties during the process of classification documents as forum pages and retrieving the information about the users and their posts. The classification and data extraction from web sites takes place in sub-component called Quantum Detector, designed to cooperate with the crawler.

Quantum Detector can be configured for processing particular formatting of the Web forum page. The configuration file is divided into two parts:

Classifiers. describing characteristic fragments of web site, which must appear in HTML in order to classify a web site into one of different models;

Elements. describing fragments of Web site which should be extracted from given document.

Elements describe precise paths in the document structure tree defined by HTML tags. Each element selects particular type of information, like name, city or e-mail address. The configuration file makes it possible to define repeatable elements, conditional elements and methods for extracting information from longer text.

The values returned from Quantum Detector are later converted to simple data objects (Forum, Section, Discussion, UserPost, User, Quotation) that can be easily stored in database.

4.3   Client Application

The client application of the system, WebSearchUI, provides the following operations:

- Choosing measures to be used in comparison process
- Choosing maximum tolerance of difference between measure vectors
- Comparing two chosen users
- Finding similar users inside given forum
- Finding user similar to chosen user on chosen forum
- Displaying detailed information about users

The presentation layer is also capable of showing detailed comparison for each measure, including highlighting parts of user post which affect considered measure.

5   Implementation and Evaluation

The implementation uses of the Java programming language and its related technologies. The persistence layer was prepared using JDBC and Hibernate to store the information about users and measures values. The Spring IoC container has been applied to configure application objects structure. There was also a graphical user interface prepared using JSF 2.0. The user interface executes the methods of application published as web service of the JAX-WS technology. The components were deployed on the JBoss application server.

SWAT and WebSearch components must be run on the same application server to ensure high performance – they exchange significant amount of data in data extraction process.

The major aim of the data extraction implementation was to provide a module enabling fast analysis of possibly wide variety of different forum pages. The persistence layer implementation impacts considerably on the application performance. It was a root cause of the decision about usage of a database in no-transaction mode.

The implementation should enable to define extensions of forum page model analysis strategy. The prepared solutions provides a possibility to add different strategies of model processing other than simple data persistence. Currently there is only one implementation of aforementioned strategy - it simply stores data directly in database. All words are stored as identifiers provided by the crawling system instead of plain text. This approach simplifies the process of searching for specific elements in user's posts.

5.1 Measures

The measures mechanism implemented in the system provides simple and easily extensible way of describing particular area of user activity inside a forum. In the current configuration the measures are calculated on user request only and are based on data stored in database, but data analysis strategy layer enables to do so during data gathering process.

All combinations of this measures can be applied to identify similar persons in virtual discussion groups. After the calculation of the measure vector for the user it's value is stored in database. During the next request of the vector evaluation it can be retrieved and used without recalculation.

All measures described in section 3 have been thoroughly tested and proved to be useful in test case scenarios.

5.2 Evaluation of Solutions

The method presented in this paper aims at finding several Internet identities of a single person. The quality of such method is very hard to evaluate. The method can give very good solution in some cases and fail in another. Several ideas for assessing the quality of solution can be proposed:

- Analyze each of the measures separately using selected set of posts. This approach can be used for tuning parameters of the measures and can prove correctness of the approach.
- Manually select or prepare set of posts with particular similarities. The results can be measured, but may be insignificant in real-world scenarios.
- Find real persons having several accounts on different forums. This method can give convincing results, however it wont prove anything.
- Divide posts of a selected user into two parts. The parts should be similar to each other, however the results may strongly depend on division.

The first of these approaches has been utilized so far. Further evaluation of the prototype system is planned in the near future.

5.3 Tests of the Prototype System

The most critical component in terms of application performance was the Quantum Detector – a module responsible for the recognition of forum pages and the retrieval of the information about the users and their posts. Two aspects of the performance can be indicated and discussed during the presentation of detection plugin. The first of them is the number of Web pages processed per second. It is influenced by time spent on searching for specified elements in HTML documents and by persistence of retrieved information. The second one is the processing efficiency determining how many of analyzed HTML documents will be classified as forum page.

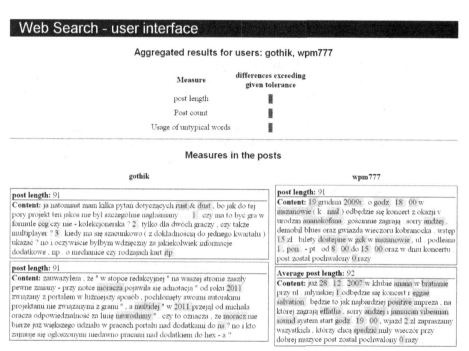

Fig. 2. Results of users comparison presented in the GUI of the system

The processing performance was verified during system development. The tests were executed using one personal computer. Random set of 30000 pages has been processed by the system, achieving up to 2.5 pages per second. Increase of recognized forums types influences negatively on the analysis performance. The experiments were executed with configuration defining structure of several very popular forum models: phpBB2, phpBB3 prosilver, phpBB3 subsilver and myBB.

Figure 2 depicts comparison of two selected users of Polish forum. Following measures have been used: average usage of untypical words by a user, average length of user's posts and post count. The upper frame shows aggregated measures comparison result – number of elements of measure difference vector that exceeds given tolerance. All posts published by pair of users are shown in the bottom part of the page. Highlighted tokens affect the first measure only: typos, words from foreign languages or abbreviations.

First two measures suggest, that the users are similar. Both have the same number of posts and similar length of posts. Third measure suggests the contrary – average number of untypical words used by user "wpm777" is far greater than those of user "gothik". Weighted aggregation of these results is a similarity factor between the users. This trivial example shows the idea of the approach. In order to achieve high quality of comparison results, more complicated measures have to be used on larger set of collected data.

6   Conclusions and Further Development

The problem of anonymity of Internet forums users is becoming more and more significant. Methods for automated identification of unreliable or illegal information are needed. The ideas and algorithms described in this paper may form the basis for creating such methods. The prototype implementation proves, that the approach can be useful for finding different Internet identities of a single person.

Current version of the system suffers from several technical issues, like poor performance and incomplete methods for parsing forums content. Further development will aim to resolve these problems.

Further research will focus on tuning existing and finding better measures for characterizing user's activities. Furthermore, better methods for results evaluation are needed.

Acknowledgments. The research leading to these results has received funding from the European Community's Seventh Framework Program (FP7/2007-2013) under grant agreement nr 218086.

References

1. Opalinski, A., Turek, W.: Information retrieval and identity analysis. In: Metody sztucznej inteligencji w dzialaniach na rzecz bezpieczenstwa publicznego, pp. 173–194 (2009) ISBN 978-83-7464-268-2
2. Chen, H., Chung, W., Xu, J.J., Wang, G., Qin, Y., Chau, M.C.L.: Crime data mining: a general framework and some examples. Computer 37(4), 50–56 (2004)
3. Chang, W., Chung, W., Chen, H., Chou, S.: An International Perspective on Fighting Cybercrime. In: Chen, H., Miranda, R., Zeng, D.D., Demchak, C.C., Schroeder, J., Madhusudan, T. (eds.) ISI 2003. LNCS, vol. 2665, pp. 379–384. Springer, Heidelberg (2003)
4. Ott, M., Choi, Y., Cardie, C., Hancock, J.T.: Finding deceptive opinion spam by any stretch of the imagination. In: Proceedings of the 49th Annual Meeting of the Association for Computational Linguistics: Human Language Technologies, HLT 2011, vol. 1, pp. 309–319. Association for Computational Linguistics, Stroudsburg (2011)
5. Jindal, N., Liu, B.: Opinion spam and analysis. In: Proceedings of the International Conference on Web Search and Web Data Mining (WSDM 2008), pp. 219–230. ACM, New York (2008)

6. Abbasi, A., Chen, H.: Writeprints: A stylometric approach to identity-level identification and similarity detection in cyberspace. ACM Trans. Inf. Syst. 26(2), Article 7 (2008)

7. Turney, P.D.: Thumbs up or thumbs down?: semantic orientation applied to unsupervised classification of reviews. In: Proceedings of the 40th Annual Meeting on Association for Computational Linguistics (ACL 2002), pp. 417–424. Association for Computational Linguistics, Stroudsburg (2002)

8. Wiebe, J., Wilson, T., Cardie, C.: Annotating expressions of opinions and emotions in language. Lang Resources and Evaluation 39, 65–210 (2005)

9. Pang, B., Lee, L.: A sentimental education: sentiment analysis using subjectivity summarization based on minimum cuts. In: Proceedings of the 42nd Annual Meeting on Association for Computational Linguistics (ACL 2004), Article 271. Association for Computational Linguistics, Stroudsburg (2004)

10. Slaninova, K., Martinovic, J., Drazdilova, P., Obadi, G., Snasel, V.: Analysis of Social Networks Extracted from Log Files. In: Handbook of Social Network Technologies and Applications, Part 1, pp. 115–146 (2010)

11. Wasserman, S., Faust, K.: Social Network Analysis: Methods and Applications. Cambridge University Press, Cambridge (1994)

Ontology Oriented Storage, Retrieval and Interpretation for a Dynamic Map System*

Igor Wojnicki, Piotr Szwed, Wojciech Chmiel, and Sebastian Ernst

AGH University of Science and Technology, Department of Automatics
{wojnicki,wch,ernst}@agh.edu.pl,
pszwed@ia.agh.edu.pl

Abstract. This paper presents the Dynamic Map system, one of the key products of the INSIGMA Project. The main focus is on the map and dynamic data storage subsystem, which utilizes a spatial database and is based on ontologies. First, the data models used are described, including the OpenStreetMap-based structure for the static map and the ontology-driven structure for dynamic parameters and events. The approach to generation of database structures from OWL is described in detail, followed by descriptions of the OSM import process, the GPS tracker module, the sensor state analyzer and the event interpreter. Finally, the planned future enhancements are outlined and discussed.

Keywords: ontology, database, map, traffic, events.

1 Introduction and Motivation

The paper discusses the main components of the Dynamic Map, one of the key subsystem being the result of the INSIGMA Project [2]. The aim of the Project is development of an Intelligent Information System for Global Monitoring, Detection and Identification of Threats. The Dynamic Map can be considered a composition of a spatial databases storing static, temporal and dynamic data relevant for urban traffic and a set of software modules responsible for data collection and interpretation.

The logical structure of the Dynamic Map is comprised of four layers, as shown in Fig. 1: (1) the digital static map which represents the road network and other map objects, (2) the traffic organization layer, (3) dynamic information about traffic conditions and (4) dynamic and temporary information about events such as traffic jams, accidents and weather conditions. Following the terminology used internally within the project, we will refer to the two lower layers of the Dynamic Map as the static map, and to the upper layers as the dynamic map.

The innovative concept behind the INSIGMA Project assumes that up-to-date information about the current urban traffic conditions is provided by a network

---

* Work has been co-financed by the European Regional Development Fund under the Innovative Economy Operational Programme, INSIGMA project no. POIG.01.01.02-00-062/09.

A. Dziech and A. Czyżewski (Eds.): MCSS 2012, CCIS 287, pp. 380–391, 2012.

of traffic surveillance sensors including: video image processors, inductive loops, acoustic arrays, weather stations and onboard GPS receivers installed in vehicles. Although the system will integrate various types of sensors, we focus on data obtained from cameras, which are treated as the primary source of traffic information. Video streams originating from cameras pointing at selected streets or crossroads are processed and analyzed on-line to calculate various traffic parameters, e.g.: average speed, length of queue of vehicles approaching a crossroad or a time necessary to make a specific maneuver [4]. Results of calculations are sent to the traffic repository, the Storage System, on a regular basis.

Availability of data describing the current traffic conditions is a key requirement, as it will be extensively used by various services provided by the system, including individual route planning (for normal traffic participants, but also privileged users as police, ambulances, fire brigades), urban-wide traffic optimization, current traffic information, traffic monitoring by external entities (e.g. police services, road administrators).

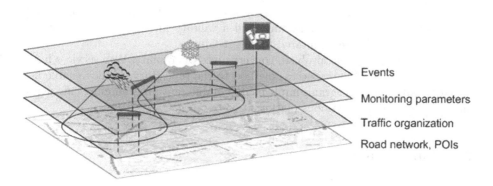

Fig. 1. The logical layers of the Dynamic Map

The repository stores traffic information that other subsystems, e.g. route planning and traffic optimization subsystems, should be aware of. The data includes measured traffic parameters, weather conditions and events (traffic jams, accidents, etc.). The data can substantially differ in datatypes, ranges and units. Moreover, it must be attached to the underlying road network structure: roads, lanes, crossroads or areas. This implies the need of semantic support which allows to differentiate various types of parameters and to verify whether the data feeds are correct with respect to assumed restrictions. System maintainability is also an important issue: while integrating new types of sensors and measured parameters, the repository should smoothly incorporate the changes without affecting the internal structure. This entails the need for discoverability: a client should be capable of querying about types of measured parameters and their values for an area of interest, i.e. to perform a route planning task.

There are several problems that needs to be faced. A static, topological information data source is not uniform. It is compiled from the road structure and additional data defining junctions, lanes, special areas etc. The compilation process needs to be precisely defined and has to take map topology changes into account.

The dynamically-changing information needs to be semantically connected both to the static information as well as concepts describing its nature, i.e. what is a sensor, its type, capabilities, what kind of data it feeds in, where it is located, what is the meaning of transmitted data, etc. This semantic connection is provided by the ontology. Such dynamic data is subject to further interpretation, generating even more semantically annotated information, i.e. traffic jam or accident detection.

Another important requirement relates to the performance. The repository should accept an assumed number of data feeds and client requests. Especially, frequent updates need to be taken into account. This leads to the use of relational or NoSQL databases, in spite of their lack of built-in semantic support, as their performance is superior to storage solutions based on RDF triple models, especially for data entry.

An issue that should also be addressed is data reliability. All data originating from sensors is assigned a timestamp and a validity interval. Sensor activities and their data feeds are monitored to detect failures.

The last, but the least, problem concerns interoperability. Web services have been chosen as the primary access method to the repository, as they enable integration with client programs implemented in different programming languages and being executed on various platforms.

2   Map Storage System

The proposed Map Storage System consists of two components: the static map and the dynamic map. The system is designed to fulfill requirements of the INSIGMA Project.

The static map provides detailed data regarding the road network and infrastructure. It serves as the basis for visualisation, route planning, and as reference for dynamic data source location. The base layer of the static map is street network data, imported from the OpenStreetMap[1] (OSM) project. As the OSM model, described in detail in Section 3.2, lacks certain elements (e.g. lanes, roadsigns, maneuvers) required in the INSIGMA Project, it has been supplemented by an additional layer. An ontology for the static map, based on map features extracted from the OSM model, has been described in Section 3.1.

The dynamic map holds dynamically-changing data, such as traffic parameters and events. To facilitate future extensions, it has also been based on an ontology (see Section 3.1). The current implementation uses a relational database (PostgreSQL).

Individual components have been described in respective sections. Interactions among them are showed in Fig. 2.

---

[1] http://www.openstreetmap.org

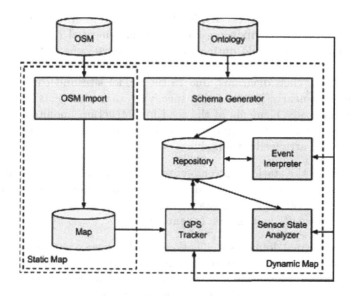

Fig. 2. Interactions among presented components

3 Data Models

This section describes the data models used for various parts of the Dynamic Map system.

3.1 Ontologies

According to a well known classification presented in [5] ontologies can be used in two phases of the software lifecycle: development and exploitation.

Ontologies finding application during the system development phase fall into two categories:

- ontologies formalizing its domain, i.e. classifying physical objects, abstract concepts and events; these ontologies apart from providing a shared reference model can be used for automatic or semi-automatic generation of database schemas or defining data exchanged between components;
- task ontologies describing software functions, services and interfaces that can be used during the system integration.

During the system exploitation ontologies can be used when the domain model or behavior description can not be fully analyzed and specified during the system development; ontologies allow to extend the basic domain model by introducing new concepts, that were not identified during the system development, classification of information is required, and the complexity of the tasks is a prerequisite for the use of advanced semantic tools (reasoners), or some kind of reasoning supporting decision making should be applied.

The system build within the INSIGMA project is at present at the development phase. Several components have been implemented (but still not integrated). The ontologies build during their development constitute a formalized domain model encoded in OWL language. We find that such representation is superior to UML class diagrams, due to the model size and complexity. The model comprises nearly 1 000 classes specifying various objects appearing on maps, types of roads, elements of the road infrastructure, monitoring parameters, sensor, events, traffic organization, types of vehicles and user preferences. These classes are arranged into modules:

1. Ontologies of the static map (`osm-core.owl`, `osm.owl`, `static-map.owl`)
2. Monitoring parameters ontology defining dynamic traffic properties delivered by sensors (`param.owl` and `sensor.owl`)
3. Ontology of events and threats (`event.owl`)

Based on these formal domain specifications the database schemas for three data repositories were semi-automatically generated: Static Map, Monitoring Parameter and Events. The repositories are implemented as PostgreSQL relational databases; the choice of relational representation was driven by the performance issues. The rules of ontology TBox translation are discussed in detail in the section @xxx. An important factor related to the translation, is that it should be revertible. Assuming that the records in the database represent individuals in ontology (ABox), the database schema should provide additional semantic information that can be used to reproduce information on individuals and their relations in form of RDF graph that is required as an input for reasoners.

3.2   Static Map

The static map is based on the OpenStreetMap (OSM) map model, which consists of the following elements:

- **nodes:** single points with geographic coordinates and unique IDs,
- **ways:** sequences of node elements, used to represent roads as well as other features, such as building shapes,
- **relations:** sets of way and/or node elements which represent a given entity (e.g. a tram line or a complex junction).

The semantics of each OSM element is defined by attached tags; for instance, every way element representing a road has a *highway* tag with a value determining the road type (motorway, local road, etc.) [1].

The OSM layer needs to be extended to fulfill the needs of the INSIGMA project. Therefore, the OSM layer has been supplemented with an extension layer, introducing the following elements: the *Crossroads* class, which defines crossroads by specifying all of its entry and exit roads; the *Turn* class, which can be used to model maneuver restrictions; the *Lane* class, used to represent individual lanes; the *SMNode* class, which allows for definition of "nodes" which do not exist in OSM and their relation to existing OSM elements.

The components are defined in the static map ontology. Part of the database schema used to store data imported from OSM is based on the OSM Sinple Schema, as described by the `pgsimple_schema_0.6.sql` file, which is a component of the Osmosis import tool (see Section 4.1). Additional components have been modeled using the static map ontology.

3.3 Dynamic Map

The Dynamic Map consists of the Repository, which is a database core component, and software modules supporting it. The Repository database schema is synthesized by the Schema Generator. The main goal of the Repository is to efficiently store and retrieve facts for given ontological classes. Other modules use the Repository, retrieving, interpreting, processing or generating new facts. They can also optionally access the Ontology.

Since there are certain performance requirements the Repository schema is synthesized from OWL in a two stage process. The first stage regards flattening desired class structure, the second stage generates actual relations for the flattened structure.

The flattening reduces number of records needed to represent single ontological fact. Other approaches such as [6], which store information about each class as separate relation require multiple records for storing a single fact. Assuming that a single fact is a set of property values of given class to be stored, each property value is stored at different relation plus some additional records need to be generated if there is a class hierarchy involved (in case of an *is-a* relationship between classes). The proposed flattening represents a tree of classes interconnected with is-a relationships as a single class. Additionally in order to preserve information about sub- and superclasses some metadata has to be stored along with the new class instances as well, so-called instance metadata.

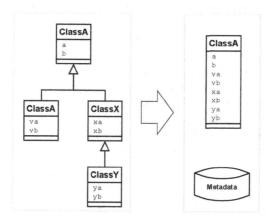

Fig. 3. The First Stage: Class Flattening

An example of the flattening operation is presented in Fig. 3. On the left side there is a class hierarchy which is transformed into a single class on the right side.

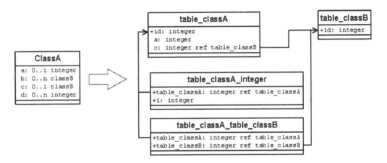

Fig. 4. The Second Stage: Relational Schemas (+ stands for Primary Key)

If there are any instances of the new class the Metadata contains information that translates them back to proper original classes.

The second stage, translating classes from the first stage into relations is given in Fig. 4. For each flattened class a relation is created. Such a relation has a primary key desined as *id*. Depending on particular property's arity and the class, the property belongs to, different actions are taken. In case of property *a*, it corresponds to an attribute (*table_classA.a*) since it regards a base datatype class of maximal arity of one. The property *c* is translated into the attribute *table_classA.c* which is a foreign key to *table_classB*, since *c* is of maximal arity of one and it is of *classB*. The properties *b* and *d* are represented with use of the auxiliary relations *table_classA_table_classB* and *table_classA_integer* respectively, since they are of any arity.

Additionally appropriate metadata allowing to recreate class instances from the relational model is also provided, so-called structure metadata. It is given in Table 1 as *meta_param* relation. Its attributes are interpreted as: *cls* – class name, *prop* – property name, *tbl* – relation name, *att* – attribute name, *ref_tbl* – name of the relation which holds property instance (optional, for properties of classes other than basic types), *key_att* – foreign key attribute name, (optional, for arities greater than 1), *ptr_tbl* – name of the relation that implements a foreign-primary key relationship (optional, for arities greater than 1).

The instance metadata is implemented as a three-attribute relation: *meta_param_pk_tbl_cls(pk, tbl, cls)*. It allows to map any record (identified by *pk*), in any relation (identified by its name: *tbl*), to be assigned to appropriate class (identified by its name: *cls*).

The proposed procedure if fully reversible. Database records can be extracted and presented as ontological facts by using the metadata. In order to find a

Table 1. Structure Metadata Relation: meta_param

cls	prop	tbl	att	ref_tbl	key_att	ptr_tbl
classA	a	table_classA	a			
classA	c	table_classA	c	table_classB		
classA	b	table_classA_ table_classB	table_classB	table_classB	table_classA	table_classA
classA	d	table_classA_	integer		table_classA	table_classA

correspondence between a record and an ontological class the following procedure has to be considered: (1) Find appropriate correspondence between the class and the relations using the structure metadata: the *meta_ param* relation, to issue a database query. (2) Find apropriate records in the relations. (3) Identify which classes or subclasses the records belong to, using the instance metadata: *meta_ param_ pk_ tbl_ cls*.

The proposed database schema and the structure metadata relation depend on the ontological classes and their relationships. The first stage metadata relation depends on class instances stored in the database. Storing a single instance of a given class requires one record for all base type class properties, one record of instance metadata and one for each property with arity greater than one. Comparing with other approaches mentioned before it gives reduction of the number of records. The number of records needed to represent an instance of a class with multiple properties is reduced proportionally to the number of properties.

The proposed data model allows efficiently store and retrieve ontological instances. They can be either accessed as plain database records or, with use of the metadata, as full ontological knowledge regarding both classes and instances. The process is fully reversible, efficient generation of appropriate RDF or OWL is possible.

4 Description of Selected Modules

This section describes selected modules of the proposed dynamic map system.

4.1 OSM Import

OpenStreetMap has been selected as the data source; therefore, a system had to be implemented for efficient import of selected map fragments. The import procedure has been presented in Fig. 5.

OSM makes its data available in the form of one huge file, `planet.osm`. Since downloading and processing of this file is unfeasible, for the time being a decision

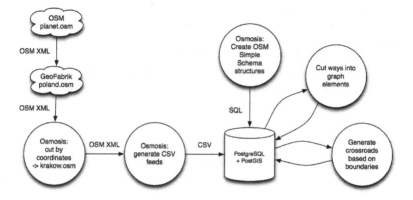

Fig. 5. Overview of the OSM import procedure

has been made to make use of ready-cut data files available at the GeoFabrik website. The import procedure is as follows:

1. Cut planet.osm to extract the map of Poland (this step is performed by GeoFabrik).
2. Download the `poland.osm` file from GeoFabrik.
3. Cut the `poland.osm` to extract the desired region (e.g. Kraków) based on coordinates; this step is performed using the Osmosis[2] tool.
4. Generate CSV files with feeds, also using Osmosis.
5. Implement the OSM Simple Schema in the database, using Osmosis accompanying SQL files.
6. Import CSV files generated in step 4 into tables.
7. Cut way elements into individual graph edges.
8. Generate crossroad data using predefined crossroad boundaries.

Boundaries of each crossroads is stored as a closed *linestring*, thus defining an area. All processing is performed using PostGIS; each node which lies within the area of a crossroads and takes part in a way which intersects with the crossroads boundary is labeled as an *input*, *output* or *input/output* node, depending on the direction of the way. All nodes which lie within the crossroads area and have not been labeled otherwise are labeled as *internal* nodes.

4.2   Schema Generator and Repository

The Generator selects particular ontological classes and provides appropriate database structures to support storing and retrieving their facts. Other modules use the schema, the facts, and optionally the Ontology to perform their actions.

Initially the Repository schemas were generated for the monitoring parameters. Currently they are being extended, to include other concepts defined by the Ontology, such as events. This extension can be easily performed by fine-tuning the Generator.

The Generator acts according to the two-stage process presented in Section 3.3. It analyzes the ontology, expressed as OWL, and generates SQL queries which, in turn, create appropriate tables. It also generates the structure metadata, providing appropriate table schemas and filling it with records. The instance metadata table is created, however no records are generated. It should be populated by other modules while storing facts in the database. The generated SQL queries comply with the SQL2 standard.

Since the process presented in Section 3.3 is formally defined with rules, the module is implemented in Prolog language. Expressing rules in Prolog is straight forward. To access and process the ontology Thea2 OWL library[3] is used.

4.3   GPS Tracker

The GPS Tracker module collects information on positions of multiple vehicles registered in the system as data sources. Periodically entered GPS coordinates

---

[2] http://wiki.openstreetmap.org/wiki/Osmosis
[3] http://www.semanticweb.gr/thea/

with accompanying timestamps are combined into vehicle trajectories, which are the basis for calculating values of several monitoring parameters related to lanes and turns.

The analysis of GPS coordinates is is a two-stage procedure. The goal of the first stage is to estimate state variables of a vehicle: location, current velocity and acceleration. The estimation is based on the position observations and uses the Kalman filtration method; In the second stage, a trajectory (a vector of state variables and their occurrence times) is projected onto the network of routes defined in the static map, then several values of monitoring parameters are calculated, assigned to lanes and turns and finally send to the repository as values of volatile parameter instances.

As we mentioned earlier, lanes and turns constitute an intermediary layer serving as a bridge between OSM data structures specifying physical course of roads and, stored in the separate repository, dynamically changing information about current traffic and traffic related events. We we adopted lazy evaluation approach, when reporting values of monitoring parameters based on the interpretation of vehicle trajectories. We start with an initial small set of lanes and turns (manually introduced during configuration of fixed sensors: cameras, inductive loops) and, if needed, dynamically create new ones when a vehicle passes successive streets and intersections.

The correctness of trajectory interpretation pose several challenges. First, it is difficult to distinguish between intentional vehicle stopping and stopping it due to a jam or other traffic obstacle. To cope with this problem, we use additional semantic information about the user profile (e.g. municipal transport bus) and the context information (e.g. location of bus stops and terminals). Second, It is assumed that information about location of lanes is unsure (in fact in OSM only numbers of lanes happens to be specified). In consequence, monitored parameters (e.g. average speed) are attributed to all lanes (with a lower confidence if several lanes exist). In some cases, the proper lane can be identified based on user profile (privileged users as buses, emergency vehicles are expected to use a separated lane). Finally, there is an issue concerning assumptions about what is known about future user behavior. If trajectory interpretation is limited to observations only, it is possible to calculate monitoring parameters based on past events. This introduces delays, which should be avoided.[4]

4.4 Sensor State Analyzer

Each value of a monitoring parameter entered to the repository is represented by a record specifying the time it was measured and the validity period. In normal

---

[4] To give an example, in the presence of a traffic jam, the time required to travel a road section and take a left turn can be calculated as equal to 30 min, after the turn is eventually made. Considering the usability of the obtained information for the route planning task, it can be stated that the information is outdated by 30 min. During this period several hundreds of vehicles could have been directed to the jammed area. However, if it is known that a user intends to make a maneuver consisting in turning left at the end of the road, the same parameter value can be estimated about 30 minutes earlier from registered speed and displacement.

operation mode it is expected, that before this time elapses, a new value will replace the old one. If a new value does not appear within the validity period, the old value can still be used, however its utility gradually diminishes until the moment, when it should be ignored by route planning algorithms as apparently outdated. From the perspective of application in route planning, it is important that such outdated values and inactive parameter instances are hidden and not visible for discovery services. This regards route planning components deployed at the server side, as well as those running in mobile devices and dynamically downloading traffic data from GSM network. Moreover, long-term absence of updated traffic data may be caused by a sensor failure and should trigger servicing and maintenance actions.

The goal of the sensor state analyzer is on-line detection of long-term inactivity observed at the input interface with respect to a monitoring parameter instance. If such inactivity is detected, the state of the instance is changed to exclude it from the discovery mechanisms. If all monitoring parameter instances that are linked to the same sensor are inactive, the analyzer module changes the sensor state and sends the notification to the human operator, who decides if the inactivity represents a system or hardware failure or is caused by a natural condition, e.g. a snow fall or a heavy rain.

4.5   Event Interpreter

The goal of the event interpreter is to identify event occurrences by analyzing the changes in the monitoring parameter values and their trends. Internally, it implements the *Observer* design pattern [3]. Several observers, e.g. TrafficJamObserver, HavyRainObserver, SnowFallObserver aimed at detecting particular event types are registered as recipients of notifications about changing values of relevant monitoring parameters. When a notification arrives, the observer makes a decision, whether a new event should be created and stored it in the event repository or properties of an existing event should be modified (changing state to terminated, duration prolonged, location extended to a wider area etc.).

To avoid immense flow of insignificant notifications, an observer during the registration indicates conditions (rules) specifying when it should be notified about the changing parameter value. The conditions are defined with use of thresholds or ranges and can be applied directly to parameter values, differences between old and new values, change directions and their speed.

5   Conclusions and Future Work

This paper introduces an outline and interactions among components of the proposed Storage System. The system handles georeferenced static data as well as dynamically changing monitoring parameters. The parameters are subject to interpretation to detect more abstract concepts, such as traffic jams, accidents etc. The proposed architecture is highly modularized. Each module regards particular functionality. There are five modules introduced. They are responsible for populating and synchronizing static data (OSM Import), generating appropriate database schemas complying with given Ontology (Schema Generator). Furthermore there are auxiliary modules interpreting raw data such as: detecting sensor conditions (Sensor State Analyzer), detecting events (Event Interpreter), or assigning monitoring parameters based on mobile sensors (GPS Tracker).

The presented solution provides a complete and uniform data source describing a dynamically changing map. The source is ontology driven. All data, while stored in a relational database, is semantically tagged. It enables RDF or OWL synthesis of gathered facts at any time.

Future work regards tweaking the Schema Generator to support broader set of ontological classes. Several other modules, interpreting gathered data are also considered. Furthermore, implementation of full RDF and OWL export capabilities, as a separate module, is also taken into account.

References

1. OpenStreetMap wiki: Map features,
   http://wiki.openstreetmap.org/wiki/Map_Features
   (last accessed: January 2012)
2. INSIGMA project, http://insigma.kt.agh.edu.pl (last accessed: March 2012)
3. Gamma, E., Helm, R., Johnson, R., Vlissides, J.: Design patterns: elements of reusable object-oriented software. Addison-Wesley Longman Publishing Co., Inc, Boston (1995)
4. Głowacz, A., Mikrut, Z., Pawlik, P.: Video detection algorithm using an optical flow calculation method. CCIS. Springer (2012)
5. Guarino, N.: Formal Ontology in Information Systems: Proceedings of the 1st International Conference, 1st edn., Trento, Italy, June 6-8. IOS Press, Amsterdam (1998)
6. LePendu, P., Dou, D., Frishkoff, G.A., Rong, J.: Ontology Database: A New Method for Semantic Modeling and an Application to Brainwave Data. In: Ludäscher, B., Mamoulis, N. (eds.) SSDBM 2008. LNCS, vol. 5069, pp. 313–330. Springer, Heidelberg (2008)

Author Index